Angsuman Sarkar, Swapnadip De, Manash Chanda, Chandan Kumar Sarkar
Low Power VLSI Design

Also of interest

Chaotic Secure Communication
K. Sun, 2016
ISBN 978-3-11-042688-5, e-ISBN 978-3-11-043406-4,
e-ISBN (EPUB) 978-3-11-043326-5, Set-ISBN 978-3-11-043407-1

Nano Devices and Sensors
J. J. Liou, S.-K. Liaw, Y.-H. Chung (Eds.), 2016
ISBN 978-1-5015-1050-2, e-ISBN 978-1-5015-0153-1,
e-ISBN (EPUB) 978-1-5015-0155-5, Set-ISBN 978-1-5015-0154-8

Systems, Automation and Control
N. Derbel (Ed.), 2016
ISBN 978-3-11-044376-9, e-ISBN 978-3-11-044843-6,
e-ISBN (EPUB) 978-3-11-044627-2, Set-ISBN 978-3-11-044844-3

Power Electrical Systems
N. Derbel (Ed.), 2016
ISBN 978-3-11-044615-9, e-ISBN 978-3-11-044841-2,
e-ISBN (EPUB) 978-3-11-044628-9, Set-ISBN 978-3-11-044842-9

Angsuman Sarkar, Swapnadip De,
Manash Chanda, Chandan Kumar Sarkar

Low Power VLSI Design

Fundamentals

DE GRUYTER
OLDENBOURG

Authors

Dr. Angsuman Sarkar
Kalyani Government Engineering College
Kalyani, India
angsumansarkar@ieee.org

Manash Chanda
Meghnad Saha Institute of Technology
Kolkata, India
manash.bst@gmail.com

Dr. Swapnadip De
Meghnad Saha Institute of Technology
Kolkata, India
swapnadipde26@yahoo.co.in

Prof. (Dr.) Chandan Kumar Sarkar
Jadavpur University
Kolkata, India
phyhod@yahoo.co.in

ISBN 978-3-11-045526-7
e-ISBN (PDF) 978-3-11-045529-8
e-ISBN (EPUB) 978-3-11-045545-8
Set-ISBN 978-3-11-045555-7

Library of Congress Cataloging-in-Publication Data
A CIP catalog record for this book has been applied for at the Library of Congress.

Bibliographic information published by the Deutsche Nationalbibliothek
The Deutsche Nationalbibliothek lists this publication in the Deutsche Nationalbibliografie; detailed bibliographic data are available on the Internet at http://dnb.dnb.de.

Cover image: MauMyHaT/iStock/thinkstock
Typesetting: Integra Software Services Pvt. Ltd.
Printing and binding: CPI books GmbH, Leck
♾ Printed on acid-free paper
Printed in Germany

www.degruyter.com

This book is dedicated to our parents and family members for their love and support.

Preface

Designing of integrated circuits (ICs) requires expertise in different areas including device physics, transistor-level design and logic-level design. An attempt has been made to cover all these topics reinforced with physical and intuitive explanations and necessary simulation analysis for better understanding of the theory. The goal of this book is to emphasize the basic physical principles and physics of transistors and behavior of circuits that can be used to explain and confront the challenging low power issues in digital and analog very-large-scale integration (VLSI) design in a comprehensive manner with a focus on predominant complementary metal–oxide–semiconductor (CMOS) technology.

An attempt has also been made to adopt a hierarchical design methodology for digital and analog energy-reduced low power VLSI design by analyzing theory and key concepts in an organized manner to provide the reader an insightful, practical guide. This book does not try to recommend specific software tools. In contrast, the main objective of this book is to demonstrate a proper path to the reader to face the challenges of low power VLSI design through a maze of names, concepts, methodologies, software tools and usage.

The purpose of this book is to present and explain in a coherent manner all the know-how of low power digital and analog VLSI design. For each chapter/topic, the book discusses several basic approaches and elaborates on the more fundamental ones. A list of complete references/bibliography has been included at the end of each chapter to provide extensive coverage of the topics by providing a list of many of the classic books/papers dealing with the findings which are considered landmarks in the VLSI industry. An attempt has been made to include all recent papers on emerging topics which are appropriate for the targeted readers of this text.

In the following, a series of capsule description of the seven chapters in the book has been introduced.

Chapter 1: Introduction to Low Power Issues in VLSI. This chapter presents an overview of the low voltage, low power digital and analog ICs and their applications, and also the context of the book and its motivations.

Chapter 2: Scaling and Short Channel Effects in MOSFET. Chapter 2 focuses on the fundamental aspects of short channel effects which is considered as a fundamental roadblock to the future downscaled devices and is of immediate relevance to the practical low power IC design. Advanced semiconductor IC fabrication technologies have increased the ability of shrinking or scaling of the devices with an intention to make them smaller, faster, less power-consuming and reliable. The aggressive scaling of the CMOS technology in the deep submicrometer regime gives rise to the detrimental short channel effects. Thus, Chapter 2 gives a brief introduction to the short channel effects and their preventions

Chapter 3: Advanced Energy-reduced CMOS Inverter Design. This chapter is concerned with the theory of inverters, known as the basic building block of the digital ICs, to set a solid foundation to understand the rest of this book by discussing basic operation and other advanced issues in both super-threshold and subthreshold regions of operation.

Chapter 4: Advanced Combinational Circuit Design. The topic on this chapter focuses on the design principles of combinational digital logic circuits with an objective to make them fast, small and reliable. After providing a preliminary review of the basic concepts of combinational digital logic design using different approaches, a number of design examples are demonstrated, followed by ultralow power implementation of combinational circuits.

Chapter 5: Advanced Energy-reduced Sequential Circuit Design. The objective of this chapter is to deal with advanced sequential circuit design techniques for clocked logic structures that are fast, reliable, energy-efficient and race-free. The chapter also discusses the energy-efficient implementation of sequential circuits with introduction to adiabatic logic.

Chapter 6: Introduction to Memory Design. This chapter considers the technological issues necessary to comprehend the design and operation of low power, high speed MOS-based memory systems. The evolution of memory design over the years has been described briefly. This chapter shows the way to design various types of ROMs and RAMs. The features and architectures of latest memories available in the market are also presented.

Chapter 7: Analog Low Power VLSI Circuit Design. The purpose of this chapter is to present the design techniques of low power analog ICs and the fundamental issues associated with the design of analog/mixed-signal system-on-chip design.

Contents

1 Introduction to Low Power Issues in VLSI

1.1 Introduction to VLSI

This chapter gives a brief description of low power digital and analog very-large-scale integration (VLSI) design. It is also treated as a review of basic principles rather than an in-depth treatment of individual advanced topics.

Over the last two decades, “wireless communication” is considered as one of the major successes of engineering, not only due to its huge technological growth, but also due to its societal and economic impact. The evolution of wireless communication is directly linked to power consumption of the devices. The gap between small-sized battery capacity and power requirement has become a critical factor to be considered for most hand-held portable wireless devices. Thus, energy efficiency and low power design has become one of the most important topics in integrated circuit (IC) design.

The chief design goal of today’s ICs can be listed as higher speed, higher accuracy and reliability and low power drain. In the design of high performance analog/digital complementary metal oxide semiconductor (CMOS) circuits, energy efficiency plays a crucial role. It has become highly difficult to maintain energy efficiency in medium-to-high accuracy circuits with highly scaled CMOS deep-submicron technologies. One of the important purposes of this book is to emphasize some of the important challenges faced by the circuit designers working with aggressively scaled devices and reviews of the state-of-the-art device/circuit-level techniques that can be employed to mitigate those challenges.

The CMOS digital ICs play a very important role in technology for modern information age. Because of their intrinsic features in low power consumption, large noise margins and ease of design, CMOS ICs have been widely used to develop random access memory (RAM) chips, microprocessor chips, digital signal processor chips and application-specific integrated circuit (ASIC) chips. The popular use of CMOS circuits continues to grow with the increasing demands for low power, low noise integrated electronic systems in the development of portable computers, portable phones and multimedia agents.

As more and more complex functions are required in various data processing and telecommunication devices, the need to integrate these functions in a small package is also increasing (International Technology Roadmap for Semiconductors, http://public.itrs.net, 2009). The level of integration is measured in terms of number of logic gates in chip. So the number of logic gates is also increasing. These advances in device manufacturing technology allow steady reduction of minimum feature size as well as channel length and power consumption per function. This also increases the speed of the device.

As process technology scales beyond 100 nm feature sizes, for functional and high-yielding silicon, the traditional design approach needs to be modified to cope with interconnect processing difficulties and other newly exacerbated physical effects. The scaling of gate oxide in the nano-CMOS regime results in a significant increase in gate direct tunneling current [1]. The effect of gate-induced drain leakage (GIDL) is felt in designs, such as DRAM and low power SRAM, where the gate voltage is driven negative with respect to the source. If these effects are not taken care of, the result will be a nonfunctional SRAM, DRAM or any other circuit that uses this technique to reduce subthreshold leakage [2].

The level of integration as measured by the number of logic gates in monolithic chip is increasing over the past three decades, mainly due to the rapid progress in process technology and interconnects technology.

- Multifunctional chips (1962) contain 2–4 logic blocks per chip.
- MSI (medium-scale integration) (1967) chips contain 20–200 logic blocks per chip.
- LSI (large-scale integration) chips contain 200–2,000 logic blocks per chip.
- VLSI (very-large-scale integration) chips contain 2,000–20,000 logic blocks per chip.
- ULSI (ultralarge-scale integration) chips contain >20,000 logic blocks per chip.

As the technology scales down with its pace determined by famous Moore's law, the supply voltage needs to be scaled down along with the transistor dimensions. This is primarily due to the fact that decreasing supply voltage helps to avoid possible breakdown of MOS transistors with ultrathin gate oxides. In an inverter cell, considered as the simplest logic gate, estimated power consumption can be expressed as

$$P = C_L \cdot V_{DD}^2 \cdot f \tag{1.1}$$

where C_L designates the load capacitor to be driven, V_{DD} is the supply voltage and f is the frequency of operation.

Therefore, if speed has to be increased, simultaneous reduction of power consumption is also necessary. Supply voltage reduction can be employed as an efficient tool to achieve it. In general, to provide sufficient lifetime to the digital circuitry and to keep power consumption at an acceptable level, downscaling is accompanied by supply voltage reduction [3–5]. It is worth mentioning that this evolution is true for digital domain, not the analog domain.

1.2 Low Power IC Design beyond Sub-20 nm Technology

Sub-20 nm process technology promises boost in performance, capacity breakthrough and significant power reduction. However, they possess several challenges to the IC design and manufacturing, requiring changes ranging from custom cell design to system-on-chip (SoC) integration. The cost of sub-20 nm is not cheap. However, despite their cost, with its power, performance and area gains (PPA), sub-20 nm process technology promises a new generation of smaller, faster and cheaper

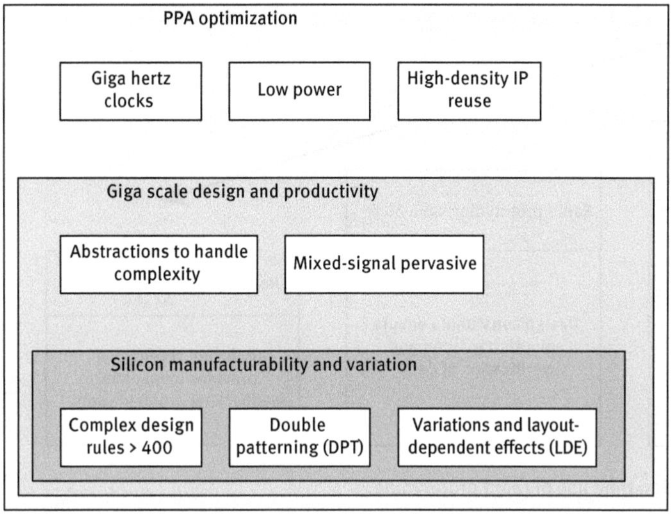

Figure 1.1: Challenges and requirements of sub-20 nm technology adaptation.

product in areas such as mobile computing, smartphones, servers, entertainment and wireless equipment. Although the projected improvements of sub-20 nm technologies are compelling, the challenges and requirements are also extremely large. The challenges and requirements for adaptation of sub-20 nm technology are shown in Figure 1.1.

1.3 Issues Related to Silicon Manufacturability and Variation

Dual pattering lithography techniques are required to pattern correctly some of the metal pitches below 80 nm, which requires extra masks and two-color layout decomposition techniques. This has an impact on every phase of IC design methodology, i.e. from standard cell development to placement, routing, extraction and physical verification. Requirement of correct-by-construction methodology needs to be employed where every tool should be “double-pattering aware” [6].

Moreover, the requirement of approximately more than 400 new design rules (some of which are very aggressive) further complicates the design process. It is expected that below 20 nm, variability will be everywhere (i.e. from channel length to channel doping). Reduction in metal pitches may cause increase in coupling between wires causing signal integrity problems. For analog circuit design, parasitics and mismatch are going to create major problems. Due to layout-dependent effects (LDE), it is expected to become difficult to model a single transistor or a cell in isolation as they are placed in close proximity in the layout structure, thus changing the behavior of the device.

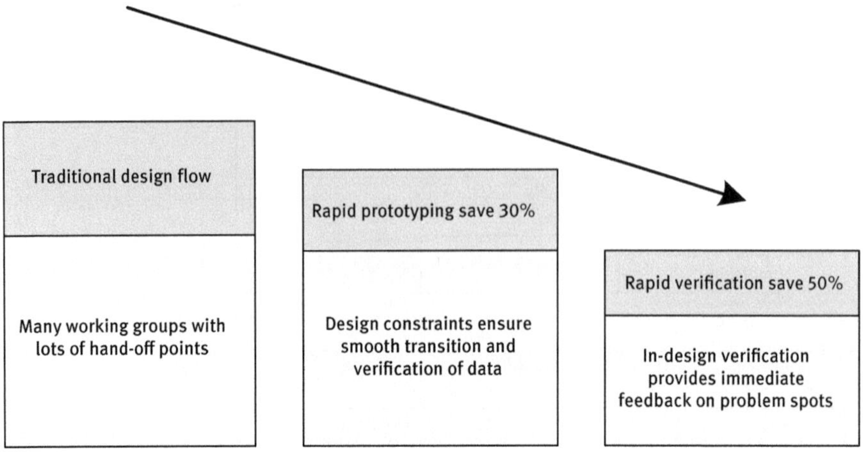

Figure 1.2: To shorten design time use of rapid prototyping.

1.4 Issues Related to Design Productivity

Sub-20 nm technology node looks awesome if the design is shipped in time in order to make the chips out of the door in time. Therefore, an accelerated and integrated end-to-end design flow is required in contrast to sequential point-tool approach. This clearly suggests the requirement of an automation tool – to achieve design cycles savings as shown in Figure 1.2.

Moreover, a flexible modeling methodology with different levels of abstraction may cause significant reduction in development time for large designs.

1.5 Limitation Faced by CMOS

As metal–oxide semiconductor field-effect transistors (MOSFETs) reach nanometer dimensions, power consumption becomes a major bottleneck for further scaling. The continued reduction of the MOSFET size is leading to an increased leakage current due to short channel effects (SCEs), such as drain-induced barrier lowering (DIBL), and the power supply voltage cannot be reduced any further because of the subthreshold slope being limited to 60 mV/decade at room temperature [7]. In this view, the exploration of alternative devices, which possibly outperform the MOSFET at these nanometer dimensions, is required.

In 1974, R. Dennard published an article [2], which has become very famous in the semiconductor device community, about how to scale a MOSFET while keeping the electric fields inside the device unchanged. He recommended that all device dimensions be scaled by $1/\kappa$; while the doping of the source and drain regions should increase by a factor of κ, applied voltages should also be scaled by $1/\kappa$. These rules have been roughly followed ever since, until rather recently.

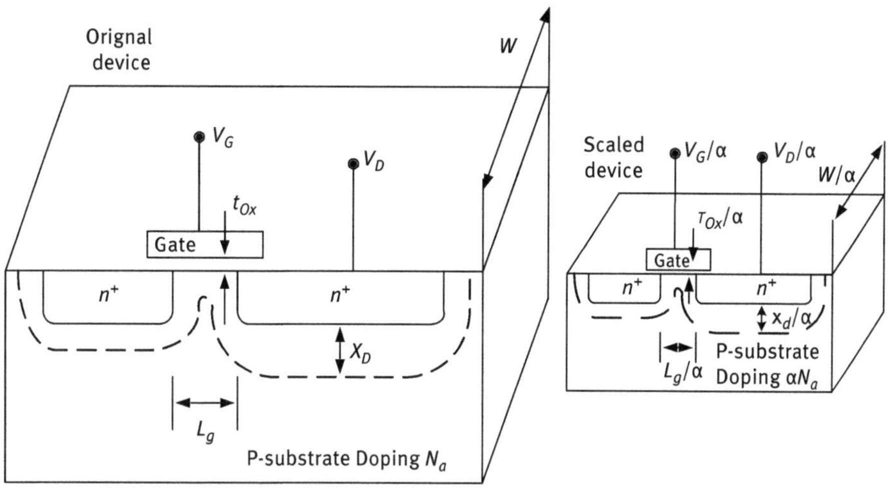

Figure 1.3: Dennard's original figures illustrating the principles of constant electric field scaling.

Figure 1.3 shows Dennard's scaling rules [2]. These scaling rules worked well for 1.4 µm node in the past. However, it no longer works for todays transistor with sizes in the deca-nanometer range. While the supply voltage V_{DD} decreased to about 20% of its original value, the threshold voltage V_{Th} only went down to approximately half of its starting value. That threshold voltage decrease did not happen as a natural result of Dennard scaling. It had to come about in other ways, such as changing the doping of the channel region under the gate. Since the electric fields inside a MOSFET stay nearly constant when the scaling rules are followed correctly, the threshold voltage stays nearly constant as well, unless other changes are made.

The most important consequence of V_{DD} reducing during device scaling while V_{Th} reduces significantly less is that the gate overdrive goes down. When gate overdrive decreases, on-current decreases, which negatively affects device performance, the I_{On}/I_{Off} ratio and dynamic speed ($C_g V_{DD}/I_{On}$). There are two possible solutions to this problem of needing a high gate overdrive: either V_{DD} can stay higher than it should with constant field scaling, or V_{Th} can be scaled down more aggressively.

1.6 International Technology Roadmap for Semiconductors

Due to the significant resources and investments required to develop the next generation of CMOS technologies, it has been necessary to identify clear goals and put collective efforts toward developing new equipment and technologies. The semiconductor roadmap represents a consensus among industry leaders and gives projected needs based on past trends. The International Technology Roadmap for Semiconductors (ITRS) [6] is the standard accepted roadmap. The gate insulator needs to be aggressively scaled down to improve the drive current and to suppress short channel

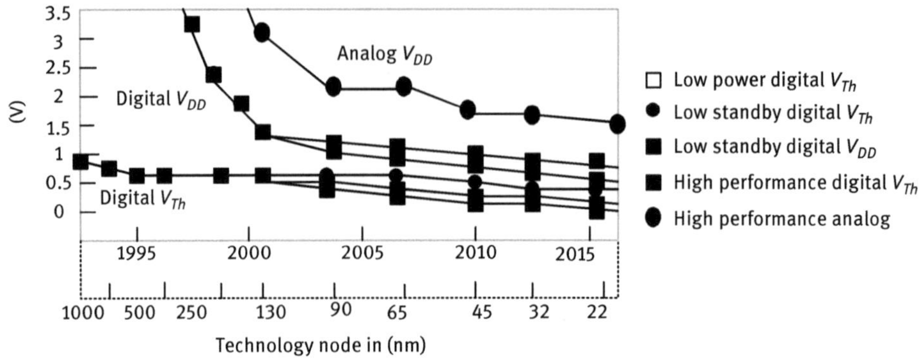

Figure 1.4: CMOS trends: Supply voltages as per ITRS 2004.

effects. Currently, in the 130 nm technology node with a 70 nm physical gate length, the physical thickness of the gate oxide is only 15 ρA, which are approximately 6 atomic layers thick. To continue the past trends in CMOS scaling, a sub-10 ρA effective oxide thickness will soon be required, which is about 4 atomic layers thick. Beyond that point, SiO_2 may lose its properties as an insulator and we may need a different material system, as will be discussed later. Figure 1.4 shows the supply voltage scaling scenario. Lower supply voltages are required due to power dissipation and reliability reasons. The roadmap distinguishes two different applications: high performance and low power circuits. High performance applications include mainstream microprocessors, and low power applications include mobile chips, where the duration of the battery power is more important than performance. On the other hand Figure 1.5

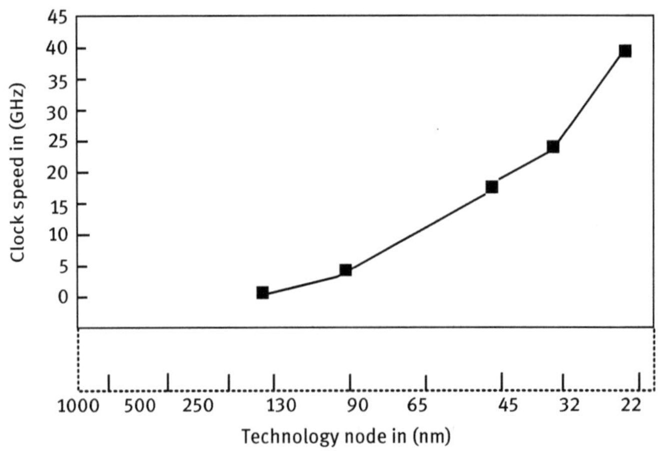

Figure 1.5: CMOS trends: On-chip clock speed as per ITRS 2004.

Table 1.1: Excerpt of 2003 ITRS technology scaling from 90 nm to 22 nm.

Year of production	2004	2007	2010	2013	2016
Technology node (nm)	90	65	45	32	22
HP physical L_g (nm)	37	25	18	13	9
EOT nm (HP/LSTP)	1.2/2.1	0.9/1.6	0.7/1.3	0.6/1.1	0.5/1.0
V_{DD} (HP/LSTP)	1.2/1.2	1.1/1.1	1.1/1.0	1.0/0.9	0.9/0.8
I_{On}/W,HP (mA/mm)	1,100	1,510	1,900	2,050	2,400
I_{Off}/W,HP (mA/mm)	0.05	0.07	0.1	0.3	0.5
I_{On}/W,LSTP (mA/mm)	440	510	760	880	860
I_{Off}/W,LSTP (mA/mm)	1e-5	1e-5	6e-5	8e-5	1e-4

HP: High performance technology, LSTP: Low standby power technology for portable applications, EOT: Equivalent Oxide Thickness.

shows the trend of on-chip clock speed variation along various technologies scaling advancement.

In summary, scaling improves cost, speed and power per function with every new technology generation. Table 1.1 shows that scaling is expected to continue.

Formerly followed scaling trends follow that $1/\kappa = 0.7$ in every 2 or 3 years which is not true for V_{DD}. In order to maintain acceptable levels of gate overdrive, V_{DD} scaling has slowed down drastically. When the supply voltage decreases along with device dimensions, the power density $I_{On} \times V_{DD}/A$ (on-current times supply voltage divided by surface area) remains constant, which means that the energy needed to drive the chip and the heat produced by the chip remain constant. This assumes that when devices scale down, we don't see chip size decreasing, but rather, more complexity and functionality are added with each generation, and chip size remains more or less constant.

When V_{DD} doesn't scale down, power density increases instead. For each MOSFET, the dynamic and static power consumption can be expressed as

$$P_{Dynamic} = fC_L V_{DD} \tag{1.2}$$

where f is the frequency and C_L is the total switched capacitive load, and

$$P_{Static} = I_{Leak} V_{DD} \tag{1.3}$$

where I_{Leak} is the sum of the leakage currents in the device when the MOSFET is in the off state.

If V_{DD} does not decrease, and yet device dimensions decrease, and more devices are added to a chip such that chip size is not significantly reduced, then it can be expected that power consumption will rise considerably.

1.7 Different Groups of MOSFETs

Moving a design from an old technology to a newer one, with smaller design rules, has always been, up to now, an interesting way to lower the power consumption and to obtain higher speed. Indeed, the overall parasitic capacitances (i.e. gates and interconnects) are decreased, the available active current per device is higher, and, consequently, the same performance can be achieved with a lower supply voltage. Moving to a new technology generation, however, induces a scale down of the power supply voltage (V_{DD}), the threshold voltage (V_{Th}) and the gate oxide thickness (t_{Ox}). Beginning with the 0.18 µm technologies, it appeared that building a transistor with a good active current (I_{On}) and a low leakage current (I_{Off}) was becoming more difficult. The following four main causes of limitations are discussed:
(1) Voltage limits and subthreshold leakage
(2) Tunneling currents
(3) Statistical dispersions
(4) Poly depletion and quantum effects

Therefore, two families of transistors were introduced: high speed transistors and low leakage transistors. The threshold voltages of the two families are tuned differently, using a different channel doping. When moving to more advanced technologies, those two families are not sufficient anymore, regarding technological constraints. The ITRS introduces three main groups of transistors:
(1) High performance (HP)
(2) Low operating power (LOP)
(3) Low standby power (LSTP)

At this stage, the channel doping not only is different, but also has the gate oxide thickness.

The HP technology uses the shortest gate lengths in order to achieve the higher drive current. A higher leakage current is also allowed in the technology. For the LOP technology, the main target is to reduce the operating power of the circuit. Compared to the HP technology, the LOP one uses a longer physical gate length, a thicker gate oxide in order to achieve a leakage current hundreds of times lower, for a given node. The main purpose of LSTP technology is to achieve transistors with a very low leakage current (roughly five orders of magnitude smaller than the HP technology). To satisfy this criterion, gate length and gate oxide scaling are relaxed, compared to both HP and LOP technologies. In addition, threshold voltage values must be significantly increased to lower the leakage current. As discussed in the following sections, many key issues, a few of which are listed below, have no ready-made available solution today:
- How to shrink the gate length and achieve good performances
- How to shrink the gate oxide thickness and match the leakage current targets
- How to reduce the supply voltage, while keeping operational circuits and low leakage current

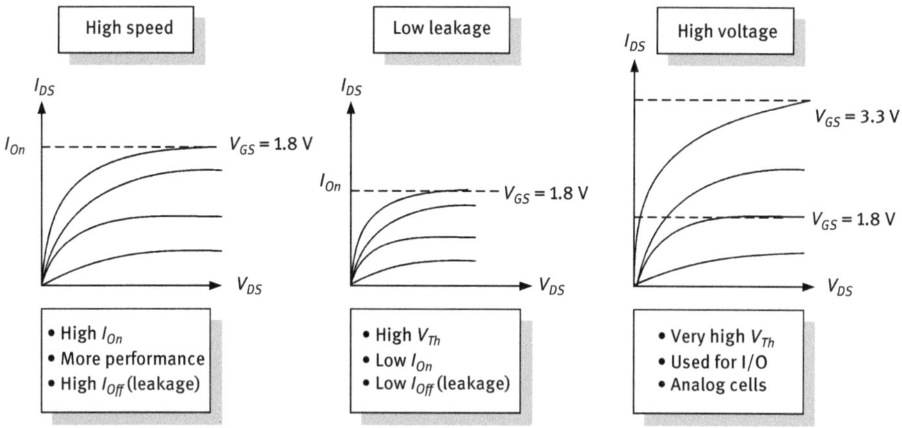

Figure 1.6: Three different types of MOSFETs introduced by ITRS roadmap.

1.8 Three MOS Types

A new kind of MOS device has been introduced in deep submicron technologies, starting with the 0.18 μm CMOS process generation. The new MOS, called "low leakage" or "High-V_{Th}" MOS device, is available as well as the normal one, called "high-speed MOS".

For I/Os operating at high voltage, specific MOS devices called "High voltage MOS" are used. We cannot use high speed or low leakage devices as their oxide is too small. A 2.5 V voltage would damage the gate oxide of a high speed MOS in 0.12 μm technology. The high voltage MOS is built using a thick oxide, two to three times thicker than the low voltage MOS, to handle high voltages as required by the I/O interfaces.

The following new kinds of MOS are introduced in deep submicron technology (0.18 μm) (shown in Figure 1.6):

- High speed: for critical path in term of speed
- Low leakage: for embedded application (less consumption)
- High voltage: for I/O which need higher voltage (oxide thicker than the other MOS)

1.9 Low Leakage MOSFET

The main drawback of the "Low leakage" MOS device is a 30% reduction of the ion current, leading to a slower switching. High speed MOS devices should be used in the case of fast operation linked to critical nodes, while low leakage MOS should be placed whenever possible, for all nodes where a maximum switching speed is not required.

There are two main reasons to keep a low voltage supply for the core of IC. The first one is low power consumption, which is of key importance for ICs used in cellular

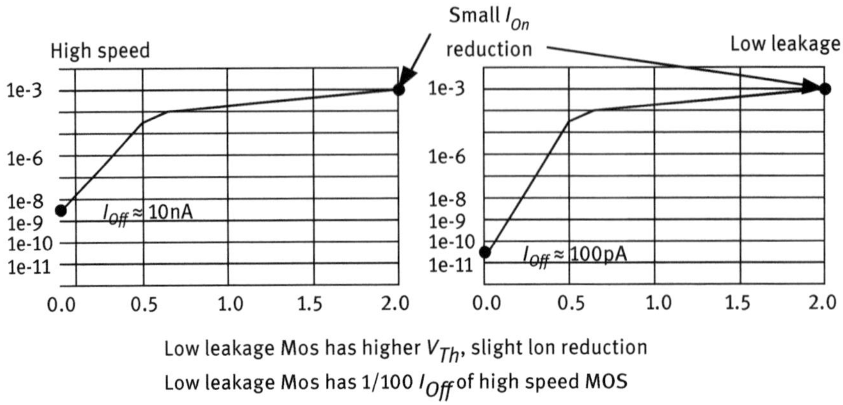

Figure 1.7: Low leakage MOSFET versus high speed MOSFET.

or any portable devices. Low supply strongly reduces power consumption by reducing the amplitude of signals, thus reducing the charge and discharge of each elementary node of the circuit. The second reason for low internal supply is the oxide breakdown. Increased switching performances have been achieved by a continuous reduction of the gate oxide thickness. In 0.12 μm technology, the MOS device has an ultrathin gate oxide, around 0.003 μm, that is 3 nm or 30 pA. The main objective is to reduce significantly the I_{Off} current which is the small current that flows between drain and source with a gate voltage 0 (supposed to be no current in first-order approximation). In Figure 1.7, the low leakage MOS device (right side) has an I_{Off} current reduced by a factor 100, thanks to a higher threshold voltage (0.45 V rather than 0.35 V).

1.10 Importance of Subthreshold Slope

While we care a lot about the "ON" behavior of MOS devices, it is equally important to know their "OFF" characteristics. One important question arises is, what is the current when $V_{GS} < V_{Th}$?

To answer this question we care about subthreshold behavior of MOSFET because it affects the operation of the dynamic circuits and it has become a significant contributor to power dissipation in high performance microprocessors (~30%).

Speed $\propto (V_{DD} - V_{Th})$ (speed is proportional to $(V_{DD} - V_{Th})$)

As the device dimensions are shrunk below 50 nm, the behavior of the device below the threshold or in the subthreshold regime becomes critical. The analysis up to now has assumed that the device turns on abruptly at a gate voltage above the threshold or that no current flows at the gate voltages below V_{Th}.

As shown in Figure 1.8, this assumption does not account for the current that flows through the channel in the region below strong inversion or in the weak inversion regime which is defined as the region where the surface band bending Φ_S is in the range

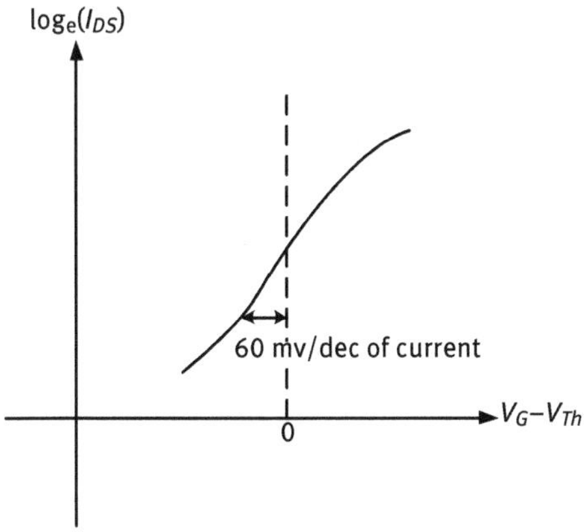

Figure 1.8: The device current does not abruptly turn the threshold off but decreases monotonically at a slope of 60 mV/decade of current.

$$\Phi_F < \Phi_S < 2\Phi_F$$

Subthreshold current is the drain-source current when the gate-source voltage is below the MOSFET threshold voltage. The threshold voltage distinguishes the conduction from nonconduction states of an MOS transistor with the basic assumption of the MOS capacitor analysis that no inversion layer charge exists below the threshold voltage. This leads to zero current below threshold. The transition from the conducting to the nonconducting state is not sharp, but continuous. This means that when the gate-source voltage increases, the charge in the channel is not created abruptly but appears gradually with V_{GS}. There are a range of gate voltages lower than V_{Th} for which there are carriers in the inversion layer that contribute to the drain current. The number of carriers that constitute the channel varies exponentially with the gate voltage. The actual subthreshold current is not zero but reduces exponentially below the threshold voltage. The subthreshold behavior is critical for dynamic circuits since one needs to ensure that no charge leaks through the transistors biased below the threshold.

An ideal I–V characteristic predicts zero drain current when $V_{GS} < V_{Th}$. Experimentally I_D is not equal to 0 when $V_{GS} < V_{Th}$. The drain current that exists for $V_{GS} < V_{Th}$ is known as the *subthreshold current.*

When the surface is in weak inversion ($-\Phi_F > \Phi_S > 0$, $V_{GS} < V_{Th}$), a conducting channel starts to form and a low level of current flows between the source and the drain as shown in Figure 1.9.

Due to that

- I_D leakage increases,
- static power increases and
- circuit instability increases.

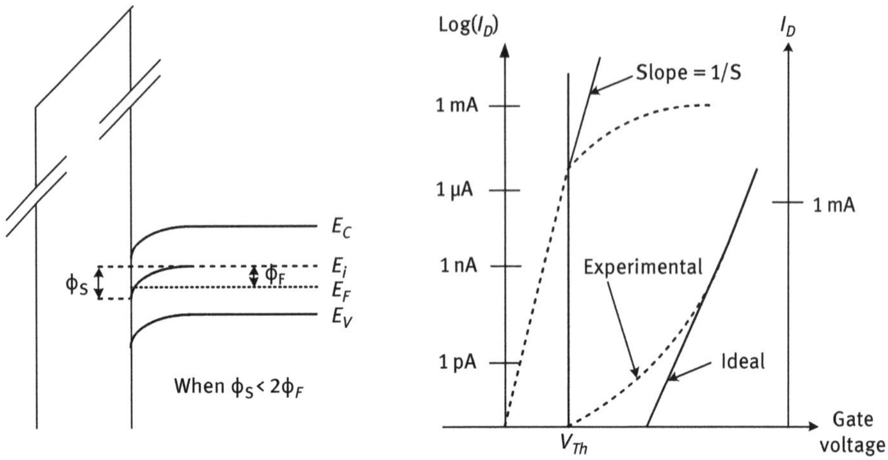

Figure 1.9: Energy band diagram of MOSFET in weak inversion.

The subthreshold current is due to weak inversion in the channel, which leads to a diffusion current from the source to the drain. Fermi level is closer to the conduction band than the valance band. So, the semiconductor surface behaves like a lightly doped *n*-type material; small conduction between *S* and *D* through this weakly inverted channel compares the barriers. In subthreshold regime, current is an exponential function of V_{GS}. For strong inversion, we lose these exponential relationships. For a very small gate voltage, subthreshold current is reduced to the leakage current of the source/drain junction which determines the off-state leakage current and causes the standby power dissipation in MOSFET.

It also tells the importance of having high quality source/drain junction so if V_{Th} is too low, the MOSFET cannot be turned off fully even at $V_G = 0$ V, causing some subthreshold leakage current.

The circuit designer must include the subthreshold current to ensure that the MOSFET is biased sufficiently below the threshold voltage in the "OFF" state. Otherwise significant power dissipation will arise as millions of MOSFETs is used in IC.

The analog circuit based on the subthreshold operation of the devices having an additional advantage of getting higher the gain due to the exponential behavior of the drain current gives rise to the higher trans-conductance factor (g_m/I_d).

We want no current when the transistor is in the off state, i.e. when the gate voltage is below the threshold voltage. If V_{Th} is large so that $|V_{GS}| < |V_{Th}|$, then the number of carriers into the channel approaches zero. However, a large V_{Th} increases the time required to switch between the on and the off conducting states, resulting in slow digital device operations. Thus we cannot make V_{DD} arbitrarily small because speed of the device is proportional to $V_{DD}-V_{Th}$.

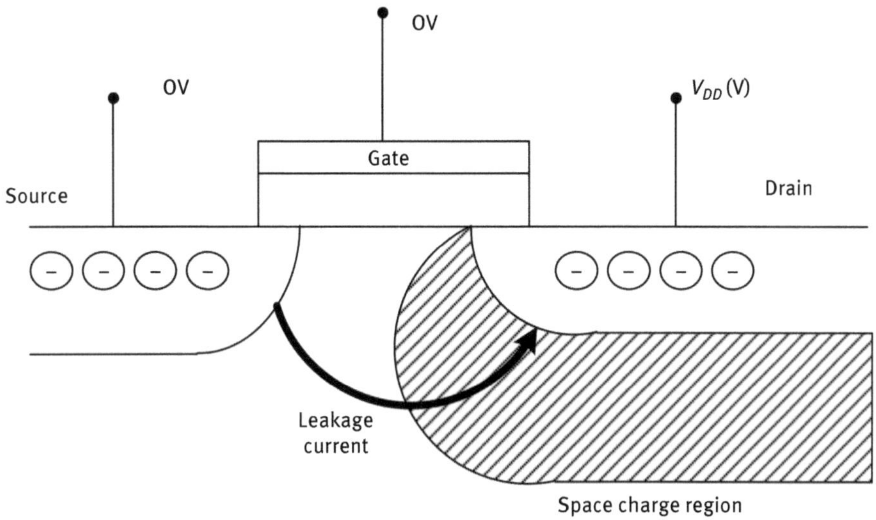

Figure 1.10: Leakage current in short channel MOSFET.

Smaller transistors require the lower operating voltages to restrict the internal electric field within reasonable limits. This in turn requires the decrease of the threshold voltage to maintain the operating speed of the device. This increases off-state leakage current. The tendency today used is to maintain circuit performance at the cost of power increase. Modern devices show a considerable current leakage even at $V_{GS} = 0$ V as shown in Figure 1.10.

1.11 Why Is Subthreshold Current Exponential in Nature?

In a MOSFET, a parasitic Bipolar Junction Transistor (BJT) is provided where we have an n-p-n sandwich with mobile minority carriers in the P region as shown in Figure 1.11. The base potential of this parasitic BJT is controlled through a capacitive divider and not directly controlled by an electrode.

We know that for a BJT

$$I_C \cong I_S \cdot e^{(qV_{BE}/KT)}.$$

In our case we have

$$I_D \cong I_0 \cdot e^{q(V_{GS}-V_{Th})/nKT},$$

where n is given by the capacitive divider given as $n = \frac{C_{Ox}+C_D}{C_{Ox}}$ where C_D is the depletion layer capacitance.

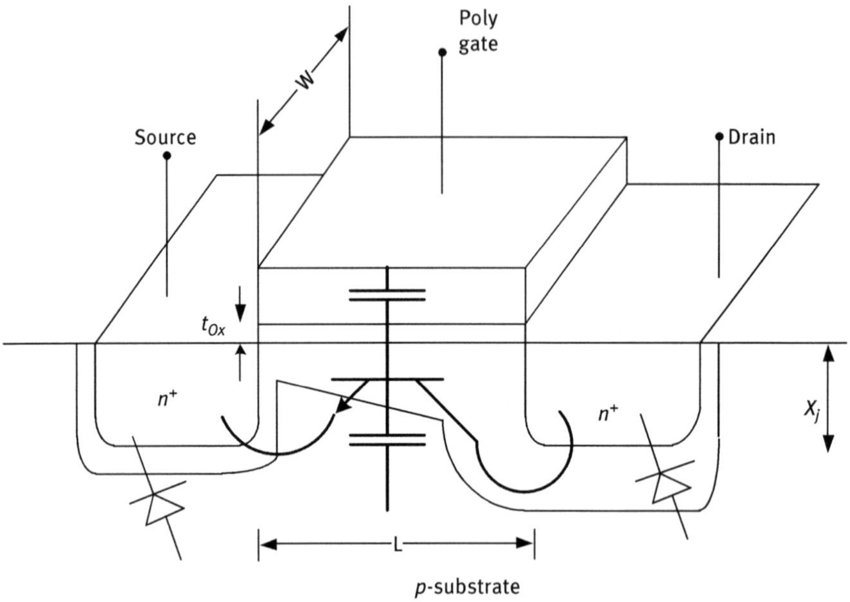

Figure 1.11: Parasitic BJT present in the MOSFET.

The value of I_0 depends on the process. For a particular technology $I_D = 0.3\ \mu A/\mu m$ which means a 1-μm-wide device will have $I_D = 0.3\ \mu A$ when $V_{GS} = V_{Th}$.

Therefore, the subthreshold current increases exponentially with surface potential. On a log plot such as Figure 1.9 the subthreshold current appears as a straight line. The inverse of the slope of that line is called "inverse subthreshold slope", "subthreshold swing" or, more simply, "subthreshold slope". It is expressed in millivolts per decade, which means "How many millivolts should the gate voltage be increased to increase the drain current by a factor 10?". The lower the value of the subthreshold slope, S, the more efficient and rapid the switching of the device from the off state to the on state.

This is an important parameter that allows quick estimation of subthreshold current in the so-called subthreshold slope factor S.

It tells us how much change in the gate control voltage (ΔV_{GS}) gives a ten times (10×) change in current. By definition the subthreshold slope is given by

$$S = \frac{dV_G}{d\log(I_D)}$$

changing the logarithm base to the natural logarithm base

$$S = \frac{\ln(10)}{\frac{d\ln(I_D)}{dV_G}}$$

since

$$I_D \propto \exp\left(\frac{qV_{GS}}{nKT}\right), \tag{1.4}$$

$$S = \frac{nkT}{q}\ln(10). \tag{1.5}$$

$n = 1$ gives $S = 60$ mV/decade (ideal BJT); $n = 1.3$ gives $S = 80$ mV/decade at room temperature. S becomes about 100 mV/decade at high temperature because kT/q increases with temperature.

For example, consider $V_{Th} = 500$ mV, $I_0 = 0.3$ μA/μm and $S = 100$ mV/decade. It indicates that current at $V_{GS} = 0$ is five decades lower, i.e. 3 pA/μm.

Now consider $V_{Th} = 100$ mV, $I_D = 0.3$ μA/μm and $S = 100$ mV/decade. It indicates that current at $V_{GS} = 0$ is 5 decades lower, i.e. 30 nA/μm.

Suppose we have 10 million transistors of width 10 μm resulting in a subthreshold current

$I_{Subthreshold} = 3$ A,

which produces huge power dissipation. This is why we cannot make V_{Th} arbitrarily small.

1.12 Subthreshold Leakage and Voltage Limits

The subthreshold current of a transistor is typically described by the following equation:

$$I_{Off} \approx a\frac{1}{L_{Eff}}\exp\left(\frac{q\left(V_G - V_{Th}\right)}{kT}\right),$$

where a is a constant, L_{Eff} is the effective gate length, V_G is the gate voltage, V_{Th} is the threshold voltage and kT/q is the thermal voltage.

In a typical scaling scenario, the electric fields are kept constant in the device by shrinking all the voltages and dimensions by the same factor. All doping levels are increased by the same scaling factor. As I_{Off} increases exponentially when V_{Th} decreases, however, static power consumption sets a lower limit to the scaling down of threshold voltages of the transistors. As the dynamic performance is directly related to the V_{DD}/V_{Th} ratio, the power supply voltage also does not scale down easily. Consequently, in the ITRS roadmap scenario, supply voltages do not shrink as rapidly as device dimensions. This results in a higher electric field in the device that has to be handled at the device level. Another consequence is the lower benefit granted to the dynamic power consumption which is proportional to V_{DD^2}.

Thus it is getting impossible to have simultaneously good active and leakage currents that several sets of transistors are required in advanced technologies.

1.13 Importance of Subthreshold Slope in Low Power Operation

Threshold voltage reduction increases the transistor leakage since a significant subthreshold current occurs during the off state of the transistor. This current has impact at the circuit level, since it is a fixed current contributing from all the devices in the off state. A subthreshold current of 10 nA at $V_{GS} = 0$ is insignificant for a single device, but in a 100 million transistor IC the impact on the overall power consumption is significant. Some technologies have MOSFETs with two different V_{Th}'s to solve the problem. The high speed devices with lower V_{Th} contribute significant higher leakage and are used in critical path where speed is important. Circuits where speed is not important are designed with higher V_{Th} transistors, reducing the overall power leakage.

Nowadays, there is a renewed interest in exploring devices that use tunneling for their on-current. In particular, there is a focus on devices which act as field-effect transistors (FETs), where a change of gate voltage turns the current on and off, but which use band-to-band tunneling in their on state, as well as in the transition between the off and on states. These devices have the potential for extremely low off current and present the possibility to lower the subthreshold swing beyond the 60 mV/decade limit of conventional MOSFETs. Therefore, they seem well adapted to be candidates for an ultimately scaled quasi-ideal switch. One such reported device is the tunnel FET (TFET) that incorporates a delta-layer of Si-Ge at the edge of the p+ region, in order to reduce the barrier width and, thereby, improve the subthreshold swing and on-current. Another is the carbon nano-tube TFET, which uses two independently controlled gates to change the energy bands in the channel.

1.14 Ultralow Voltage Operation

Power consumption is a critical issue in modern-day IC design. Ideal technology scaling reduces energy in the third order as shown in eq. (1.6):

$$\text{Delay} = \frac{1}{f} = \frac{C_S V_{DD}}{I_{Dsat}} \propto \frac{V_{DD}}{(V_{DD} - V_{Th})^{1.3}}; \text{ Power} \propto f V_{DD^2}; \text{ Energy} \propto C_S V_{DD^2}. \tag{1.6}$$

However, it is important to remember that saving energy by scaling supply voltage has not proven to be useful since (a) in modern-day IC's switching energy is no longer the chief contributor of total energy consumption due to large increase of subthreshold and gate leakage component and (b) supply voltage scales in a slow linear manner.

As supply voltage scales, quadratic to exponential savings in switching, subthreshold and gate leakage energy is theoretically expected. However, it is important to remember that supply voltage scaling affects the performance of the circuit as shown in eq. (1.6). Therefore, supply voltage is lowered to a certain value where circuit can finish their work within a stipulated deadline – a technique known as dynamic

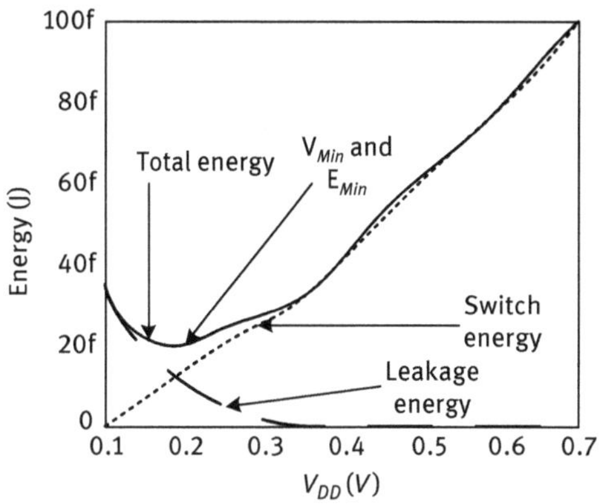

Figure 1.12: V_{Min}/E_{Min} curve.

scaling. This lower limit of the supply voltage usually lies well above the threshold voltage [8, 9].

However, in an ultralow voltage operation, supply voltage is further scaled down to threshold voltage to maximize the energy efficiency as CMOS gates can fulfill their functionality at these voltages [10]. Recently, many researchers have demonstrated successful operation of CMOS ICs with several hundreds of millivolt by achieving several orders of energy efficiency [11–13].

A peak energy-efficient point is required to be defined for ultralow power operation. Zhai et al. [14] and Calhoun and Chandrakasan [15] have shown that energy efficiency diminishes if supply voltage is scaled to a very low value. This can be primarily attributed to the fact that higher leakage energy is consumed by increasingly slow circuits, which offsets savings achieved in switching energy. As a result, total energy consumption starts to increase after a minimum is reached, referred to as V_{Min}, and corresponds to a minimum energy of E_{Min} as shown in Figure 1.12.

1.15 Low Power Analog Circuit Design

Low power analog circuit design is a complex task involving multiple trade-offs between power consumption, speed, linearity, transistor dimension, etc. For advanced deep submicron devices, increasing SCEs further complicates the design process. With aggressive downscaling of MOS devices, as SCEs becomes more prominent, the quest for accurate device models intensifies. In digital domain, the most powerful commercial circuit simulators rely on sophisticated compact models for analysis and design of digital circuits. However, in analog domain, due to the lack of available compact model, even the most experienced analog designer still relies on the hand calculation, prior to the simulation. However, for bulk MOSFETs, BSIM6 and other device compact models are available for designing low power analog circuitry.

The design of CMOS analog circuits like OP-AMPs takes advantage of working in the subthreshold or weak inversion region to compromise bandwidth, size and gain. For instance when a transistor operates in saturation region, it consumes more power to meet the specification. Therefore, the region of operation is another important design aspect.

There are several advantages of subthreshold region of operation, which can be listed as follows:

(1) Possibilities of achieving higher gain [16–19]
(2) Guaranteed low power consumption
(3) Reduced distortion and improvement in linearity than saturation region [17]
(4) Increased output resistance

However, the major difficulty with subthreshold region of operation is the reduction in the circuit bandwidth and resulting limited frequency of operation. However, it is worth mentioning that by optimizing the device structure to reduce intrinsic gate capacitances (C_{Gs} and C_{Gd}) frequency of operation can be increased.

1.16 Fundamental Consequence of Lowering Supply Voltage

From the views of plain physics, power consumption in analog circuits is proportional to the signal integrity (signal-to-noise ratio). In simple words, in an analog circuit, high power is invested to result in higher performance. It is important to remember that for a given power budget, performance degrades as supply scales down. In Figure 1.13 the power consumption of a unity-gain buffer is plotted against supply voltage for different technology nodes. Figure 1.13 indicates that minimum power consumption increases with decreasing supply voltage for performance to be remain constant, although it is worth mentioning that at a constant supply voltage, migration to new technology node lowers the power consumption.

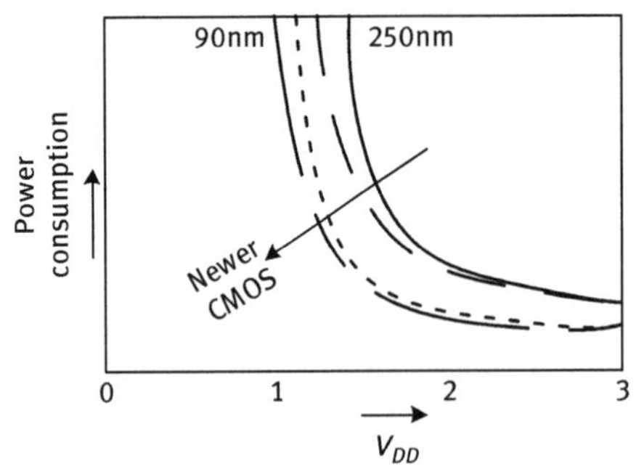

Figure 1.13: Minimum power consumption for a unity-gain buffer analog circuit with fixed topology and constant performance as a function of the supply voltage, for four technologies.

As CMOS transistors are scaled down in the nanometer range, various issues known as SCEs are causing degradation of transistor performance. These effects need to be mitigated in order to continue the historical cadence of downscaling. Proper circuit-level design with new circuit techniques was reviewed to provide leverage for improving energy efficiency of future downscaled analog circuits.

The drain-to-source saturation voltage and threshold voltage do not scale in the same manner as the supply voltage. As a result, analog designers face difficulties due to available limited voltage headroom. Some high performance analog circuits normally working under high voltage lost their performance validity in low voltage operation. Low voltage operation has also laid down several limitations on the sampled-data circuits such as switched-capacitor operation.

1.17 Analog MOS Transistor Performance Parameters

(i) Cutoff frequency is considered as one of the most important MOS device performance parameters, which is given by

$$f_t = \frac{g_m}{2\pi\left(C_{Gs} + C_{Ds}\right)}, \tag{1.7}$$

where g_m represents transconductance and C_{Gs} and C_{Ds} represent gate-to-source and drain-to-source capacitances, respectively.

This is one of the few parameters that improves with scaling. The maximum cutoff frequency is given by [20]

$$f_{t,Max} = \frac{v_{Sat}}{2\pi\, L_{Eff}}. \tag{1.8}$$

The maximum cutoff frequency is limited by saturation velocity of the channel carriers and effective distance between the source and the drain terminal.

However, it is worth mentioning that transistors generally operate much below than $f_{t,Max}$, due to the presence of parasitic capacitances.

(ii) Transconductance Generation Factor (TGF) is considered as the figure of merit to measure the efficiency to translate current into transconductance. A lower TGF indicates reduced input device ability and higher power dissipation [21]. It characterizes the device current efficiency and to obtain a certain value of conductance is given by

$$\mathrm{TGF} = g_m/I_d. \tag{1.9}$$

(iii) Transistor intrinsic gain is given by

$$A_V = \frac{g_m}{g_{Ds}}, \tag{1.10}$$

where g_{Ds} is the channel conductance. This is one factor that degrades with scaling and is considered as a design challenge for future analog circuit design with downscaled CMOS technology.

(iv) The product of g_m/I_D and f_T represents a trade-off between power and bandwidth and is utilized in moderate to high speed designs. The intrinsic gain (g_m/g_{Ds}) is a valuable figure of merit for operational transconductance amplifier. To comprise these aspects of analog/RF circuit design, a unique figure of merit, the gain transconductance frequency product, is proposed and given by [22]

$$\text{GTFP} = \left[\frac{g_m}{g_{Ds}} \cdot \frac{g_m}{I_{Ds}} \cdot f_T\right]. \tag{1.11}$$

(v) Nonlinearity of a CMOS analog circuit depends on the nonlinearity of the device drain current. Linearity is an essential requirement in all RF system in order to ensure minimal intermodulation and higher order harmonics at the output of RF front-end stages [23]. Traditional method of achieving linearity involves complex circuit design methods [24] and/or requires operation of the device in the velocity saturation regime [25], which implies high supply voltages and large power consumption – a scenario that is not ideal for portable and low power applications. For a MOSFET, transconductance and output conductance are major causes of nonlinearity. Linearity is directly proportional to transconductance and is inversely proportional to the second derivative of the transconductance [23] which indicates that devices with constant transconductance versus gate-voltage curves, and small variations over a specific voltage range, are more linear. In the following analysis of linearity for MOSFET, g_{m1}, g_{m2} and g_{m3} are given by

$$g_{m1} = \frac{\partial I_{Ds}}{\partial V_{GS}}, \quad g_{m2} = \frac{\partial^2 I_{Ds}}{\partial V_{GS^2}}, \quad g_{m3} = \frac{\partial^3 I_{Ds}}{\partial V_{GS^3}}. \tag{1.12}$$

VIP_2 and VIP_3 represent the extrapolated gate-voltage amplitudes at which the second- and third-order harmonics, respectively, become equal to the fundamental tone in the device drain current (I_D). These are the suitable FOMs, which can properly determine the distortion characteristics from DC parameters; to achieve high linearity and low distortion operations, these should be as high as possible. VIP_2 and VIP_3 represent the extrapolated gate-voltage amplitudes at which the second- and third-order harmonics, respectively, become equal to the fundamental tone in the device drain current (I_D) and are given by [26, 27]

$$\text{VIP}_2 = \left[4 \cdot \left(\frac{g_{m1}}{g_{m2}}\right)\right]_{V_{DS}=\text{Constant}}; \tag{1.13}$$

$$\mathrm{VIP}_3 = \left[\sqrt{24 \cdot \left(\frac{g_{m1}}{g_{m3}}\right)}\right]_{V_{DS}=\mathrm{Constant}}. \tag{1.14}$$

IMD_3 determines the distortion performance of a device, which should be low for minimization of distortion and is given by [22]

$$\mathrm{IMD}_3 = R_S \cdot \left\{4.5 \cdot (\mathrm{VIP}_3)^3 \cdot g_{m3}\right\}^2. \tag{1.15}$$

IIP3 is another important FOM which evaluates the linearity performance and is given by [22, 27, 28]

$$\mathrm{IIP}_3 = \frac{2 \cdot g_{m1}}{3 \cdot g_{m3} \cdot R_S} = \frac{2 \cdot \left(\frac{\partial I_D}{\partial V_{GS}}\right)}{3 \cdot R_S \cdot \left(\frac{\partial^3 I_D}{\partial V_{GS}}^3\right)}. \tag{1.16}$$

1 dB compression point is considered as a reliable measure of linearity evaluation at the onset of distortion and is given by [27, 28]

$$1\ \mathrm{dB\ compression\ point} = 0.22\sqrt{\frac{g_{m1}}{g_{m3}}}. \tag{1.17}$$

The 1 dB compression point indicates the power level that causes the gain to drop by 1 dB from its small signal value. This parameter is important for an amplifier circuit as it gives an idea about the maximum input power that the circuit can handle by providing a fixed amount of gain.

Summary

Physical dimensions of MOSFET devices are being continuously scaled down over the past four decades. This rapid cadence of MOSFET downscaling is accelerating introduction of new technologies to extend MOS scaling beyond the 100 nm node. This acceleration simultaneously requires an intense study of SCEs and their remedies in order to improve the performance and to sustain the historical cadence of miniaturization. The emphasis of this dissertation is on incorporating the recent advances in unconventional MOS device structures to determine the minimum acceptable channel length and to circumvent SCEs, considered as the most daunting roadblock for sub-100 nm MOSFET scaling. This chapter provides the incentive and guide for further research and experimental exploration of the unique features of MOSFET scaling beyond 100 nm and demonstrates a new way of engineering deep submicron MOSFETs with the focus on uncovering the potential of novel unconventional MOSFET structures in the context of design of digital logic or RF/analog ICs. This provides an incentive for circuit simulation involving unconventional device structures for next-generation ULSI circuits.

References

[1] "1965 – 'Moore's Law' Predicts the Future of Integrated Circuits". Computer History Museum. http://www.computerhistory.org/semiconductor/timeline/1965-Moore.html

[2] R. Dennard, "Design of ion-implanted MOSFETs with very small dimensions." IEEE Journal of Solid State Circuits, Vol. 87, Issue 4 (1974): 256.

[3] Y. Taur, "CMOS design near the limit of scaling." IBM Journal of Research and Development, 46:2/3 (2002): 213–222.

[4] D.D. Buss, "Technology in the internet age." IEEE International Solid-State Circuits Conference (ISSCC) Digest of Technical Papers, 2002, pp. 18–21.

[5] A.J. Annema, B. Nauta, R. van Langevelde and H. Tuinhout, "Designing outside rails constraints." IEEE International Solid-State Circuits Conference (ISSCC) Digest of Technical Papers, 2004, pp. 134–135.

[6] 20 nm Design – How this advanced technology node will transform SoCs and EDA, white paper, http://www.cadence.com/cadence/Pages/downloads.aspx?cid=2

[7] Kin P. Cheung, "On the 60 mV/dec@ 300 K limit for MOSFET subthreshold swing." IEEE International Symposium on VLSI Technology Systems and Applications (VLSI-TSA), 2010, 2010.

[8] R.J. Widlar, New developments in IC voltage regulators. IEEE Journal of Solid-State Circuits, 6.1 (1971): 27.

[9] IBM PowerPC. IBM PC. http://www.chips.ibm.com/products/powerpc.

[10] J.D. Meindl and J.A. Davis, The fundamental limit on binary switching energy for terascale integration (tsi). IEEE Journal of Solid-State Circuits, 35.10 (2000): 1515–1516.

[11] A. Wang and A. Chandrakasan, A 180 mV FFT processor using subthreshold circuit techniques. IEEE International Solid-State Circuits Conference, 2004. Digest of Technical Papers. ISSCC, Vol. 1, 2004, pp. 292.

[12] S. Hanson, Bo Zhai, Mingoo Seok, B. Cline, K. Zhou, M. Singhal, M. Minuth, J. Olson, L. Nazhandali, T. Austin, D. Sylvester and D. Blaauw. Performance and variability optimization strategies in a sub-200 mv, 3.5 pj/inst, 11 nw sub-threshold processor. 2007 IEEE Symposium on VLSI Circuits, June 2007, pp. 152–153.

[13] Joyce Kwong, Yogesh Ramadass, Naveen Verma, Markus Koesler, Korbinian Huber, Hans Moormann and Anantha Chandrakasan, A 65 nm sub-vt micro-controller with integrated SRAM and switched-capacitor dc-dc converter. IEEE International Solid-State Circuits Conference, 2008, p. 318.

[14] Bo Zhai, David Blaauw, Dennis Sylvester and Krisztian Flautner. Theoretical and practical limits of dynamic voltage scaling. Proceedings of the 41st annual conference on design automation (DAC '04), 2004, pp. 868–873.

[15] B.H. Calhoun and A. Chandrakasan, Characterizing and modeling minimum energy operation for subthreshold circuits. Proceedings of the 2004 International Symposium on Low Power Electronics and Design, 2004 (ISLPED '04), 2004, pp. 90–95.

[16] Phillip E. Allen and Douglas R. Holberg, CMOS analog circuit design, Oxford University Press, United Kingdom, 2002.

[17] David J. Comer and Donald T. Comer, "Using the weak inversion region to optimize input stage design of CMOS op amps." IEEE Transactions on Circuits and Systems II: Express Briefs 51.1 (2004): 8–14.

[18] Paul R. Gray, Paul J. Hurst, Stephen H. Lewis, Robert G. Meyer *Analysis and Design of Analog Integrated Circuits*. Wiley, New York, 2001.

[19] Yannis Tsividis, *Mixed Analog-Digital VLSI Devices and Technology*. World Scientific, Singapore, 2002.

[20] C.T. Kirk Jr, "A theory of transistor cutoff frequency (fT) falloff at high current densities." IRE Transactions on Electron Devices, 9.2 (1962): 164–174.

[21] N. Mohankumar, B. Syamal and C.K. Sarkar, "Influence of channel and gate engineering on the analog and RF performance of DG MOSFETs." IEEE Transactions on Electron Devices, 57 (2010): 820.

[22] S. Kaya and W. Ma, "Optimization of RF linearity in DG MOSFET." IEEE Electron Device Letters, 25.5 (2004): 308–310.

[23] Behzad Razavi and Razavi Behzad. *RF Microelectronics*. Vol. 2. Prentice Hall, New Jersey, 1998.

[24] J. Chen and B. Shi, "Novel constant transconductance references and comparisons with the traditional approach." Southwest Symposium on Mixed-Signal Design, 2003, pp. 104–107.

[25] H. Wong, K.F. Man and M.C. Poon, "Modeling of saturation transconductance for short-channel MOSFETs." Solid-State Electronics, 39.9 (1996): 1401–1404

[26] P.H. Woerlee, R. van Langevelde, A.H. Montree, D.B.M. Klaassen, L.F. Tiemeijer and P.W.H. de Vreede, "RF-CMOS performance trends." Proceedings of the 30th European Solid-State Device Research Conference, 2000, 11–13 September 2000, pp. 576, 579.

[27] P. Ghosh, S. Haldar, R.S. Gupta and M. Gupta, "An investigation of linearity performance and intermodulation distortion of GME CGT MOSFET for RFIC design." IEEE Transactions on Electron Devices, 59.12 (2012): 3263, 3268.

[28] Wei Ma and Savas Kaya, Impact of device physics on DG and SOI MOSFET linearity. Solid-State Electronics, 48.10–11 (2004): 1741–1746.

2 Scaling and Short Channel Effects in MOSFET

2.1 MOSFET Scaling

Scaling is the process of reducing the device dimensions while keeping the electrical characteristics constant [1]. The main problem with miniaturization is the direct dependence of electrical characteristics on physical parameters. As a result many nonideal effects hinder the performance characteristics of devices. In case of constant field scaling, the depletion region, internal fields, currents, capacitances and all dimensions are scaled by a factor k. The main drawback of this scaling scheme is that it is often not possible to scale parameters in the required proportions. Constant field scaling is only approximated and not followed exactly.

In the constant electric field scaling, the source voltages are decreased by a scale factor k. Constant voltage scaling is a more practical application of the more ideal method of constant electric field scaling [2–5]. An important drawback of this technique is tehat by not scaling the supply voltage, higher fields are created in the device. As a result, mobility degradation, hot carrier effects and other reliability problems gain prominence.

A third technique is constant electrostatic scaling in which device dimensions are reduced by the same factor k, but potentials are reduced by a different factor $\lambda = k^{0.5}$ [5]. In Ref. [6], off-current scaling is more complex in practice than the other techniques. In this case the doping profile characteristics are changed as in Refs [4–6]. All scaling methods replicate long channel behavior in a short channel device. No scaling technique provides an exact solution, and designing a device requires many iterations and experience on the part of the designer. The best technique may be the combining of one of the first three methods with one or more of the latter two.

2.2 International Technology Roadmap for Semiconductors

ITRS stands for International Technology Roadmap for Semiconductors. Semiconductor industry has been improving by reduced dimensions, leading to more transistors per chip and faster functioning [7]. These scaling trends are enabled and preceded by various research and development programs. National Technology Roadmap for Semiconductors was started by Semiconductor Industry Association (SIA). In 1998 Europe, Japan, Korea and Taiwan joined with SIA and modified it to ITRS.

Europe, Japan, Korea, Taiwan and the United States of America are the five main semiconductor chip manufacturing industries in the world. The top manufacturing organizations that sponsor ITRS are the European Semiconductor Industry Association (ESIA), the Japan Electronics and Information Technology Industries Association (JEITA), the Korean Semiconductor Industry Association (KSIA), the Taiwan

Semiconductor Industry Association (TSIA) and the United States Semiconductor Industry Association (USSIA).

So ITRS is a set of documents that are produced by the group of semiconductor industry experts mentioned above. The documents cover factory integration, assembly and packaging, system drivers design, modeling and simulation, microelectromechanical systems (MEMS), emerging research devices and so on.

2.3 Gate Oxide Scaling

Reduction of device dimensions is called scaling [8, 9]. There are two types of scaling: constant field scaling and constant voltage scaling. In case of the latter, the terminal voltages are kept constant but the device dimensions are reduced. So the gate lengths, oxide thickness, width of the device and so on are reduced by same amount in case of constant voltage scaling. As the oxide thickness is reduced, the drain current increases. Also hot carrier effect is reduced as a result.

2.4 Gate Leakage Current

As the oxide thickness is decreased, the leakage current through gate oxide increases as in Figure 2.1. This causes increase in power dissipation. The cell phones require lower current for longer battery lifetime. DRAM can be used for lower oxide leakage current. However for DRAM, the oxide thickness cannot be below 3 nm, and below 0.1 nm scaling is difficult.

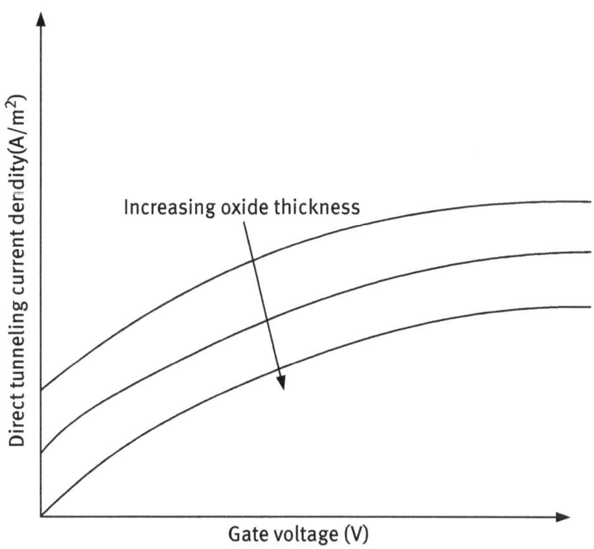

Figure 2.1: Change in tunneling current density with varying oxide thickness.

2.5 Mobility

When the oxide thickness is decreased, the vertical electric field due to gate bias increases and carriers are pulled toward the oxide–silicon interface. The mobility of carriers reduces as a result of increased surface scattering. The scattering reduces the carrier transport efficiency.

2.6 High-*k* Gate Dielectrics

As the SiO_2 thickness is reduced with scaling, the gate tunneling current increases and the undesirable hot carrier effect increases. To reduce the gate leakage current high-*k* gate dielectrics are replacing the SiO_2 layer [10]. As the dielectric constant of high-*k* materials is high, in order to keep same gate capacitance, the thickness of the material needs to be increased (Figure 2.2). So the leakage current decreases using high-*k* materials.

2.7 Key Guidelines for Selecting an Alternative Gate Dielectric

The choice of high-*k* material to replace SiO_2 as gate dielectric is based on the following key factors:
(1) Leakage current through the material
(2) Reliability of the material
(3) Quality of Si/material interface
(4) Thermodynamic stability with respect to silicon
(5) Material compatibility

2.8 Materials

The research is on for a suitable high-*k* material to replace SiO_2 as gate dielectric. It is seen that both HfO_2 and ZrO_2 are stable with silicon substrate and hence they replace

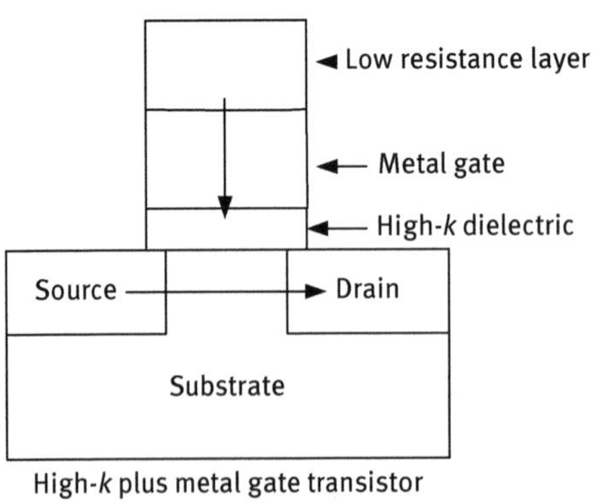

Figure 2.2: MOS transistor with high-*k* gate oxide and metal gate.

TiO_2 and Ta_2O_5 as the dielectric material [11, 12]. Recently amorphous dielectric replaces the polycrystalline form based on the uniformity argument. However, for the amorphous dielectric, $HfAlO_2$ replaces HfO_2 since HfO_2 crystallizes at relatively lower temperature.

2.9 Gate Tunneling Current

As the oxide thickness is reduced due to scaling of MOSFET, the current tunneling through the oxide increases as a result of increased vertical electric field for a constant gate bias [13, 14]. The gate current in Figure 2.3 increases exponentially with decrease of oxide thickness. The gate tunneling current is very important for designing VLSI circuits.

2.10 Gate Length Scaling

Reduction of device dimension is very important for increasing the number of transistors within a particular chip area. In case of gate length scaling, parameters related to gate length are changed. If k be the scaling factor (Figure 2.4) such that $k < 1$, for the gate length scaling, length of gate reduces from L_g to kL_g after scaling. Also the gate width Z reduces to kZ.

2.11 Introduction to Short Channel Effect in MOSFET

A MOSFET device is called short when the length of the channel is of the same order as the depletion layer width at the source and the drain ends respectively. The various short channel effects are as follows:

(1) Reduction in the threshold voltage
(2) Drain-induced barrier lowering (DIBL)

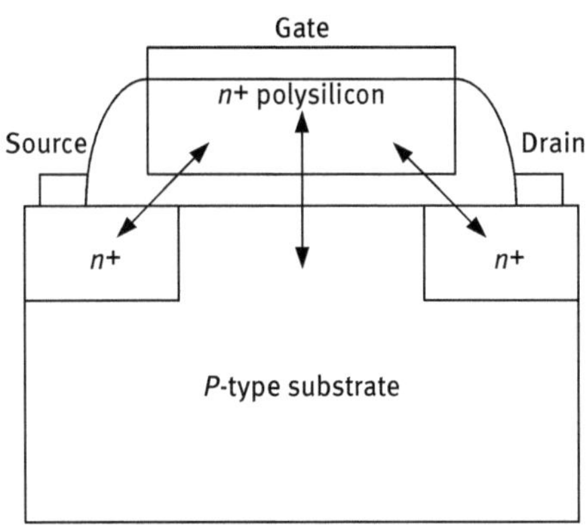

Figure 2.3: The gate tunneling current through the gate oxide layer.

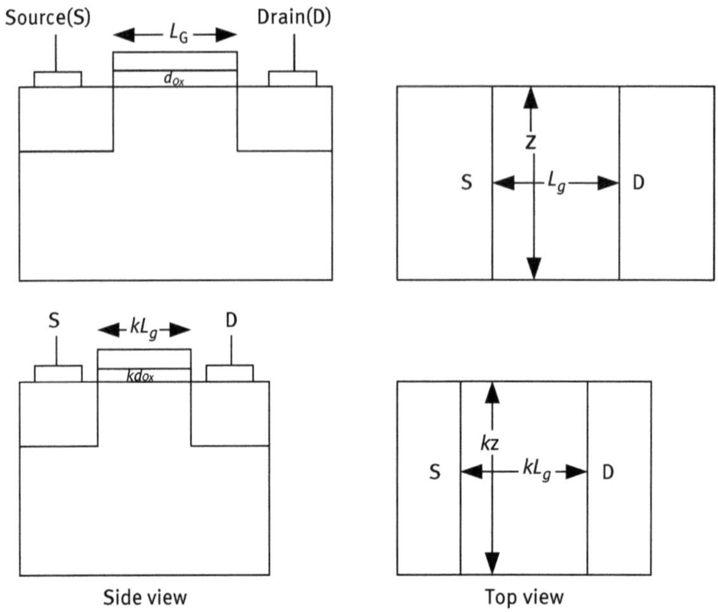

Figure 2.4: The side view and top view of gate length and width scaling.

(3) Mobility degradation and surface scattering
(4) Hot carrier effect
(5) Punch-through effect
(6) Increase in the off-state leakage current

2.11.1 Reduction of Effective Threshold Voltage

The threshold voltage stands for the minimum value of the gate-to-source voltage for creating the inversion layer in the channel region under the gate. As the channel length is decreased with scaling, the number of electrons from the $n+$ source and drain regions accumulating under the gate increases and hence the effective threshold voltage reduces with scaling [15]. This effect can be explained by a charge sharing mechanism as in Figure 2.5. Due to this mechanism it is easier for the gate to deplete the channel of mobile charges. The charge sharing effect is more important as the channel length is decreased.

2.11.2 Drain-induced Barrier Lowering

It is seen that with reduction of channel length, the threshold voltage reduces with scaling. Also as the drain-to-source voltage increases, the off-state current increases as in Figure 2.6. This effect is called DIBL [16].

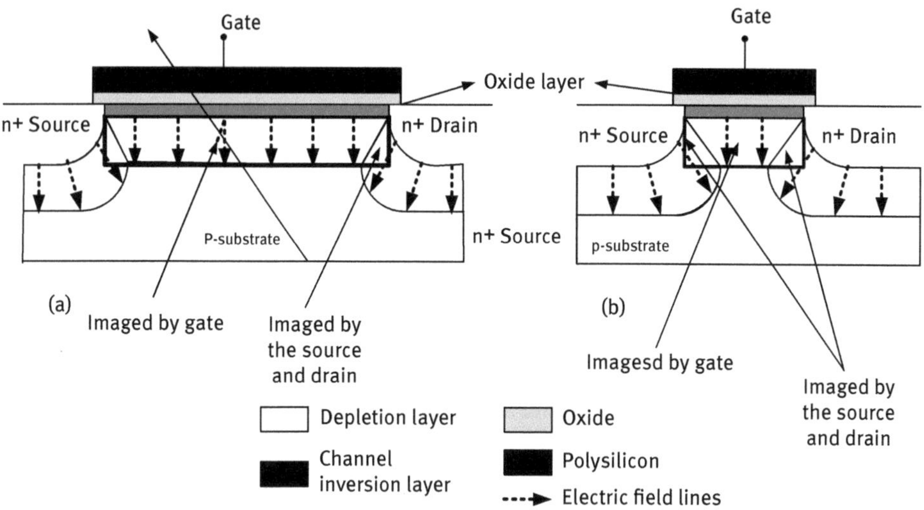

Figure 2.5: Illustration of the threshold voltage related to short channel effects and charge sharing between the source/drain depletion regions and the channel depletion region.

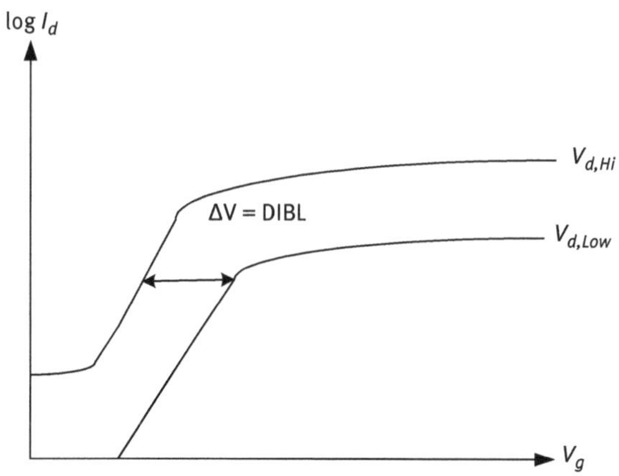

Figure 2.6: Current versus gate voltage plot at low and high drain biases, demonstrating DIBL.

When a drain voltage is applied, the DIBL effect is caused as a result of decrease of the barrier potential at the source end as in Figure 2.7.

A measure of DIBL is

$$\text{DIBL} = \frac{V_{Th,Lin} - V_{Th,Sat}}{V_{DD} - V_{d,Lin}}$$

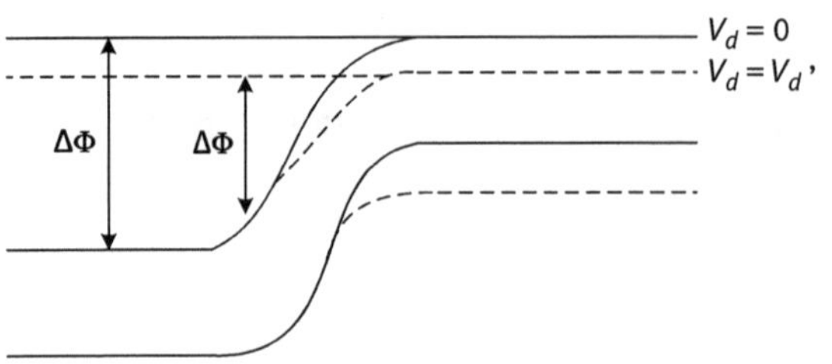

Figure 2.7: The energy band diagram at the source end of an nMOS device with and without an applied drain bias.

where V_{DD} is the supply voltage, $V_{d,Lin}$ is the linear drain voltage and $V_{Th,Lin}$ and $V_{Th,Sat}$ are the threshold voltages in the linear and the saturated operations, respectively.

2.11.3 Mobility Degradation and Surface Scattering

Mobility of a carrier is defined as the average drift velocity of the carrier per electric field.

Let the drift velocity = V_d and electric field = E.

So the mobility $u = V_d/E$

The reduction of mobility of a MOSFET is due to the following two electric fields:

(a) Vertical electric field
(b) Horizontal electric field

2.11.3.1 Vertical Electric Field Mobility Degradation

When a positive gate voltage is applied to an *n*-channel MOSFET, a vertical electric field is created as in Figure 2.8. Due to a positive drain-to-source voltage, electrons move from the source to the drain end. Due to the vertical electric field, electrons are attracted toward the oxide–semiconductor interface which is rough. As a result carriers lose mobility. The electrons moving from the source to the drain end undergo surface scattering as a result of which mobility is degraded to a large extent. The smaller the channel, the more is the surface scattering and lower is the mobility.

2.11.3.2 Horizontal Electric Field Mobility Degradation

The drain voltage induced electric field E_y plays a significant role in mobility degradation when compared to the gate voltage induced electric field, as a result of velocity saturation [17–20].

Carrier velocity $v \alpha E_y$

So $v = \mu_s E_y$ where μ_s = Mobility of surface electrons (Figure 2.9). When V_{DS} is small, velocity saturation of carriers takes place for short channel MOSFETs. Horizontal mobility degradation takes place with horizontal electric field.

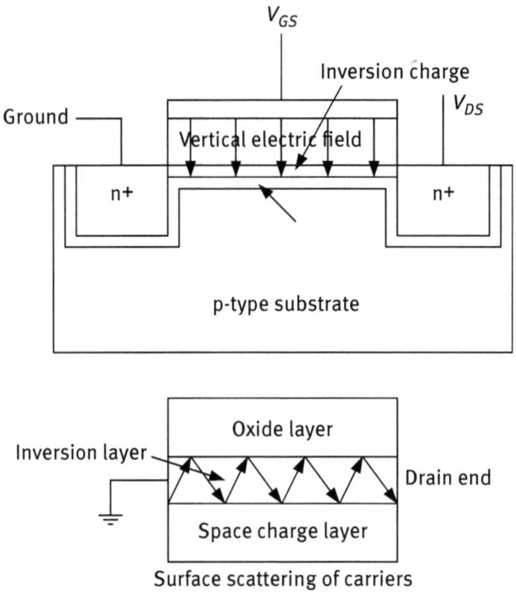

Figure 2.8: Vertical electric field in a short channel MOSFET and due to that surface scattering.

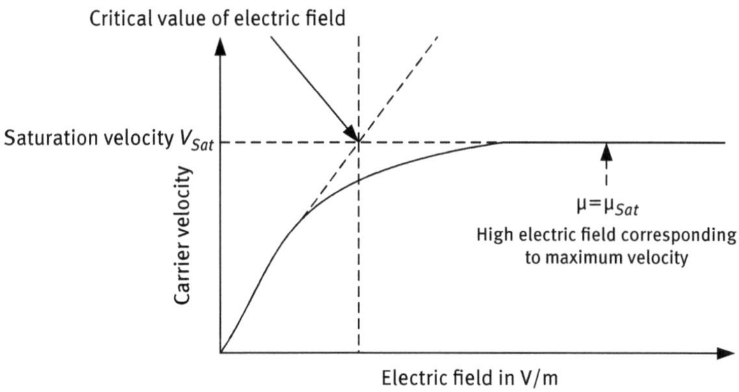

Figure 2.9: Electric field versus carrier velocity.

A model for the horizontal mobility μ_H is given as

$$\begin{aligned} u_H &= u_0/[1+\{V_{DS}/(L_{Effecrit})\}] \\ &= u_0/[1+(m_1 V_{DS})] \end{aligned}$$

$1/L_{Effecrti}$ = the drain bias mobility reduction parameter and often denoted as m_1. The electric field E_{Crit} is shown in Figure 2.9. For a large transistor, $m_1 < 1$ and $u_H = u_0$.

2.11.4 Surface Scattering

As the channel length is reduced in case of constant voltage scaling, the horizontal and vertical components of the electric field increase and hence the surface scattering effect increases. This causes reduced value of mobility of electrons in the inversion layer.

2.11.5 Hot Carrier Effect

Hot carrier stands for high energy electrons or holes that are accelerated due to the high horizontal and vertical electric fields due to the gate and drain bias [17–20]. In case of constant voltage scaling, as the channel length is reduced, the oxide thickness also decreases, keeping the terminal voltages constant. So the vertical electric field increases due to scaling (Figure 2.10). The electrons are attracted toward the oxide–silicon interface. Some of these high energy electrons overcome the oxide-silicon potential barrier and get trapped in the oxide. These electrons are called hot electrons and they degrade the performance of oxide (Figures 2.11 and 2.12).

2.11.6 Punch-through Effect

As the channel length is decreased with scaling, the depletion layers under the source and drain overlap to form a single depletion layer. This causes a very large current to flow from the source to the drain with increasing drain bias (Figure 2.13).

2.11.7 Velocity Saturation Effect

As the dimension of MOSFET is reduced, the channel length and the oxide thickness decrease. As a result, the horizontal and vertical electric fields increase. The drift velocity of electrons being proportional to the electric field should increase. However,

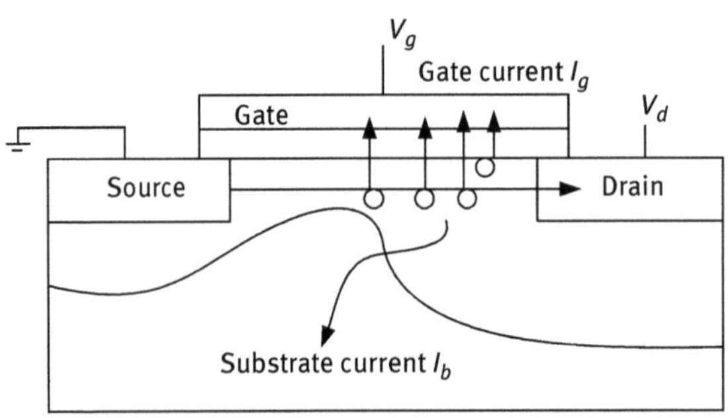

Figure 2.10: Hot carrier effect.

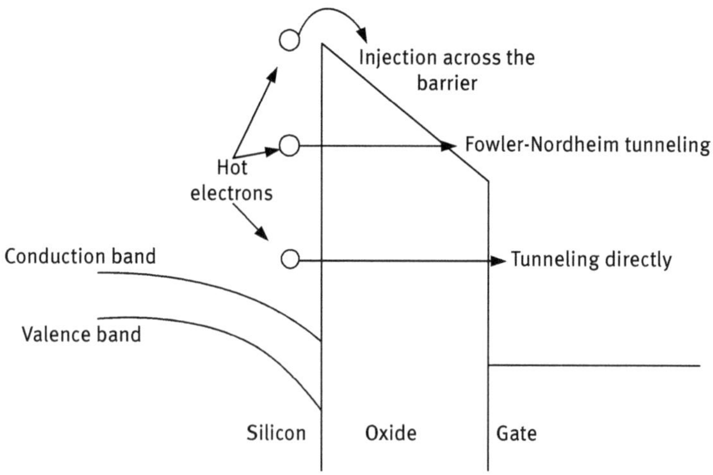

Figure 2.11: Three different types of carrier injection into the gate resulting in hot carrier effects.

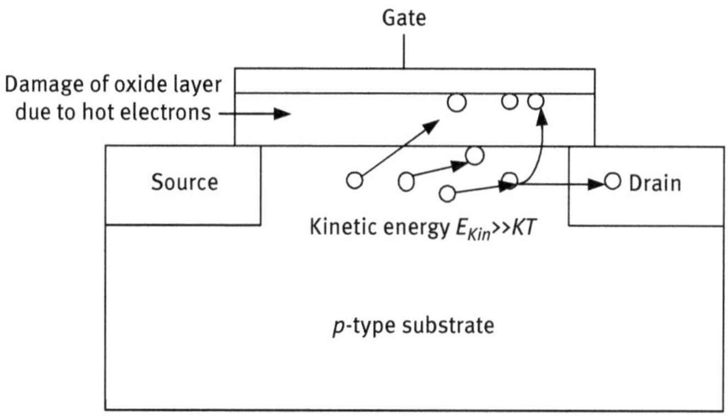

Figure 2.12: Damage of oxide due to hot carriers in short channel MOSFET.

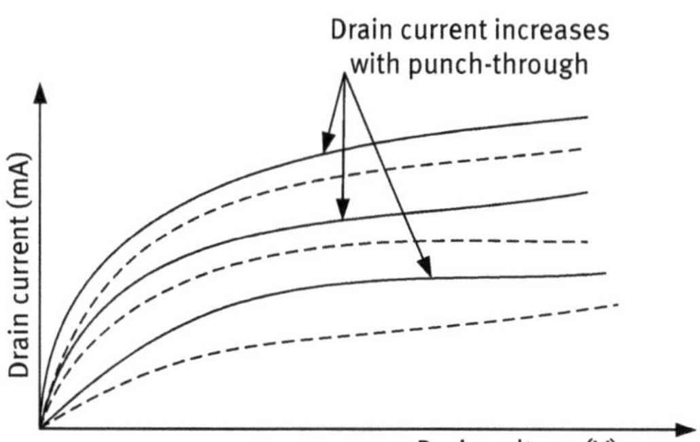

Figure 2.13: Current–voltage characteristics with and without punch-through effect.

with increase of electric fields the mobility degradation increases as in Refs [13, 15]. Due to this opposite effect, the velocity of electrons saturates at a high value.

The saturation drain current is given by

$$I_{Ds(Sat)} = W \cdot vd_{(Sat)} \cdot \int_0^{leff} q \cdot n(x)dx = W \cdot vd_{(Sat)} \cdot |Q_I|.$$

So

$$I_{Ds(Sat)} = W \cdot vd_{(Sat)} \cdot C_{Ox} \cdot V_{Dsat}.$$

2.11.8 Increase in Off-state Leakage Current

For long channel device, as distance between the source and the drain is very high, positive V_{DS} voltage is not able to attract carriers from the source under subthreshold condition. But for short channel MOSFET, positive drain-to-source voltage attracts more carriers from the source; so "OFF" state leakage current increases.

Due to the reduction of the threshold voltage for short channel MOSFET, the transistor leakage increases since a significant subthreshold current occurs during the off state of the transistor (Figure 2.14). This current has impact at the circuit level, since it is a fixed current contributing from all the devices in the off state. A subthreshold current of 10 nA at $V_{GS} = 0$ is negligibly small for a single device, but in a 100 million transistor IC the impact on the overall power consumption is huge.

2.12 Motivation for Present Research

In short channel devices, some problems (commonly called short channel effects) affect the performance of the devices. When the length of the channel decreases, the

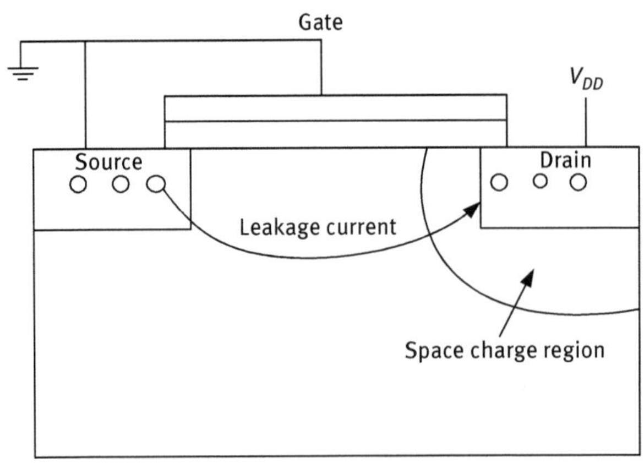

Figure 2.14: Leakage current in case of short channel MOSFET.

fraction of charge in the channel region decreases. With increase of the drain bias, the reverse biased space charge region at the drain end extends further into the channel and the gate has less charge control. In case of a short channel MOSFET, the $n+$ source and the drain regions induce an appreciable amount of the depletion charge. For a short channel device, the channel charge is shared by the gate, substrate, source and drain, called charge sharing. As a result, the source and drain depletion regions come very close to one another.

For a long channel MOSFET the threshold voltage expression thus overestimates the depletion charge supported by the gate voltage. So the gate voltage required to offset the depletion charge will be less. Hence the estimated threshold voltage value of long channel MOSFET will be greater than the actual value for a short channel device [21]. As the depletion depth increases, the surface potential increases, making the channel more attractive for electrons. So the device conduction current increases. This effect causes reduction of the threshold voltage V_{Th} as the drain current is a function of (V_{GS}-V_{Th}) where V_{GS} is the applied gate bias.

As the gate-oxide thickness is reduced, the oxide breakdown and the oxide reliability become a matter of concern. Many oxides having higher permittivity called high-k dielectrics have been considered as alternate oxides to replace silicon dioxide as gate dielectric. Avalanche breakdown is an undesirable short channel effect which occurs due to the high velocity of electrons in the presence of a large longitudinal electric field, generating electron–hole pairs by impact ionization and subsequent ionization. When the device length is reduced, the drain region moves closer to the source, and its electric field affects the whole channel. This effect is termed as drain-induced barrier lowering or DIBL, since the drain lowers the potential barrier for the flow of carriers from the source end. The threshold voltage gets lowered as a result of this. Punch-through occurs when the source and the drain depletion regions merge into a single depletion region.

Velocity saturation occurs due to the mobility reduction and is important in the submicron devices. The performance of short channel MOSFET is also affected by the velocity saturation, leading to lowering of the transconductance in the saturation mode. To sustain the continued density and performance increases, much of the attention has turned to the modified device structures and to new materials – among them high permittivity ("high-k") gate dielectrics such as hafnium oxide (HfO_2) to replace SiO_2.

2.12.1 Lightly Doped Drain Structure

It is a kind of graded structure commonly used as source/drain in submicron devices. In this case, a lightly doped n-region (n-) is first created by low energy P or As implantation and then the oxide spacers are formed at the polysilicon gate sidewall. The oxide spacers serve as a mask for the standard $n+$ As implant. The $n+$ implants diffuse under the spacers to the edges of the gate.

In an LDD structure, the channel side is doped less heavily to reduce the lateral field for the consideration of hot carrier effect. The channel extension has shallower junction depth to reduce the detrimental short channel effects. With the introduction of an *n*-region between the drain and the channel, the peak channel electric field is not only shifted toward the drain end but is also decreased to about 80% of the value for a conventional device. Due to the reduction and shift of the peak electric field inside the drain region, carrier injection into the oxide is reduced resulting in a more reliable device. This structure results in a higher breakdown voltage and lower substrate current. It is to be noted that the overlap capacitance is also reduced, resulting in a lower gate capacitance and higher speed. The raised source/drain structure is an advanced design where a heavily doped epitaxial layer is grown over the source/drain regions. The purpose is to minimize junction depth to control the short channel effects.

2.12.2 Channel Engineering Technique

The pocket doping (symmetrical and asymmetrical) at the two ends of a MOSFET can suppress the short channel effects [22–25]. The structure of single halo MOSFET is shown in Figure 2.15.

Some papers are already published focusing on the subthreshold behavior of pocket implanted MOSFET [26–29]. In Ref. [6], models for the subthreshold and the super-threshold currents in sub-100 nm pocket n-MOSFETs for low voltage applications are calculated based on the diffusion current transport equation. The double halo (DH) and the dual material gate (DMG) devices suppress the short channel effects quite efficiently [30, 31]. In case of the DH an extra pocket region at the drain end is added (Figure 2.16) along with that at the source end as in the case of single halo (LAC) devices. Both the pockets are symmetrical [32, 33]. Most significant advantage is that the heavy doping at the drain end restricts the drain field from penetration into

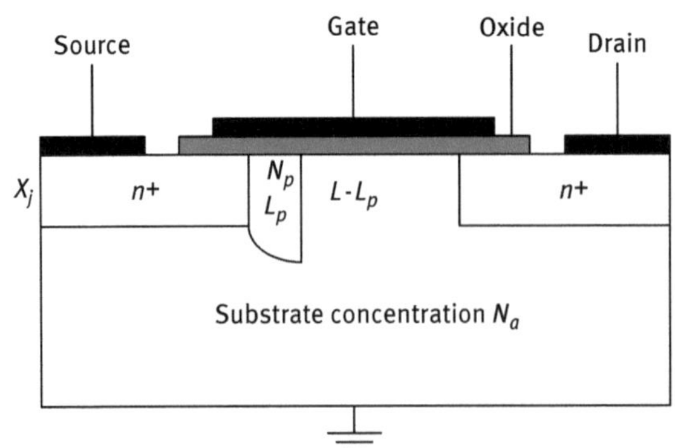

Figure 2.15: Single halo MOSFET.

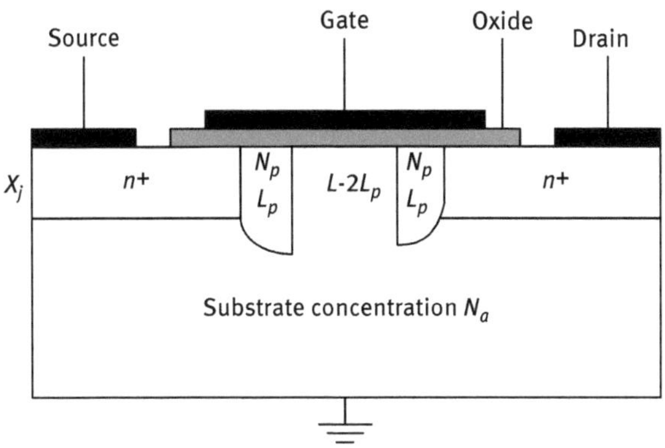

Figure 2.16: Double halo MOSFET.

the source region. Due to heavy pocket doping it can absorb more electric field lines from the drain region, as a result of which the short channel effects are reduced.

The DIBL effect can be reduced by increasing the substrate doping concentration at the edges of the source and the drain junctions, which are called halos. With decrease of channel length for halo devices, the average channel doping concentration increases [34]. So the threshold voltage increases when the gate length is reduced, called the "reverse short-channel effect". When the gate length is reduced, the short channel effect becomes dominant and the threshold voltage falls off [35]. This can be reversed by locally increasing the channel doping next to the drain or drain/source junctions. Experiments are carried out with halo structures over the last decade. It is seen that V_{Th} of all devices decreases as V_{DS} increases, but the amount of V_{Th} drop as a result of DIBL effect is very low in the pocket-implanted devices. This phenomenon is very important for a short channel device, in which the short channel behavior is unacceptable without lateral channel engineering [36–38].

2.12.3 Gate Engineering Technique

A DMG MOSFET is one in which the gate material is made of two different metals with different work functions. A step potential profile is formed in the channel and the electron drift velocity is increased as a result. However, the various short channel effects are reduced considerably [39] in this case as the control of gate voltage increases compared to the drain one [26, 27, 40, 41].

2.12.4 Single Halo Dual Material Gate MOSFET

In this case a small region with very high positive doping concentration is considered at the source end along with the gate material, made of two different metals with

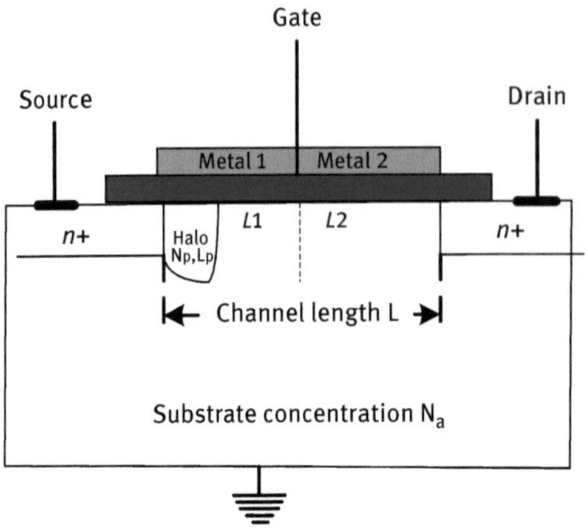

Figure 2.17: Single halo dual material gate MOSFET.

different work functions (Figure 2.17). The work function corresponding to the first metal is greater than that corresponding to the second. So although the small positively doped pocket leads to reduced drain current, the DMG structure increases it. The net result is that the drain current is increased in case of single halo dual material gate (SHDMG) MOSFET. However, the control of gate voltage is increased and the short channel effects are reduced using this structure [40–42]. It is seen that such structures work well for channel lengths up to 40 nm.

2.12.5 Double Halo Dual Material Gate MOSFET

In case of DH MOSFET, two small heavy pocket doping regions are considered at the two ends. These regions absorb a large number of electrons, and hence, the drain current is significantly reduced considering two halo regions. However, in case of Double Halo Dual Material Gate (DHDMG) MOSFET, the gate material is consisted of two different metals with different work functions [42]. This creates a step profile in the channel and the drain current is increased as compared to DH case (Figure 2.18). In subthreshold regime, such a structure is very effective in suppressing short channel effects.

2.12.6 Double Gate MOSFET

In this case there is a very thin Si layer for a channel, with two gates electrically connected, one on each side of the channel. Short channel effects are greatly suppressed in such a structure since the two gates very effectively terminate the drain electric field lines, preventing the effect of the drain potential from being felt at the source end.

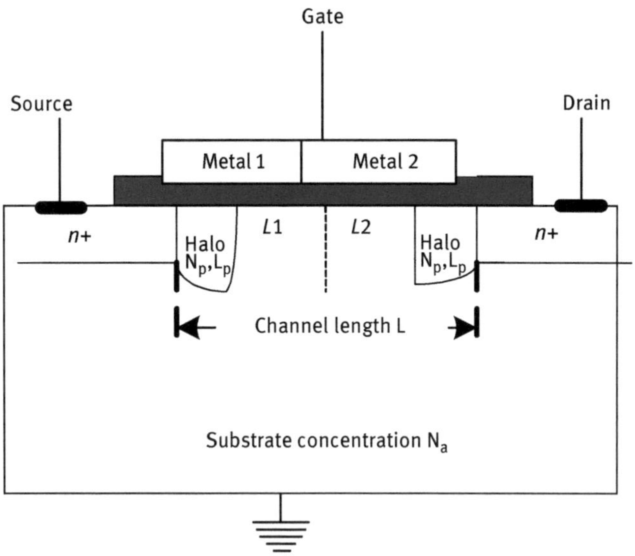

Figure 2.18: Double halo dual material gate MOSFET.

The variation of the threshold voltage with the drain potential and the gate length of a double gate MOSFET is much less than that of a conventional single gate structure of the same channel length [43–46].

The double gate MOSFET structure as given in Figure 2.19 shows a second gate electrode at the opposite side of the thin silicon body. It has two channels at the top and the bottom of the device and is reported to be relatively free of short channel effects.

Due to the two gates in Figure 2.20, the gate control of the channel increases. However, the double gate MOSFET structure is yet to be used in the manufacture of IC due to its significant fabrication problems. Simulations have been carried out for a 30 nm channel length double gate nMOSFET in Refs [43–46]. Very fast switching times of the order of 1–2 ps are observed in existing research works.

2.12.6.1 Advantages of Double Gate MOSFET

(1) Increased scalability
(2) Lower junction capacitance
(3) Possibility of light doping
(4) Larger drive current

2.12.6.2 Disadvantages of Double Gate MOSFET

(1) Higher series source and drain resistance. Raised source-to-drain structure is required to lower it.
(2) Difficulty in fabrication.

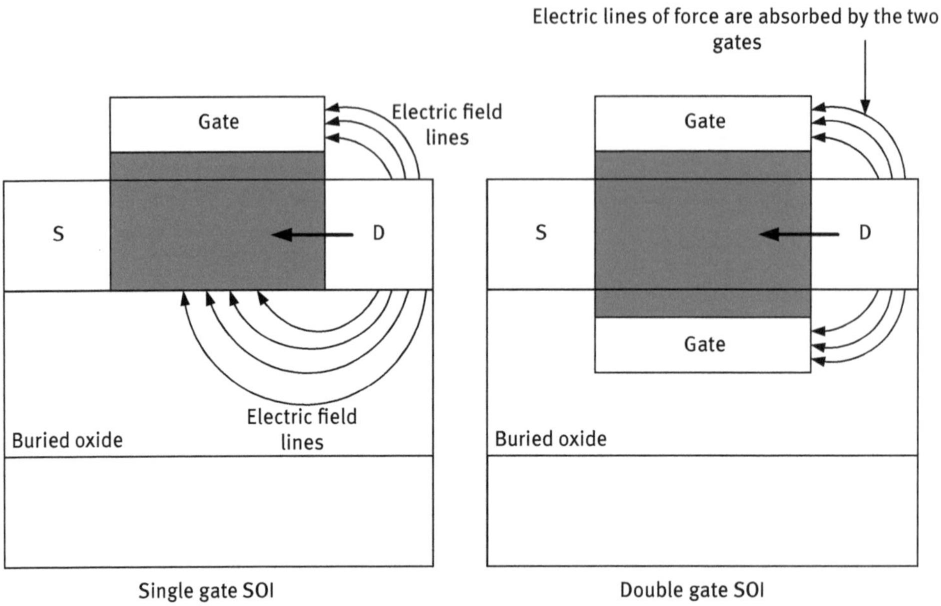

Figure 2.19: Field lines for single gate and double gate SOI MOSFETs.

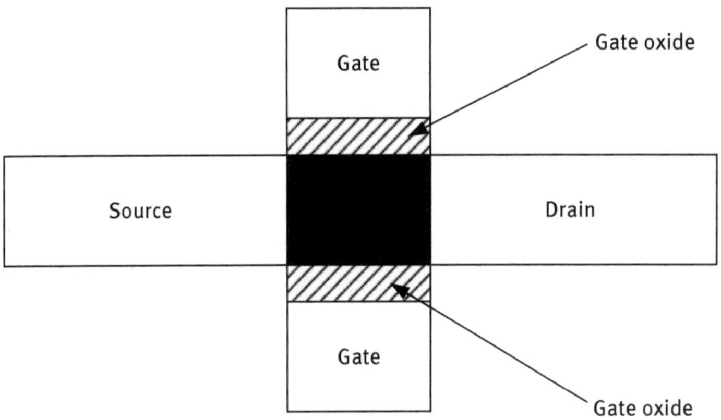

Figure 2.20: Double gate MOSFET structure.

The challenges are in the fabrication of double gate MOSFETs with both gates self-aligned to the source and the drain.

2.12.7 Dual Material Double Gate MOSFET

In this case gate engineering is applied to a double gate MOSFET [47] as in Figure 2.21. The gate material is made of two metals with different work functions. The work function of the first metal is higher than that of the second. So a step potential profile is

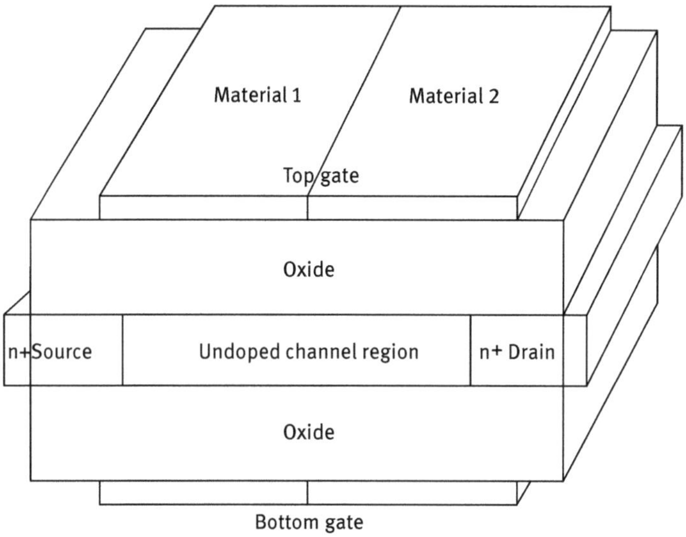

Figure 2.21: Dual material double gate MOSFET.

formed and the electrons have to overcome it for reaching the drain. The two gates effectively terminate the electric field lines from the drain as a result of the drain bias. So the effect of drain bias on the electrons at the source end is greatly reduced. The short channel effects are greatly suppressed as a result [48, 49].

2.12.8 Triple Material Double Gate MOSFET

In this case the gate material is made of three metals with different work functions [50–53]. The work function of the first metal near the source end is highest while that near the drain end is least. The work function of the metal in between is intermediate between the high value and the low one as in Figure 2.22. Two step potential profiles are created as a result. This structure is expected to provide further improvement in the performance over double gate and the DMDG MOSFET structures due to the two steps present in the surface potential profile. It is seen that the triple-material double gate (TMDG) MOSFET with a large ratio of L_1:L_2:L_3 can provide better suppression of SCEs. Therefore, it is concluded that a TMDG structure outperforms other gate-engineered SOI MOSFET device structures in suppressing the undesirable short channel effects.

2.12.9 FinFET

The FinFET consists of a channel formed in a vertical silicon fin controlled by a self-aligned double gate as shown in Figure 2.23. When the fins are viewed from the top, they are made thin enough such that the two gates control the entire fully depleted channel. The FinFET is similar to the conventional planar MOSFET in the layout and

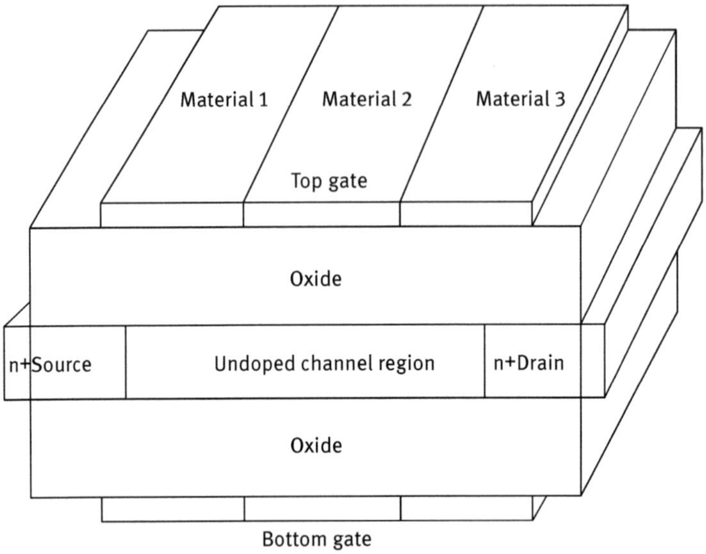

Figure 2.22: Structure of triple material double gate MOSFET.

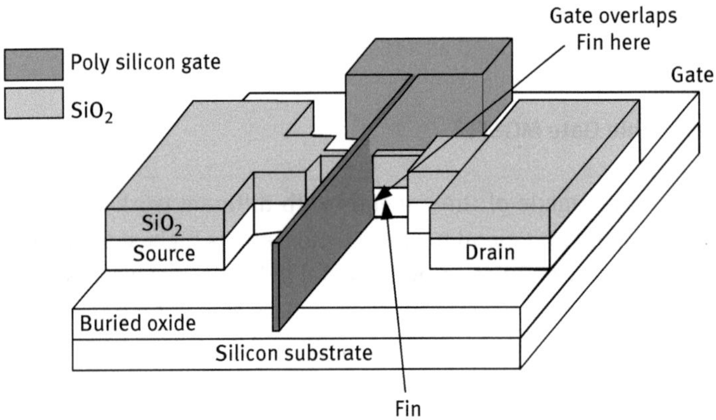

Figure 2.23: Perspective view of FinFET.

the fabrication. It provides a range of channel lengths, CMOS compatibility and large packing density compared to other double gate structures. n-channel FinFETs show good short channel performance down to a gate length of 17 nm. These results indicate that the FinFET is a promising device structure for future CMOS technology [54, 55].

Advantage of FinFET is that the fabrication in this case is a less complex process than double gate MOSFET and is very compatible with the current techniques.

2.12.10 Triple Gate MOSFET

Such structures have three effective gates. Intel first introduced them to increase transistor switching performance and decrease leakage power dissipation. They showed that using triple gate transistors increases the operating speed by 37% and reduces the power dissipation by 50% of the earlier values of transistor.

Such structures are basically FinFETs with active top gate. A single gate controls two vertical sections and one horizontal in the channel. The gate dielectric must be equally thin on the three sides of the body. The magnitude of the current can be changed by altering the fin width.

Such a triple gate structure allows thrice the surface area for electrons to travel. The additional gate control increases the amount of current flow during the on state and reduce the current to as close to zero as possible during the off state. So the overall transient response of the device improves. The reduced leakage current for such types of devices leads to reduction of overall power consumption.

2.12.11 Gate-all-around MOSFET

The Gate-all-around (GAA) structure in Ref. [56] is similar to the vertical surrounding gate (SRG) MOSFET. They are thin, fully depleted SOI MOSFETs consisting of two gates. These devices possess several advantages like low leakage current, minimum threshold voltage shift and better characteristics when compared with bulk and SOI MOSFETs.

The gate control over the channel increases for such devices and the overall device performance improves. In such devices the gate oxide and the gate are wrapped around the channel. First a silicon rod is patterned on the buried oxide layer for fabricating GAA. Then the undercut is formed underneath the rod by wet etching using HF. Then the polysilicon gate wraps the silicon channel. Finally, the gate is patterned and the S/D regions are using ion implantation.

These devices show increased transconductance due to volume inversion. The threshold voltage variation with temperature is least, allowing the device to operate at higher temperatures. The leakage current is far less compared to other MOSFETs resulting in excellent output swing. Such devices are quite immune against radiation effect. In terms of the packing density, the GAA MOSFET is less efficient when compared with SRG or VRG (vertical surrounding gate) MOSFETs.

2.12.12 Surrounding Gate MOSFET

As semiconductor devices are scaled down to the sub-0.1 μm regime, gate oxide tunneling and several lithographic issues become paramount in further scaling. SRG MOSFETs fabricated on SOI substrates have shown promise in overcoming the

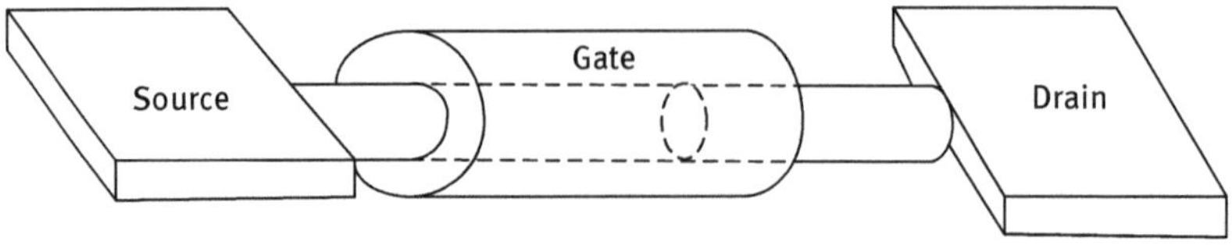

Figure 2.24: Cylindrical surrounding gate MOSFET.

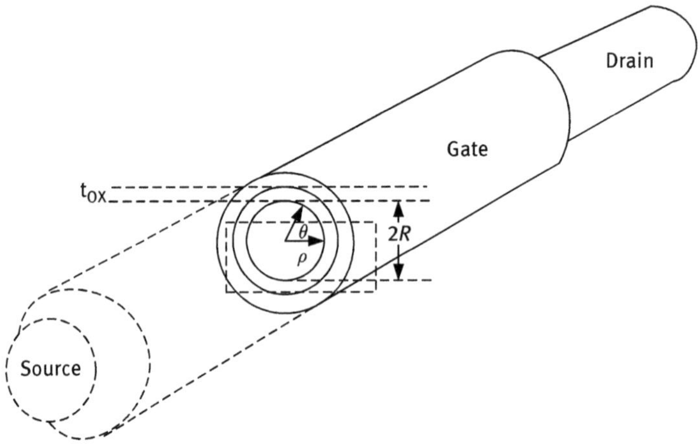

Figure 2.25: Cylindrical gate MOSFET with dimension.

short channel effects [57, 58]. The physical structure of SRG MOSFETs is shown in Figure 2.24.

Electrical lines of force from drain end are totally absorbed by the SRG structure. So drain control is totally eliminated and such structures provide best control over the channel from all around.

Figure 2.25 (ρ, θ) represents the cylindrical coordinate. The threshold voltage of a SRG MOSFET is governed by the oxide thickness (t_{Ox}) and the radius of the cylinder (R). A MOSFET will have larger threshold voltage for a thinner oxide or smaller radius. To obtain a desired threshold voltage we have to choose the right material for the gate [57, 58].

2.12.13 Silicon Nanowires

An alternative to planar MOSFET structure that is being explored is the surrounding gate nanowire MOSFET. The SRG nanowire transistor gives better performance than single gate device. A detailed study of SCEs for SRG nanaowire MOSFETs using device modeling is highly essential to obtain a deeper understanding of various physical effects. It is also important to consider quantum effects in SRG MOSFET models as the quantum confinement effects start to become significant when silicon film thickness is smaller than 10 nm [59–62].

It is very important to accurately predict the RF performance of SRG nanowires with downscaling.

2.13 Fringing-induced Barrier Lowering

Increased gate leakage is one major limiting factor on aggressive scaling of gate dielectric for deep-submicron CMOS technology [63–65]. Search has been on for a suitable high-*k* dielectric, which can replace the silicon dioxide. However, this gives rise to significant fringing capacitances, consisting of gate-dielectric fringing capacitance between gate electrode bottom and the source/drain surface and the gate electrode fringing capacitance between the gate electrode edge and the source/drain surface. Both these effects affect the electrical characteristics and degrade the short channel performance of MOSFET as they increase the fringing field from the gate to the source/drain regions. Recently, conformal mapping transformation was employed to model the fringing parasitic capacitance [63–65]. The fringing-induced barrier lowering (FIBL) is demonstrated through the effect of charges induced by the fringing capacitances. The results show that the fringing capacitances increase with the *k*-value of the gate dielectric or thickness of the gate electrode. The fringing capacitances can also impact the threshold voltage of the device. The effect of fringing capacitances increases as the device is continuously scaled down. So these fields need to be incorporated in the analytical models in order to get a perfect estimation of the characteristic parameters like surface potential, threshold voltage and drain current of the deep submicron devices.

2.14 Silicon-on-insulator MOSFETs

In this case a thin layer of silicon is fabricated on top of an insulator. The thin layer of silicon is separated from the *p*-type substrate by a thick layer of buried SiO_2 [66]. The thickness of the silicon film typically ranges between 50 and 200 nm, while the buried oxide thickness is 80–400 nm. For thin film silicon, the depletion region under the gate extends through the buried oxide and the device is called "*fully depleted*". Otherwise the transistor is said to be "*partially depleted*".

The advantages of SOI MOSFETs over the conventional ones are as follows:

(1) SOI circuits consist of single-device islands dielectrically isolated from each other and from the underlying substrate. The lateral isolation offers more compact design and the simplified technology than in the bulk silicon.
(2) The current in SOI MOSFETs is higher than that of bulk devices.
(3) The advantages of the SOI substrate include the improved MOSFET scaling due to its thin body.

(4) The buried oxide layer serves as good isolation to reduce capacitance to the substrate, giving rise to higher speed.
(5) Device isolation is much easier, simply by removing the surrounding "*thin film*". This can significantly improve the circuit density.
(6) The limited extension of drain and source regions allows SOI devices to be less affected by short channel effects, originated from "*charge sharing*" between gate and junctions.

The disadvantages of SOI are higher wafer cost, the kink effect and worse heat conduction because of the oxide layer.

2.15 Nonconventional Double Gate MOSFETs

VLSI technology focuses on miniaturization of device dimensions in order to incorporate large number of devices on a single chip. With further development and enhancement of this technology, the need for shrinking device dimensions will substantially increase. However, this shrinking of device dimensions leads to undesirable short channel effects like hot electron effect, DIBL and so on. To solve these effects and to facilitate the production of very large-scale integrated circuits, engineering techniques like channel engineering and gate engineering are used. In channel engineering technique halo implants like single halo and DH are induced in the MOSFETs that help negate the adverse short channel effects. It helps achieve better control over threshold voltage. Further gate engineering techniques are used to amalgamate two different metals M1 and M2 with work functions φ_1 and φ_2, laterally in a gate to reduce DIBL.

The combination of these techniques can be further utilized to lower hot electron effects. And finally, the use of double gates helps to achieve better control over the potential barrier between the source and drain terminals. Previous research work have included the 2-D modeling of subthreshold surface potential for short channel single gate dual material double halo (SGDMDH) MOSFETs and double gate dual material double halo (DGDMDH) MOSFET by solving pseudo 2D Poisson's equation.

However, the focus has been on modeling and design of double gate single material double halo (DGSMDH) MOSFET, double gate dual material single halo (DGDMSH) MOSFET and double gate single material single halo (DGSMSH) MOSFET based on the study and previous research work done on DGDMDH MOSFET. In recent work the pseudo 2-D Poisson's equation for DGDMDH MOSFET has been modified to suit the model description for the DGSMDH, DGDMSH and DGSMSH MOSFETs. The comparative study and analysis of these models on surface potential versus channel length can go a long way in determining which model serves the best to reduce the short channel effects for high density integration in the VLSI industry. Moreover, variation of parameters like V_{DS} (drain-to-source voltage), N_a (acceptor concentration) and work function have been carried out and plots have been simulated to simplify and efficiently focus on the characteristics that need to be taken care of before device designing. The effect of using multiple gates, multiple gate materials and multiple halo implants and their efficiency in reducing the undesirable short channel effect is the major theme of our analysis.

DGSMDH MOSFET

Here we discuss and compare the models and calculate the surface potentials for four devices namely double halo double material double gate (DHDMDG) MOSFET, double halo single material double gate (DHSMDG) MOSFET, single halo double material double gate (SHDMDG) MOSFET and single halo single material double gate (SHSMDG) MOSFET. This discussion with their varied changes will help to compare and analyze the potential and other features of the respective devices (Figure 2.26).

Double Gate Dual Material Double Halo MOSFET

The cross-sectional area diagram for a DGDMDH MOSFET is as shown in Ref. [67]. The surface potential can be found as in Refs [67, 68] (Figure 2.27).

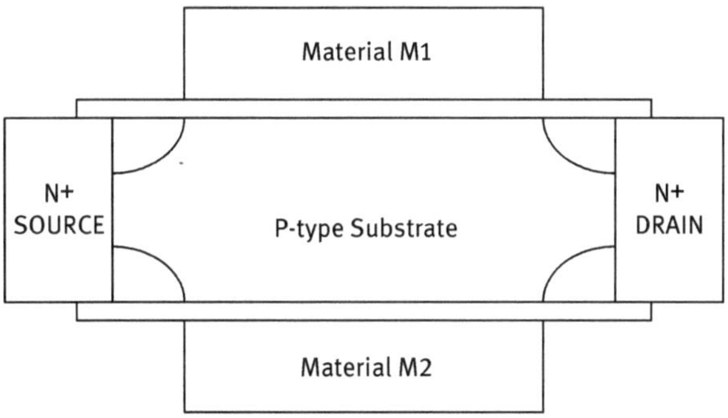

Figure 2.26: Structure of DGSMDH MOSFET.

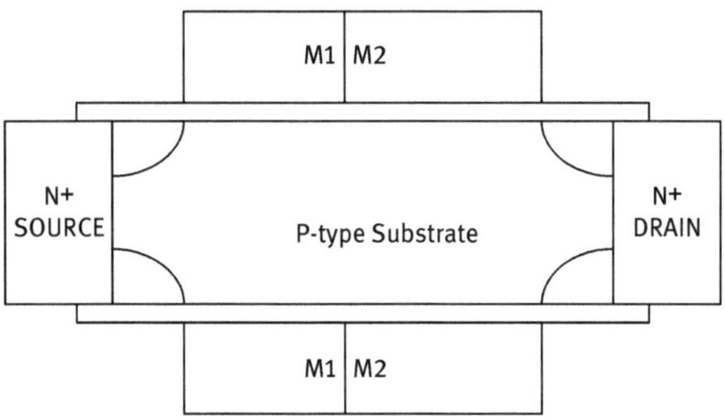

Figure 2.27: Structure of DGDMDH MOSFET.

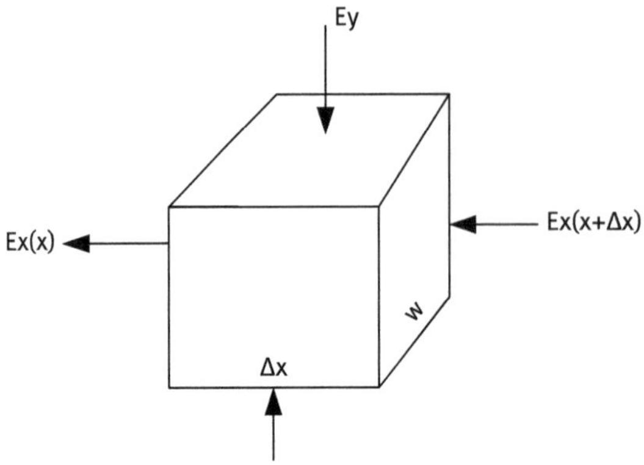

Figure 2.28: Elementary Gaussian rectangular box.

Let us consider an elementary rectangular Gaussian surface. The rectangular box of length Δx, at a given x, covers the entire depletion region width y_d and w is the gate width as shown in Figure 2.28. We use a pseudo 2D analysis because the analytical solution of 2D Poisson's equation, though accurate, is highly complicated. Since the device has two gates, the vertical electric field components E_{y1} on the top or front side and E_{y2} on the bottom or back side are both non-zero. So we can obtain E_{y1} and E_{y2} from the potential balance equations as follows:

$$V_{Gbf} = V_{Fbf} + \psi_S + \psi_{Ox} \quad \text{for the front side and} \tag{2.1}$$

$$V_{Gbb} = V_{Fbb} + \psi_S + \psi_{Ox} \quad \text{for the back side.} \tag{2.2}$$

We define surface potential Ψs as the potential drop from the surface to the bulk outside the depletion region, Ψ_{Ox} as potential drop across oxide layer, V_{Gbf} as the front gate bias voltage, V_{Gbb} as back gate bias voltage, V_{Fbf} and V_{Fbb} are the flat band voltages under front and back gates, respectively.

The flux lines terminating on the interface change per unit area Q_0 will not contribute the same since it is modeled as effective interface charge which resides on the oxide side of the interface. Thus only flux lines terminating on Q_c will contribute the same and for an oxide thickness of t_{Ox}, the corresponding vertical field can be obtained as

$$E_{y1} = \frac{V_{Gbf} - V_{Fbf} - \psi_S}{t_{Ox}}, \tag{2.3}$$

$$E_{y2} = \frac{V_{Gbb} - V_{Fbb} - \psi_S}{t_{Ox}}. \tag{2.4}$$

Equation (2.3) is obtained from eq. (2.1) and eq. (2.4) is obtained from eq. (2.2).

In weak inversion, the inversion layer charge can be ignored. For a substrate doping of N_a, the total depletion charge due to the ionized acceptor atoms within the Gaussian surface is given by $-qN_a\Delta x y_d W$, where q is the electronic charge. If the dielectric permittivity of the medium is ε and the mobile charge carriers are ignored, after applying of Gauss's law to the said surface, we get

$$\varepsilon \oint_{surface} \vec{E}\cdot\overrightarrow{Ds} = -qN_a y_d \Delta x W. \tag{2.5}$$

The left, right, top and bottom surfaces of the Gaussian box will have non-zero contribution to the left-hand side of eq. (2.5). If ε_{Si} and ε_{Ox} are the dielectric permittivity of the silicon and SiO_2, respectively, eq. (2.5) becomes

$$-\varepsilon_{Ox}E_{y1}\Delta x W + \varepsilon_{Ox}E_{y2}\Delta x W + \varepsilon_{Si}\left\{-E_x\left(x+\Delta x\right) + E_x\left(x\right)\right\} y_d W = -qN_a y_d \Delta x. \tag{2.6}$$

By replacing $E_x = \left(\frac{d\psi_s}{dx}\right)$ and solving eq. (2.6) we get eq. (2.7).

$$-\varepsilon_{Si}\left(\frac{d^2\psi_s}{dx^2}\right) - \left(\frac{C_{Ox}}{y_d}\right)\psi_s = qN_a - \left(\frac{C_{Ox}}{y_d}\right)\left(V_{GSf} - V_{GSb}\right), \tag{2.7}$$

$$V'_{GSf} = V_{GSf} + V_{Sb} - V_{Fbf}, \tag{2.8}$$

$$V'_{GSb} = V_{GSb} + V_{Sb} - V_{Fbb}, \tag{2.9}$$

where
- V_{GSf}/V_{GSb} = front/back gate-to-source voltage
- V_{Sb} = source bias voltage
- V_{Fbf}/V_{Fbb} = front and back flat band voltage under the front and back gate of MOSFET.
- *For single metal gate MOSFET*
- ϕ_M = work function of metal

$$\phi_t = \frac{kT}{q}, \tag{2.10}$$

where ϕ_t = thermal voltage.

Halo region flat band voltage

For front,

$$V_{Fbfp} = -\left(\frac{E_g}{2q}\right) - \phi_t \ln\frac{N_p}{n_i}. \tag{2.11}$$

For back,

$$V_{Fbbp} = -\left(\frac{E_g}{2q}\right) - \phi_t \ln\frac{N_p}{n_i}, \tag{2.12}$$

where
- N_p = halo doping concentration
- n_i = intrinsic carrier density
- E_g = bang gap in silicon = 1.1 eV
- Fermi potential of n-type substrate

$$\phi_{Fn} = \phi_t \ln \frac{N_a}{n_i}, \tag{2.13}$$

where
- N_a = p-type substrate doping concentration
- *Flat band voltage under metal of gate*

For front,

$$V_{Fbf1} = (\phi_M - \phi_s)/q. \tag{2.14}$$

For back,

$$V_{Fbb1} = (\phi_M - \phi_s)/q. \tag{2.15}$$

Since there is single material,

$$V_{Fbf1} = V_{Fbf2}, \tag{2.16}$$

$$V_{Fbb1} = V_{Fbb2}. \tag{2.17}$$

Work function of silicon substrate

$$\phi_s = \left(\frac{E_g}{2q}\right) + \phi_{Fn} + \chi, \tag{2.18}$$

where
χ =electron affinity of silicon

$$C_{Ox} = \frac{\varepsilon_{Ox}}{t_{Ox}}, \tag{2.19}$$

where
- C_{Ox} = oxide capacitance per unit gate area
- t_{Ox} = gate oxide thickness
- ε_{Ox} = dielectric permittivity of SiO_2

$$y_d = \sqrt{\frac{2\varepsilon_{Si}\psi_s}{qN_a}}, \tag{2.20}$$

where,

- y_d =depletion layer depth under gate
- ε_{Si} =dielectric permittivity of Si

and

$$\psi_s = \left(\frac{y}{2} + \sqrt{\frac{y^2}{4} + V_{Gb} - V_{Fb}}\right)^2, \tag{2.21}$$

where

ψ_s = gate controlled subthreshold surface potential for long channel MOSFET

and

$$y = \frac{\sqrt{2q\varepsilon_{Si}N_a}}{C_{Ox}}, \tag{2.22}$$

where y = body effect coefficient.

For short DHSMDG MOSFET, there is nonuniformity in y_d due to the two junctions, metal and the two halo regions [68–71]. The depletion layer depth around the source and the drain junctions is a complex function of substrate doping, drain and source bias voltages and junction depth. The surface potential depends on depletion layer thickness, which is not constant. So $y_d(x)$ is modeled first for predicting the surface potential accurately [68, 69, 71]. If the channel length is not too small and a reasonable amount of voltage is applied to the source or drain, $y_d(x)$ varies with x (Figure 2.29).

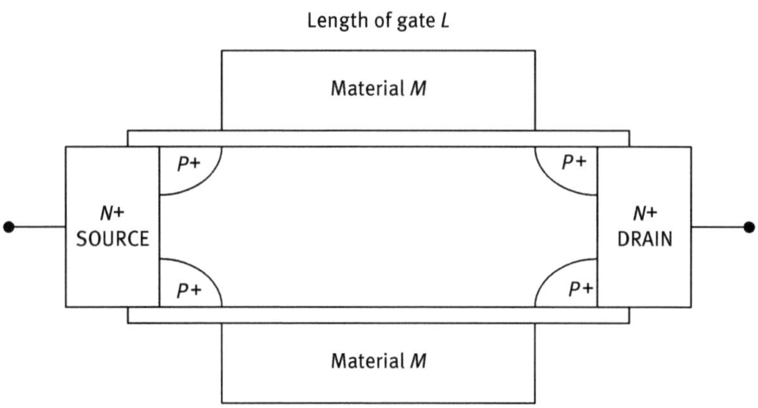

Figure 2.29: Structure of DGSMDH MOSFET showing different substrate and doping materials.

We consider an approximate model as $y_{d(x)} = (ax+b)^2$.
Putting $y_{d(x)} = (ax+b)^2$ in eq. (2.7), we get,

$$(ax+b)^2 \frac{d^2\psi_s}{dx^2} - \frac{C_{Ox}}{\varepsilon_{Si}}\psi_s = \frac{qN_a}{\varepsilon_{Si}}(ax+b)^2 - \frac{C_{Ox}}{\varepsilon_{Si}}\left(V_{GSf} - V_{GSb}\right). \tag{2.23}$$

The channel is divided into five different regions.
The boundary values of the potential are as follows:
At $x_1 = 0$, position of channel length voltage,

$$V_1 = V_{Bi} + V_{Sb}. \tag{2.24}$$

At $x_7 = 0$, position of channel length voltage,

$$V_7 = V_{Bi} + V_{Db}, \tag{2.25}$$

where

$$V_{Bi} = \frac{E_g}{2q} + \phi_{Fp}. \tag{2.26}$$

V_{Bi} =built-in potential

$$\phi_{Fp} = \phi_t \ln \frac{N_p}{n_i}. \tag{2.27}$$

ϕ_{Fp} = Fermi potential of p-type substrate.
Putting $t = \ln(ax+b)$, $\frac{dt}{dx} = \frac{a}{ax+b}$
ψ_s can be obtained as

$$(ax+b)^2 \frac{d^2\psi_s}{dx^2} = a^2 D(D-1)\psi \tag{2.28}$$

$$\mathrm{D} = \frac{d}{dt},$$

where

$$d = \sqrt{\frac{1}{4} + \frac{C_{Ox}}{\varepsilon_{Si}a^2}}, \tag{2.29}$$

and

$$\beta = \frac{qN_a}{2\varepsilon_{Si}a^2 - C_{Ox}}. \tag{2.30}$$

So complementary function $CF = e^{\frac{t}{2}}\left(C_1e^{dt} + C_2e^{-dt}\right)$, (2.31) where C_1 and C_2 are arbitrary constants.

$$\text{Partial fraction } PI = \beta e^{2t} + \left(V'_{GSf} - V'_{GSb}\right). \tag{2.32}$$

So front gate flat band voltage under halo and material is

$$V_{Fbfp1} = V_{Fbfp2} = \left(V_{Fbfp} - V_{Fbf1}\right). \tag{2.33}$$

Back gate flat band voltage under halo and material is

$$V_{Fbbp1} = V_{Fbbp2} = \left(V_{Fbbp} - V_{Fbb1}\right). \tag{2.34}$$

For region 1,

$$x_1 = 0 < x \le x_2 = L_p \text{ (halo length)}$$

$$y_1 = \left(x_j + x_{Rs}\right) = \left(x_j + \sqrt{\frac{2\varepsilon_{Si} V_1}{qN_p}}\right) / \xi_s, \tag{2.35}$$

where

$$x_{rs} = \sqrt{\frac{2\varepsilon_{Si} V_1}{qN_p}}. \tag{2.36}$$

And ξ_s = fitting parameter for the source side

$$\xi_s = \frac{4V_1}{V_{Bi}} \tag{2.37}$$

$$y_2 = \frac{\sqrt{2\varepsilon_{Si}\psi_{s1}}}{qN_p}, \tag{2.38}$$

where

$$\psi_{s1} = \left(-\frac{y_p}{2} + \sqrt{\frac{y_p^2}{2} + \left(V_{Gbf} - V_{Fbfp1}\right) - \left(V_{Gbb} - V_{Fbbp1}\right)}\right)^2 \tag{2.39}$$

and

$$y_p = \frac{\sqrt{2q\varepsilon_{Si} N_p}}{C_{Ox}}. \tag{2.40}$$

For region 2,

$$x_3 = x_2, x_4 = x_c;$$

$$y_3 = y_2 = \sqrt{\frac{2\varepsilon_{Si}\psi_{s1}}{qN_p}}; \tag{2.41}$$

$$y_4 = \sqrt{\frac{2\varepsilon_{Si}\psi_{s2}}{qN_p}}, \tag{2.42}$$

where

$$\psi_{s2} = \left(-\frac{y_p}{2} + \sqrt{\frac{y_p^2}{2} + \left(V_{Gbf} - V_{Fbfp}\right) - \left(V_{Gbb} - V_{Fbbp}\right)}\right)^2. \tag{2.43}$$

For region 3,

$$x_5 = x_4, x_6 = L_1 \quad x_7 = x_6, x_8 = L - x_c$$

$$y_5 = y_4 = \sqrt{\frac{2\varepsilon_{Si}\psi_{s2}}{qN_p}}; \tag{2.44}$$

$$y_6 = \sqrt{\frac{2\varepsilon_{Si}\left(\left(\psi_{s2} + \psi_{s3}\right)/2\right)}{qN_a}}; \tag{2.45}$$

$$y_8 = y_6 = \sqrt{\frac{2\varepsilon_{Si}\left(\left(\psi_{s2} + \psi_{s3}\right)/2\right)}{qN_a}}; \tag{2.46}$$

$$y_5 = y_7 = \sqrt{\frac{2\varepsilon_{Si}\psi_{s2}}{qN_p}}, \tag{2.47}$$

where

$$\psi_{s3} = \left(-\frac{y_a}{2} + \sqrt{\frac{y_a^2}{2} + \left(V_{Gbf} - V_{Fbf1}\right) - \left(V_{Gbb} - V_{Fbb1}\right)}\right)^2 \tag{2.48}$$

and

$$y_a = \frac{\sqrt{2q\varepsilon_{Si}N_a}}{C_{Ox}}. \tag{2.49}$$

For region 4,

$$x_9 = x_8, x_{10} = L - L_p;$$

$$y_9 = y_8 = \sqrt{\frac{2\varepsilon_{Si}\left(\psi_{s2+\psi_{s3}}\right)}{qN_a}}; \tag{2.50}$$

$$y_{10} = \sqrt{\frac{2\varepsilon_{Si}\psi_{s4}}{qN_a}}, \tag{2.51}$$

where

$$\psi_{s4} = \left(-\frac{y_p}{2} + \sqrt{\frac{y_p^2}{2} + \left(V_{Gbf} - V_{Fbfp2}\right) - \left(V_{Gbb} - V_{Fbbp2}\right)}\right)^2. \tag{2.52}$$

For region 5,

$$x_{11} = x_{10},\ x_{12} = L$$

$$y_{11} = y_{10} = \sqrt{\frac{2\varepsilon_{Si}\psi_{s4}}{qN_a}}; \tag{2.53}$$

$$y_{12} = \left(x_j + x_{Rd}\right) = \left(x_j + \sqrt{\frac{2\varepsilon_{Si}V_7}{qN_p}}\right)/\xi_d, \tag{2.54}$$

where

$$\xi_d = \frac{4V_7}{V_{Bi}} \tag{2.55}$$

and ξ_d = fitting parameter for drain side.
For region 1,

$$t_1 < t < t_2;$$

$$t_1 = \ln\left(a_1x_1 + b_1\right); \tag{2.56}$$

$$t_2 = \ln\left(a_1x_2 + b_1\right); \tag{2.57}$$

$$a_1 = \frac{\sqrt{y_2} - \sqrt{y_1}}{(x_2 - x_1)}; \tag{2.58}$$

$$b_1 = \frac{x_2\sqrt{y_1} - x_1\sqrt{y_2}}{(x_2 - x_1)}; \tag{2.59}$$

$$d_1 = \sqrt{\frac{1}{4} + \frac{C_{Ox}}{\varepsilon_{Si}a_1^2}}; \tag{2.60}$$

$$\beta_1 = \frac{qN_p}{2\varepsilon_{Si}a_1^2 - C_{Ox}}; \tag{2.61}$$

$$V_{GSf1} = V_{GSf} + V_{Sb} - V_{Fbfp1}; \tag{2.62}$$

$$V_{GSb1} = V_{GSb} + V_{Sb} - V_{Fbbp1}; \tag{2.63}$$

$$V_{GS1} = V_{GSf1} - V_{GSb1}; \tag{2.64}$$

$$\begin{aligned}\psi_{s1}(t) = \frac{1}{\sinh(d_1t_2-d_1t_1)} \Big\{ \left(V_2 - V_{GS1} - \beta_1 e^{2t_2}\right) e^{\frac{t-t_2}{2}} \sinh(d_1t - d_1t_1) \\ - \left(V_1 - V_{GS1} - \beta_1 e^{2t_1}\right) e^{\frac{t-t_1}{2}} \sinh(d_1t - d_1t_2) \Big\} + \beta_1 e^{2t} + V_{GS1}.\end{aligned} \tag{2.65}$$

For region 2,

$$t_3 < t < t_4;$$

$$t_4 = \ln(a_2x_3 + b_2); \tag{2.66}$$

$$t_4 = \ln(a_2x_4 + b_2); \tag{2.67}$$

$$a_2 = \frac{\sqrt{y_4} - \sqrt{3}}{(x_4 - x_3)}; \tag{2.68}$$

$$b_2 = \frac{x_4\sqrt{y_3} - x_3\sqrt{y_5}}{(x_4 - x_3)}; \tag{2.69}$$

$$d_2 = \sqrt{\frac{1}{4} + \frac{C_{Ox}}{\varepsilon_{Si} a_2^2}}; \tag{2.70}$$

$$\beta_2 = \frac{qN_p}{2\varepsilon_{Si} a_2^2 - C_{Ox}}; \tag{2.71}$$

$$V_{GSf2} = V_{GSf} + V_{Sb} - V_{Fbfp}; \tag{2.72}$$

$$V_{GSb2} = V_{GSb} + V_{Sb} - V_{Fbbp}; \tag{2.73}$$

$$V_{GS2} = V_{GSf2} - V_{GSb2}; \tag{2.74}$$

$$\begin{aligned}\psi_{s2}(t) = \frac{1}{\sinh(d_2t_4 - d_2t_3)} \Big\{ \left(V_3 - V_{GS2} - \beta_2 e^{2t_4}\right) e^{\frac{t-t_4}{2}} \sinh(d_2t - d_2t_3) \\ - \left(V_2 - V_{GS2} - \beta_2 e^{2t_3}\right) e^{\frac{t-t_3}{2}} \sinh(d_2t - d_2t_4) \Big\} + \beta_2 e^{2t} + V_{GS2}.\end{aligned} \tag{2.75}$$

For region 3,

$$t_5 < t < t_8;$$

$$t_5 = \ln(a_3x_5 + b_3); \tag{2.76}$$

$$t_6 = \ln(a_3x_6 + b_3); \tag{2.77}$$

$$t_7 = \ln(a_4x_7 + b_4); \tag{2.78}$$

$$t_8 = \ln\left(a_4 x_8 + b_4\right); \tag{2.79}$$

$$a_3 = \frac{\sqrt{y_6} - \sqrt{y_5}}{(x_6 - x_5)}; \tag{2.80}$$

$$b_3 = \frac{x_6\sqrt{y_5} - x_5\sqrt{y_6}}{(x_6 - x_5)}; \tag{2.81}$$

$$a_4 = \frac{\sqrt{y_8} - \sqrt{y_7}}{(x_8 - x_7)}; \tag{2.82}$$

$$b_4 = \frac{x_8\sqrt{y_7} - x_7\sqrt{y_8}}{(x_8 - x_7)}; \tag{2.83}$$

$$d_3 = \sqrt{\frac{1}{4} + \frac{C_{Ox}}{\varepsilon_{Si} a_3^2}}; \tag{2.84}$$

$$d_4 = \sqrt{\frac{1}{4} + \frac{C_{Ox}}{\varepsilon_{Si} a_4^2}}; \tag{2.85}$$

$$\beta_3 = \frac{qN_p}{2\varepsilon_{Si} a_3^2 - C_{Ox}}; \tag{2.86}$$

$$\beta_4 = \frac{qN_p}{2\varepsilon_{Si} a_4^2 - C_{Ox}}. \tag{2.87}$$

Now, as we use a single material, we have

$$V_{GSf3} = V_{GSf4} = V_{GSf} + V_{Sb} - V_{Fbf1}, \tag{2.88}$$

$$V_{GSb3} = V_{GSb4} = V_{GSb} + V_{Sb} - V_{Fbb1}, \tag{2.89}$$

$$V_{GS3} = V_{GS4} = V_{GSf3} - V_{GSb3}, \tag{2.90}$$

$$\psi_{s3}(t) = \frac{1}{\sinh(d_3 t_6 - d_3 t_5)} \left\{ \left(V_4 - V_{GS3} - \beta_3 e^{2t_6}\right) e^{\frac{t-t_6}{2}} \sinh(d_3 t - d_3 t_5) - \left(V_3 - V_{GS3} - \beta_3 e^{2t_5}\right) e^{\frac{t-t_5}{2}} \sinh(d_3 t - d_3 t_6) \right\} + \beta_3 e^{2t} + V_{GS3} \tag{2.91}$$

$$\psi_{s4}(t) = \frac{1}{\sinh(d_4 t_8 - d_4 t_7)} \left\{ \left(V_5 - V_{GS4} - \beta_4 e^{2t_8}\right) e^{\frac{t-t_8}{2}} \sinh(d_4 t - d_4 t_7) - \left(V_4 - V_{GS4} - \beta_4 e^{2t_7}\right) e^{\frac{t-t_7}{2}} \sinh(d_4 t - d_4 t_8) \right\} + \beta_4 e^{2t} + V_{GS4}. \tag{2.92}$$

For region 4,

$$t_9 < t < t_{10};$$

$$t_9 = \ln\left(a_5 x_9 + b_5\right); \tag{2.93}$$

$$t_{10} = \ln\left(a_5 x_{10} + b_5\right); \tag{2.94}$$

$$a_5 = \frac{\sqrt{y_{10}} - \sqrt{y_9}}{(x_{10} - x_9)}; \tag{2.95}$$

$$b_5 = \frac{x_{10}\sqrt{y_9} - x_9\sqrt{y_{10}}}{(x_{10} - x_9)}; \tag{2.96}$$

$$d_5 = \sqrt{\frac{1}{4} + \frac{C_{Ox}}{\varepsilon_{Si} a_5^2}}; \tag{2.97}$$

$$\beta_5 = \frac{qN_p}{2\varepsilon_{Si} a_5^2 - C_{Ox}}; \tag{2.98}$$

$$V_{GSf5} = V_{GSf} + V_{Sb} - V_{Fbfp}; \tag{2.99}$$

$$V_{GSb5} = V_{GSb} + V_{Sb} - V_{Fbbp}; \tag{2.100}$$

$$V_{GS5} = V_{GSf5} - V_{GSb5}; \tag{2.101}$$

$$\psi_{s5}(t) = \frac{1}{\sinh(d_5 t_{10} - d_5 t_9)} \left\{ \left(V_6 - V_{GS5} - \beta_5 e^{2t_{10}}\right) e^{\frac{t-t_{10}}{2}} \sinh(d_5 t - d_5 t_9) - \left(V_5 - V_{GS5} - \beta_5 e^{2t_9}\right) e^{\frac{t-t_9}{2}} \sinh(d_5 t - d_5 t_{10}) \right\} + \beta_5 e^{2t} + V_{GS5}. \tag{2.102}$$

For region 5,

$$t_{11} < t < t_{12};$$

$$t_{11} = \ln(a_6 x_{11} + b_6); \tag{2.103}$$

$$t_{12} = \ln(a_6 x_{12} + b_6); \tag{2.104}$$

$$a_6 = \frac{\sqrt{y_{12}} - \sqrt{y_{11}}}{(x_{12} - x_{11})}; \tag{2.105}$$

$$b_6 = \frac{x_{12}\sqrt{y_{11}} - x_{11}\sqrt{y_{12}}}{(x_{12} - x_{11})}; \tag{2.106}$$

$$d_6 = \sqrt{\frac{1}{4} + \frac{C_{Ox}}{\varepsilon_{Si} a_6^2}}; \tag{2.107}$$

$$\beta_6 = \frac{qN_p}{2\varepsilon_{Si} a_6^2 - C_{Ox}}; \tag{2.108}$$

$$V_{GSf6} = V_{GSf} + V_{Sb} - V_{Fbfp2}; \tag{2.109}$$

$$V_{GSb5} = V_{GSb} + V_{Sb} - V_{Fbbp2}; \tag{2.110}$$

$$V_{GS6} = V_{GSf6} - V_{GSb6}; \tag{2.111}$$

$$\psi_{s6}(t) = \frac{1}{\sinh(d_6 t_{12} - d_6 t_{11})} \left\{ \left(V_7 - V_{GS6} - \beta_6 e^{2t_{12}}\right) e^{\frac{t-t_{12}}{2}} \sinh(d_6 t - d_6 t_{11}) - \left(V_6 - V_{GS6} - \beta_6 e^{2t_{11}}\right) e^{\frac{t-t_{11}}{2}} \sinh(d_6 t - d_6 t_{12}) \right\} + \beta_6 e^{2t} + V_{GS6}. \tag{2.112}$$

The unknown potentials V_2, V_3, V_4, V_5 and V_6 are found by solving the following simultaneous equations obtained by applying the continuity of derivative of potential at the boundaries between two regions. Though there are five regions in the channel, we take V_5 as the voltage in region 5 for ease of calculation.

Interface between region 1 and region 2:

$$a_{11}V_2 + a_{12}V_3 + a_{13}V_4 + a_{14}V_5 + a_{15}V_6 = A_1. \tag{2.113}$$

Interface between region 2 and region 3:

$$a_{21}V_2 + a_{22}V_3 + a_{23}V_4 + a_{24}V_5 + a_{25}V_6 = A_2. \tag{2.114}$$

Interface between region 3 and region 4:

$$a_{31}V_2 + a_{32}V_3 + a_{33}V_4 + a_{34}V_5 + a_{35}V_6 = A_3; \tag{2.115}$$

$$a_{41}V_2 + a_{42}V_3 + a_{43}V_4 + a_{44}V_5 + a_{45}V_6 = A_4. \tag{2.116}$$

Interface between region 4 and region 5:

$$a_{51}V_2 + a_{52}V_3 + a_{53}V_4 + a_{54}V_5 + a_{55}V_6 = A_5, \tag{2.117}$$

where

$$a_{11} = \left[\left(d_1a_1\coth(d_1t_2 - d_1t_1) + \frac{a_1}{2}\right)e^{-t_2} + \left(d_2a_2\coth(d_2t_4 - d_2t_3) - \frac{a_1}{2}\right)e^{-t_3}\right]; \tag{2.118}$$

$$a_{12} = -d_2a_2\operatorname{csch}(d_2t_4 - d_2t_3)\,e^{\frac{-t_3-t_4}{2}}; \tag{2.119}$$

$$a_{13} = 0, a_{31} = 0. \tag{2.120}$$

$$A_1 = \begin{array}{l}\left[d_1a_1\left(V_1 - V_{GS1} - \beta_1e^{2t_1}\right)\operatorname{csch}(d_1t_2 - d_1t_1)\,e^{\frac{-t_2-t_1}{2}}\right] \\ -\left[d_2a_2\left(V_{GS2} + \beta_2e^{2t_4}\right)\operatorname{csch}(d_2t_4 - d_2t_3)\,e^{\frac{-t_3-t_4}{2}}\right] \\ +\left[\left(V_{GS2} + \beta_2e^{2t_3}\right)\left(d_2a_2\coth(d_2t_4 - d_2t_3) - \frac{a_2}{2}\right)e^{-t_3} + 2a_2\beta_2e^{t_3}\right] \\ +\left[\left(V_{GS1} + \beta_1 + e^{2t_2}\right)\left(d_1a_1\coth(d_1t_2 - d_1t_1) + \frac{a_1}{2}\right)e^{-t_2} + 2a_1\beta_1e^{t_2}\right]\end{array}; \tag{2.121}$$

$$a_{21} = -d_2a_2\operatorname{csch}(d_2t_4 - d_2t_3)\,e^{\frac{-t_3-t_4}{2}} = a_{12}; \tag{2.122}$$

$$a_{22} = \left[\left(d_2a_2\coth(d_2t_4 - d_2t_3) + \frac{a_2}{2}\right)e^{t_4} + \left(d_3a_3\coth(d_3t_6 - d_3t_5) - \frac{a_3}{2}\right)e^{-t_5}\right]; \tag{2.123}$$

$$a_{23} = (-1)\operatorname{csch}(d_3t_6 - d_3t_5)\left(d_3a_3e^{\frac{-t_5-t_6}{2}}\right); \tag{2.124}$$

$$a_{32} = a_{23}; \tag{2.125}$$

$$\begin{aligned} A_2 = & \left[-\left(V_{GS3}+\beta_3 e^{2t_6}\right)\left(\operatorname{csch}\left(d_3t_6-d_3t_5\right)\left(d_3a_3e^{\frac{-t_2-t_1}{2}}\right)\right)\right] \\ & +\left[\left(V_{GS2}+\beta_2 e^{2t_3}\right)\left(d_2a_2\operatorname{csch}\left(d_2t_4-d_2t_3\right)-e^{\frac{-t_3-t_4}{2}}\right)-2a_2\beta_2e^{t_4}\right] \\ & +\left[\left(V_{GS3}+\beta_3 e^{2t_5}\right)\left(d_3a_3\coth\left(d_3t_6-d_3t_5\right)-\tfrac{a_3}{2}\right)e^{-t_5}+2a_2\beta_2e^{t_3}\right] \\ & +\left[\left(V_{GS2}+\beta_2+e^{2t_4}\right)\left(d_2a_2\coth\left(d_2t_4-d_2t_3\right)+\tfrac{a_2}{2}\right)e^{-t_4}+2a_3\beta_3e^{t_5}\right] \end{aligned} ; \quad (2.126)$$

$$a_{13}=0, a_{31}=0; \quad (2.127)$$

$$a_{33}=\left[\left(d_3a_3\coth\left(d_3t_6-d_3t_5\right)+\frac{a_3}{2}\right)e^{-t_6}+\left(d_4a_4\coth\left(d_4t_8-d_4t_7\right)-\frac{a_4}{2}\right)e^{-t_7}\right]; \quad (2.128)$$

$$\begin{aligned} A_3 = & \left[-\left(V_{GS4}+\beta_4 e^{2t_8}\right)\left(d_4a_4\operatorname{csch}\left(d_4t_8-d_4t_7\right)\left(e^{\frac{-t_7-t_8}{2}}\right)\right)\right] \\ & -\left[\left(V_{GS3}+\beta_3 e^{2t_5}\right)\left(d_3a_3\operatorname{csch}\left(d_3t_6-d_3t_5\right)e^{\frac{-t_5-t_6}{2}}\right)-2a_4\beta_4e^{t_7}\right] \\ & +\left[\left(V_{GS4}+\beta_4 e^{2t_7}\right)\left(d_4a_4\coth\left(d_4t_8-d_4t_7\right)-\tfrac{a_4}{2}\right)e^{-t_7}\right] \\ & +\left[\left(V_{GS3}+\beta_3+e^{2t_6}\right)\left(d_3a_3\coth\left(d_3t_6-d_3t_5\right)+\tfrac{a_3}{2}\right)e^{-t_6}-2a_3\beta_3e^{t_6}\right] \end{aligned} ; \quad (2.129)$$

$$a_{44}=\left[\left(d_4a_4\coth\left(d_4t_8-d_4t_7\right)+\frac{a_4}{2}\right)e^{-t_8}+\left(d_5a_5\coth\left(d_5t_{10}-d_5t_9\right)-\frac{a_5}{2}\right)e^{-t_9}\right]; \quad (2.130)$$

$$a_{55}=\left[\left(d_5a_5\coth\left(d_5t_{10}-d_5t_9\right)+\frac{a_5}{2}\right)e^{-t_{10}}+\left(d_6a_6\coth\left(d_6t_{12}-d_6t_{11}\right)-\frac{a_6}{2}\right)e^{-t_{11}}\right]; \quad (2.131)$$

$$a_{45}=-\operatorname{csch}\left(d_5t_{10}-d_5t_9\right)\left(d_5a_5e^{\frac{-t_9-t_{10}}{2}}\right); \quad (2.132)$$

$$a_{54}=a_{45}; \quad (2.133)$$

$$a_{43}=-\operatorname{csch}\left(d_4t_8-d_4t_7\right)\left(d_4a_4e^{\frac{-t_7-t_8}{2}}\right); \quad (2.134)$$

$$a_{43}=a_{34}; \quad (2.135)$$

$$a_{35}=a_{53}=0, a_{14}=a_{41}=0, a_{24}=a_{42}=0, a_{15}=a_{51}=0, a_{25}=a_{52}=0; \quad (2.136)$$

$$\begin{aligned} A_4 = & \left[-\left(V_{GS5}+\beta_5 e^{2t_{10}}\right)\left(d_5a_5\operatorname{csch}\left(d_5t_{10}-d_5t_9\right)\left(e^{\frac{-t_{10}-t_9}{2}}\right)\right)\right] \\ & -\left[\left(V_{GS4}+\beta_4 e^{2t_7}\right)\left(d_4a_4\operatorname{csch}\left(d_4t_8-d_4t_7\right)e^{\frac{-t_8-t_7}{2}}\right)-2a_4\beta_4e^{t_8}\right] \\ & +\left[\left(V_{GS5}+\beta_5 e^{2t_9}\right)\left(d_5a_5\coth\left(d_5t_{10}-d_5t_9\right)-\tfrac{a_5}{2}\right)e^{-t_9}\right] \\ & +\left[\left(V_{GS4}+\beta_4 e^{2t_8}\right)\left(d_4a_4\coth\left(d_4t_8-d_4t_7\right)+\tfrac{a_4}{2}\right)e^{-t_8}-2a_3\beta_3e^{t_6}\right] \end{aligned} ; \quad (2.137)$$

$$\begin{aligned} A_5 = & \left[\left(V_7-V_{GS6}+\beta_6 e^{2t_{12}}\right)\left(d_6a_6\operatorname{csch}\left(d_6t_{12}-d_6t_{11}\right)\left(e^{\frac{-t_{12}-t_{11}}{2}}\right)\right)\right] \\ & -\left[\left(V_{GS5}+\beta_5+e^{2t_9}\right)\left(d_4a_4\operatorname{csch}\left(d_5t_{10}-d_5t_9\right)e^{\frac{-t_9-t_{10}}{2}}\right)-2a_5\beta_5e^{t_{10}}\right] \\ & +\left[\left(V_{GS6}+\beta_6 e^{2t_{11}}\right)\left(d_6a_6\coth\left(d_6t_{12}-d_6t_{11}\right)-\tfrac{a_6}{2}\right)e^{-t_{11}}\right] \\ & +\left[\left(V_{GS5}+\beta_5 e^{2t_{10}}\right)\left(d_5a_5\coth\left(d_5t_{10}-d_5t_9\right)+\tfrac{a_5}{2}\right)e^{-t_{10}}+2a_6\beta_6e^{t_{11}}\right] \end{aligned} . \quad (2.138)$$

Surface potential can be easily calculated by substituting the values of V_2, V_3, V_4, V_5 and V_6.

Comparative study and various MATLAB7.0.1 simulations have been presented below to focus on the improvement over the previously existing models and choose the best out of the newly designed ones. Varying device parameters and biasing conditions have been used for these models to portray the surface potential profiles and make the comparative analysis easier.

The plot in Figure 2.30 compares the surface potential for DHDMDG, SHDMDG, SMDGDH and SHSMDG MOSFETs [71]. The value of Na is 4×10^{17}, V_{GS} is 0 V and the length of the device L is taken as 30×10^{-7}. The plots show that the surface potential decreases from SMDHDG to SHSMDG to DMDHDG to SHDMDG.

The plot in Figure 2.31 shows the variation of V_{DS} for values 0.5 V, 1 V and 1.5 V for DGDMDH MOSFET. The value of Na is 4×10^{17}, V_{GS} is 0 V and the length of the device L is taken as 30×10^{-7}. With this variation, the surface potential decreases from V_{DS} = 1.5 V to V_{DS} = 1.0 V to V_{DS} = 0.5 V.

The plot in Figure 2.32 shows the variation of V_{DS} for values 0.5 V, 1 V and 1.5 V for DGSMDH MOSFET. The value of Na is 4×10^{17}, V_{GS} is 0 V and the length of the device L is taken as 30×10^{-7}. With this variation, the surface potential decreases from V_{DS} = 1.5 V to V_{DS} = 1.0 V to V_{DS} = 0.5 V.

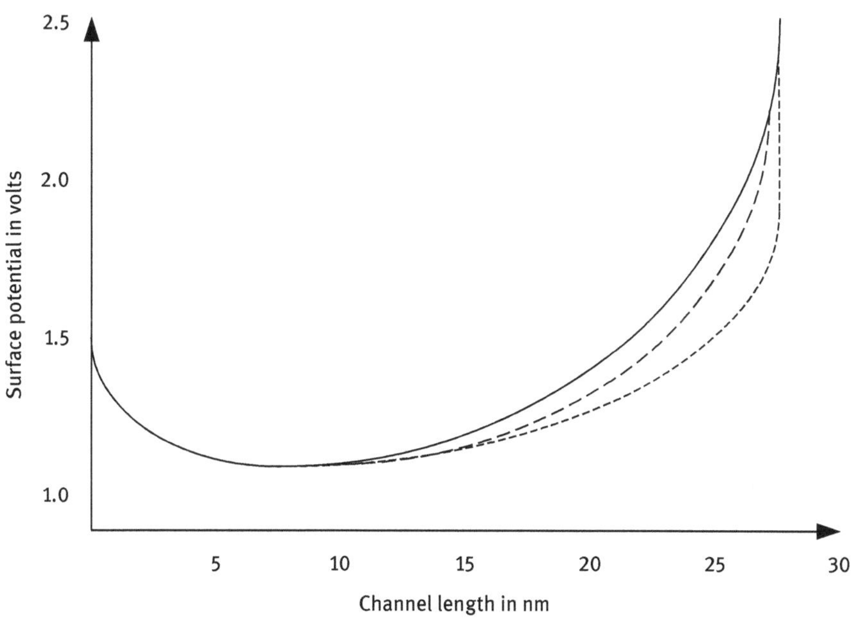

Figure 2.30: Plot of subthreshold surface potential vs. channel length for different channel and gate-engineered double gate structures.

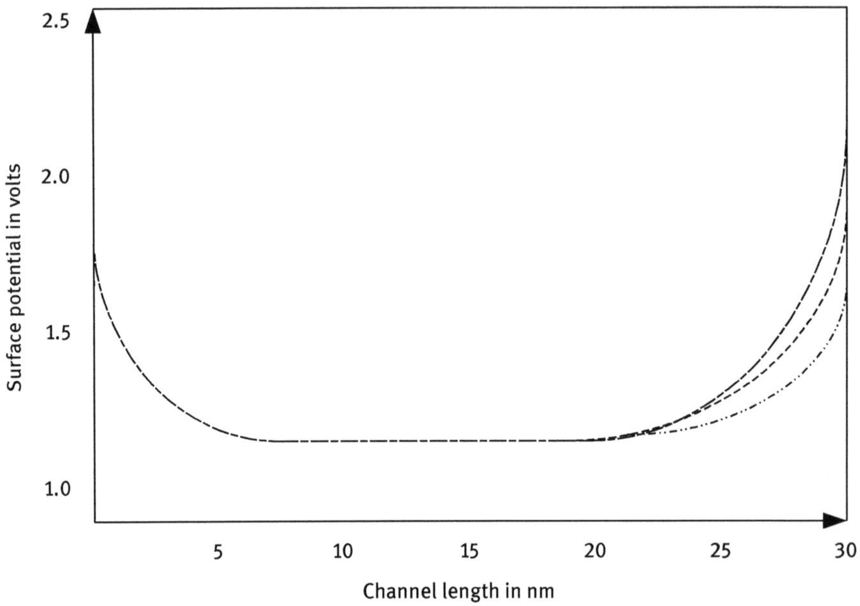

Figure 2.31: Plots of subthreshold surface potential vs. channel length plots for DGDMDH MOSFET for 3 different source-to-drain voltages.

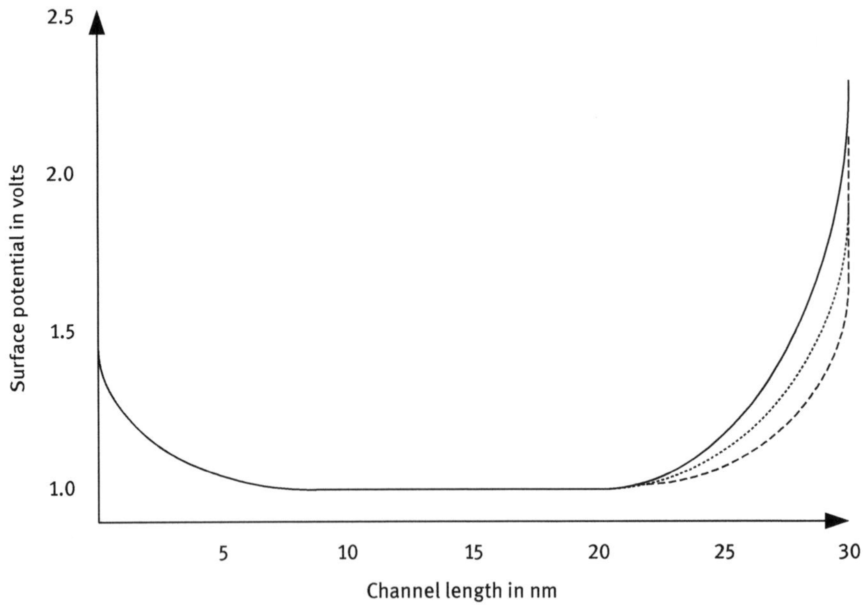

Figure 2.32: Plots of subthreshold surface potential vs. channel length for DGSMDH MOSFET for three different values of drain-to-source voltages.

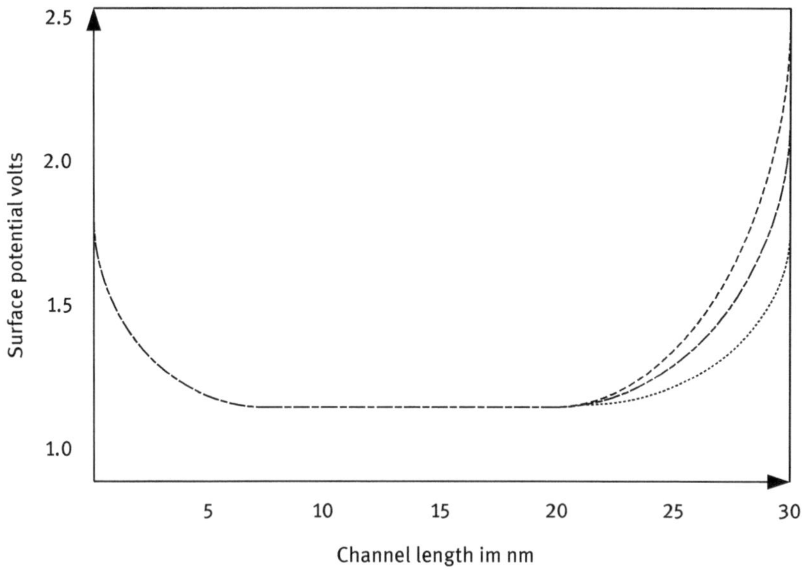

Figure 2.33: Plots of subthreshold surface potential vs. channel length for DGDMSH MOSFET for 3 different values of drain-to-source voltages.

The plot in Figure 2.33 shows the variation of V_{DS} for values 0.5 V, 1 V and 1.5 V for DGDMSH MOSFET. The value of Na is 4×10^{17}, V_{GS} is 0 V and the length of the device L is taken as 30×10^{-7}. With this variation, the surface potential decreases from V_{DS} = 1.5 V to V_{DS} = 1.0 V to V_{DS}= 0.5 V.

The plot in Figure 2.34 shows the variation of Na for values 2×10^{15}, $4 \times 10e^{17}$ and $9 \times 10e^{18}$ for DGDMDH MOSFET. The value of V_{GS} is 0 V, and the length of the device L is taken as 30×10^{-7}. With this variation, the surface potential decreases from Na = $9 \times 10e^{18}$ to Na = $4 \times 10e^{17}$ to $2 \times 10e^{15}$.

In this work the main focus has been on modeling the surface potential of various short channel MOSFETs. The characteristics are affected by the depletion layer formation near the source and the drain regions. The analytical modeling has been done by solving the pseudo-2D Poisson's equation and applying Gauss's law to the same, and thereby solving it. The same procedure has been applied for various structures of double gate MOSFETs like Double Gate Dual Material Double Halo (DGDMDH), DGSMDH, DGDMSH and DGSMSH MOSFETs.

2.16 Tunnel Field-effect Transistor

The tunnel field-effect transistor (TFET) has structure almost similar to that of MOSFET. However, the switching mechanism of this transistor is different from that of conventional MOSFET, making this device candidate for low power electronics. In this

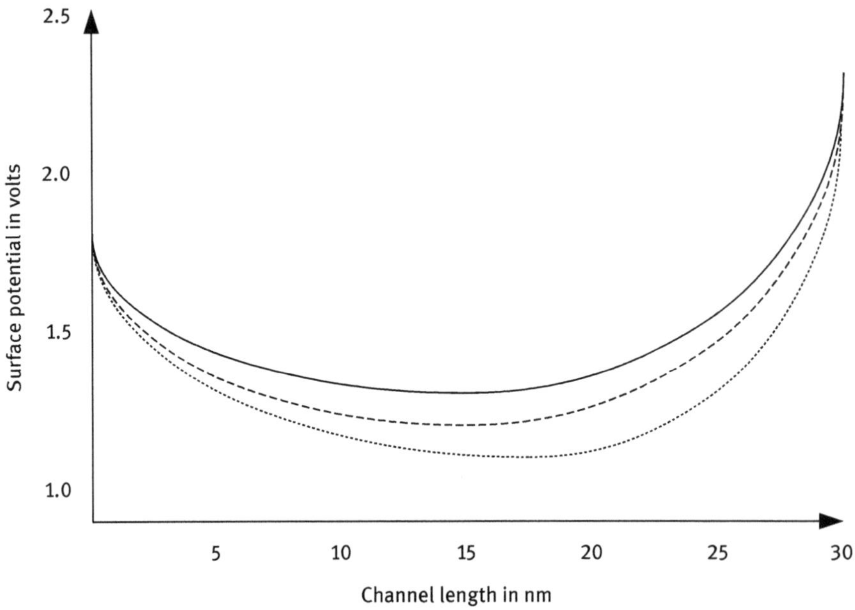

Figure 2.34: Plots of subthreshold surface potential vs. channel length for DGDMDH MOSFET for 3 different values of acceptor ion concentrations.

case, the basic principle is that of quantum tunneling through a potential barrier, unlike in case of conventional MOSFET, where thermionic emission over the potential barrier is the driving force behind electron transmission. As a result, subthreshold swing of TFETs goes below 60 mV/dec at room temperature [72–74]. Joerg Appenzeller in 2004 reported that a TFET with channel made of carbon nanotube can give subthreshold swing as low as 40 mV per decade. It is also observed that use of low power TFETs can lead to considerable saving of power when compared to conventional MOSFETs. The basic structure of TFET is shown below:

It is seen from Figure 2.35 that the source and drain regions are oppositely doped [75–78]. The gate terminal controls the flow of electrons. The higher the on-current, the more is the speed of transistor. The lowering of threshold voltage is very important for constant field scaling. For constant voltage scaling, the supply voltage cannot be scaled down with lowering of device dimensions. This leads to reduction of processor speed with scaling. However, TFET with a slope far below 63 mV/dec permits scaling of device dimensions with increase in processor frequency. In TFETs with increase in voltage, accumulation of electron occurs in the intrinsic region of the device. When the valence band of the *p*-region and the conduction band of the intrinsic region get aligned due to the band tunneling effect in TFET, electrons flow from the valence band of *p*-region to the conduction band, leading to flow of current in the device. When the gate voltage is reduced, the two bands become misaligned and current cannot

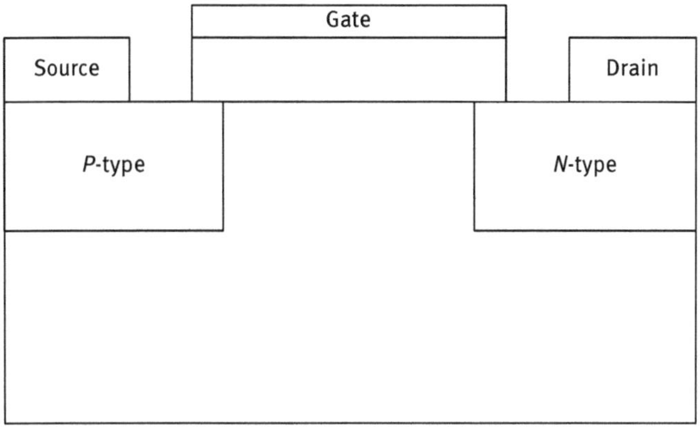

Figure 2.35: Structure of TFET.

flow through the device. Recently double gate TFET structures have been proposed by researchers to overcome the limitations of lateral TFET structures like requirement of very sharp doping profile, large gate leakage, etc.

2.17 IMOS Device

The conventional MOSFETs suffer from the limitation of achieving low subthreshold slope. This is due to the inherent property of the conventional MOS devices that the transport phenomenon is based on the diffusion process. The need of achieving steeper SS for the high switching applications become imminent, unlike in case of the conventional devices. In this device [79, 80] the carrier transport phenomena is impact ionization based. IMOS can achieve SS as low as 5 mV/decade and I_{On}/I_{Off} of 10^7. Due to impact-ionization-based avalanche breakdown, the device can easily switch from on state to off state. The off-state leakage current also decreases and this leads to improvement in suppression of the SCEs.

Till now, different versions of the device like depletion-IMOS, IMOS transistor with lower breakdown voltage and higher impact ionization area, and so on have been reported in various literatures [81, 82]. The device is basically a gated PIN diode. In this case, the depletion region extends up to the *i* region; therefore, the electric field along the horizontal direction will be lower. Therefore the avalanche multiplication process will dominate and desirable SS characteristics can be obtained.

2.18 Summary

It is seen that for continuing the scaling of MOSFET as per the ITRS, conventional device structures cannot be used due to the short channel effects. This chapter deals with the challenges of device scaling. The halo and DMG structures can suppress short channel effects in the 70 nm gate length regime. It is also seen that the combination of halo and DMG can suppress SCEs effectively up to a gate length of around 40 nm. However, when the device is scaled down further the nonconventional structures like double gate, GAA, FinFET, SRG and so on can be used. It is found that the introduction of the DMG structure in a fully depleted SOI double gate MOSFET leads to reduced SCEs due to a step function in the channel potential profile, thereby improving device performance and enhancing device scalability some steps further. Among all gate-engineered structures, TM-DG MOSFET structure outperforms other gate-engineered structure due to its enhanced gate transport efficiency resulting from the two steps present in the surface potential profile. It is seen that the SRG transistor exhibited improved performance than traditional single gate devices. Therefore, it is envisioned as a future replacement of silicon planar MOSFETs in digital circuits and is a potential candidate for providing long-term solutions to continue scaling CMOS beyond the 100 nm technology node. On the other hand, the double gate MOS transistor with the triple material gate technology with its higher carrier transport efficiency and SRG MOSFET with its best electrostatic gate control over the channel shows improved analog/RF performance parameters compared to that of the conventional MOSFET architecture and thus may be suitable to realize the low cost and high performance analog/RF mixed signal SoC applications in the sub-100 nm regime.

References

[1] G. Baccarani, M. Wordeman and R. Dennard, "Generalized scaling theory and its application to 1/4 micrometer MOSFET design." IEEE Transactions on Electron Devices, ED-31.4 (1984): 452.

[2] Robert H. Dennard, Fritz H. Gaensslen, Hwa-Nien Yu, V. Leo Rideout, Ernest Bassous and Andre R. Leblanc, "Design of ion-implanted MOSFETs with very small physical dimensions." IEEE Journal of Solid-State Circuits, SC-9 (1974): 256–258.

[3] N. Weste and K. Eshragian, *Principles of CMOS VLSI Design: A Systems Perspective*. Addison Wesley Publishing Company, USA, 1993.

[4] International Technology Roadmap for Semiconductors, http://www.sematech.org, December 4, 2001.

[5] B. Streetman, *Solid State Electronic Devices*. Pearson Prentice Hall, 1995.

[6] S. Pang and J.R. Brews, "Models for subthreshold and above subthreshold currents in 0.1 μm pocket n-MOSFETs for low voltage applications." IEEE Transactions on Electron Devices, 49 (2002): 832–839).

[7] International Technology Roadmaps for Semiconductor (ITRS), 1999 and 2005 edition.

[8] Claudio Fiegna, "The effect of scaling on the performance of small-signal MOS amplifiers." Proceedings of ISCAS 2000, May 2000, pp. 733–736.

[9] R.H. Dennard, F.H. Gaensslen, L. Kuhn and H.N. Yu, "Design of micron MOS switching devices." IEDM Digital Technical Papers, 1972, pp. 344.

[10] B.Cheng, M.Cao, R.Rao, A. Inani, P.V. Voorde, W.M. Greene, et.al., "The impact of high gate dielectrics and metal gate electrodes on sub-100 nm MOSFETs." IEEE Transactions on Electron Devices, 46.7 (1999): 1537–1544.

[11] G.V. Reddy and M.J. Kumar, "A new dual-material double-gate (DMDG) nanoscale SOI MOSFET—Two-dimensional analytical modeling and simulation." IEEE Transactions on Nanotechnology 4 (2005): 260–268.

[12] U.K. Mishra, A.S. Brown and S.E. Rosenbaum, "DC and RF performance of 0.1-μm gate length Al0.48As/Ga0.47In0.53As pseudomorphic HEMT." IEDM Technical Digest, 1988, pp.180–183.
[13] R.-H. Yan, A. Ourmazd, K.F. Lee, D.Y. Jeon, C.S. Rafferty and M.R. Pinto, "Scaling the Si metal-oxide-semiconductor field effect transistor into the 0.1 pn regime using vertical doping engineering." Applied Physics Letters, 59 (1991): 3315, and R.-H. Yan, A. Ourmazd and K.F. Lee, "Scaling the Si MOSFET From bulk to SO1 to bulk." Applied Physics Letters, 59 (1991): 1704.
[14] D.A. Antoniadis and J.E. Chung, "Physics and technology of ultra short channel MOSFET devices." IEDM Technical Digest, 1991, p. 21.
[15] Y.P. Tsividis, *Operation and Modeling of the MOS Transistor*. Mcgraw Hill, New York, 1999.
[16] A. Chaudhary and M.J. Kumar, "Controlling short-channel effects in deep submicron SOI MOSFETs for improved reliability: A review." IEEE Transactions on Device and Materials Reliability, 4.1 (2004): 99–109.
[17] S.M. Sze, *Physics of Semiconductor Device*, 2nd edn. John Wiley & Sons, Inc., New York, 1981.
[18] B. Razavi, *Design of Analog CMOS Integrated Circuit.* TMH, 2001.
[19] N. Arora, "MOSFET models for VLSI circuit simulation: theory and practice." World Scientific Publishing Company, 2007. Reprinted from Springer-Verlag, 1993.
[20] Gordon Moore, *Cramming More Components onto Integrated Circuits*, 1965.
[21] Z.-H. Liu, C. Hu, J.-H. Huang, T.-Y. Chan, Threshold voltage model for deep-submicrometer MOSFETs. IEEE Transactions on Electron Devices, 40.1 (1993): 86–95.
[22] K.Y. Lim and X. Zhou, "Modeling of threshold voltage with non-uniform substrate doping." Proceedings of the IEEE International Conference on Semiconductor Electronics (ICSE '98), Malaysia, 1998, pp. 27–31.
[23] B. Yu, C.H. Wann, E.D. Nowak, K. Noda and C. Hu, "Short channel effect improved by lateral channel engineering in deep-submicrometer MOSFETs." IEEE Transactions on Electron Devices, 44 (1997): 627–633.
[24] B. Yu, H. Wang, O. Millic, Q. Xiang, W. Wang, J.X. An and M.R. Lin, "50 nm gate length CMOS transistor with super-halo: Design, process and reliability." IEDM Technical Digest, 1999, pp. 653–656.
[25] K.M. Cao, W. Liu, X. Jin, K. Vasant, K. Green, J. Krick, T. Vrotsos and C. Hu, "Modeling of pocket implanted MOSFETs for anomalous analog behavior." IEEE IEDM Technical Digest, 1999, pp. 171–174.
[26] S. Baishya, A. Mallik and C.K. Sarkar, "A subthreshold surface potential and drain current model for lateral asymmetric channel (LAC) MOSFETs." IETE Journal of Research, 52 (2006): 379–390.
[27] S. Baishya, A. Mallik and C.K. Sarkar, "Subthreshold surface potential and drain current models for short-channel pocket implanted MOSFETs." Microelectronics Engineering. Available online.
[28] J.-G. Su, C.-T. Huang, S.-C. Wong, C.-C. Cheng, C.-C. Wang, H.-L. Shiang and B.-Y. Tsui, "Tilt angle effect on optimizing HALO PMOS and NMOS performance." Proceedings of IEEE IEDM, 1997, pp. 11–14.
[29] H.S. Shin, C. Lee, S.W. Hwang, B.G. Park and H.S. Min, "Channel length independent subthreshold characteristics in submicron MOSFETs." IEEE Electron Device Letters, 19 (1998): 137–139.
[30] Ali Khakifirooz and Dimitri A. Antoniadis, "MOSFET performance scaling—part I: Historical trend." IEEE Transactions on Electron Devices, 55.6 (2008): 1391–1400.
[31] Ali Khakifirooz, and Dimitri A. Antoniadis, "MOSFET performance scaling—Part II: Future directions." IEEE Transactions on Electron Devices, 55.6 (2008): 1401–1408.
[32] S. Kang and Y. Leblebici, *CMOS Digital Integrated Circuits, Analysis and Design*. TMH Edition. McGraw-Hill Higher Education.
[33] S. Borkar, "Design challenges of technology scaling." IEEE Micro, 19.4 (1999): 2–29.

[34] Saxena M, Haldar S, Gupta M, Gupta RS, "Design considerations for novel device architecture: hetero-material double-gate (HEM-DG) MOSFET with sub-100 nm gate length." Solid-State Electronics, 48.7 (2004): 1167–1174.

[35] W. Long, H. Ou, J. M. Kuo, and K. K. Chin, "Dual material gate (DMG) field effect transistor." IEEE Transactions on Electron Devices, 46.5 (1999): 865–870.

[36] X. Zhou and W. Long, "A novel hetero-material gate (HMG) MOSFET for deep-submicron ULSI technology." IEEE Transactions on Electron Devices, 45.11 (1998): 2546–2548.

[37] A. Chaudhry and M.J. Kumar, "Investigation of the novel attributes of a fully depleted dual-material gate SOI MOSFET." IEEE Transactions on Electron Devices, 51.9 (2004): 1463–1467.

[38] S. Baishya, A. Mallik and C.K. Sarkar, "A pseudo two-dimensional subthreshold surface potential model for dual-material gate MOSFETs." IEEE Transactions on Electron Devices, 54 (2007): 2520–2525.

[39] S.Baishya, A.Mallik and C.K. Sarkar, "A subthreshold surface potential model for short-channel MOSFET taking into account the varying depth of channel depletion layer due to source and drain junctions." IEEE Transactions on Electron Devices, 53 (2006): 507–514.

[40] S. De, A. Sarkar and , C.K. Sarkar, "Effect of fringing field in modeling of subthreshold surface potential in dual material gate (DMG) MOSFETS." ICECE 2008 (Vol. 1), 20–22 December, pp. 148–151, available in IEEE Xplore, 2008.

[41] A. Sarkar, S. De, M. Nagarajan, C.K. Sarkar and S. Baishya "Effect of fringing fields on subthreshold surface potential of channel engineered short channel MOSFETs." Tencon 2008, IEEE Region 10 Conference, 19–21 November, pp. 1–6, available in IEEE Xplore, 2008.

[42] Swapnadip De, Angsuman Sarkar and Chandan Kumar Sarkar, "Modelling of parameters for asymmetric halo and symmetric DHDMG n-MOSFETs." International Journal of Electronics, 98.10 (2011): 1365–1381, Taylor & Francis Group.

[43] C.C. Tsai, Y.J. Lee, J.L. Wang, K.F. Wei, I.-C. Lee, C.-C. Chen and H.-C. Cheng, "High-performance top and bottom double-gate low temperature poly-silicon thin film transistors fabricated by excimer laser crystallization." Solid State Electron, 52 (2008): 365–371.

[44] Leland Chang, Stephen Tang, Tsu-Jae King, Jeffrey Bokor, and Chenming Hu, "Gate length scaling and threshold voltage control of double-gate MOSFETs." IEDM, 2000, pp. 719–722.

[45] T. Ernst, S. Cristoloveanu, G. Ghibaudo, T. Ouisse, S. Horiguchi, Y. Ono, Y. Takahashi and K. Murase, "Ultimate thin double-gate SOI MOSFETs." IEEE Transactions on Electron Devices 50 (2003): 830–838.

[46] Andrew R. Brown, Jeremy R. Watling, Asen Asenov, "A 3-D atomistic study of archetypal double gate MOSFET structures." Journal of Computational Electronics, 1 (2002): 165–169.

[47] G.V. Reddy and M.J. Kumar, "A new dual-material double-gate (DMDG) nanoscale SOI MOSFET—Two-dimensional analytical modeling and simulation." IEEE Transactions on Nanotechnology 4 (2005): 260–268.

[48] M. Jagadesh Kumar and G. Venkateshwar Reddy, "Diminished short channel effects in nanoscale double-gate silicon-on-insulator metal-oxide-semiconductor field-effect-transistors due to induced back-gate step potential." Japanese Journal of Applied Physics, 44.9A (2005): 6508–6509, 2005.

[49] T.K. Chiang and M.L. Chen, "A new two-dimensional analytical model for short-channel symmetrical dual-material double-gate metal-oxide-semiconductor field effect transistors." Japan Society of Applied Physics, 46.6A (2007): pp. 3283–3290.

[50] P.K. Tiwari, S. Dubey, M. Singh and S. Jit, "A two-dimensional analytical model for threshold voltage of short channel triple-material double-gate metal-oxide-semiconductor field effect transistors." Journal of Applied Physics 108 (2010): 074508– 074508–8.

[51] P. Razavi and A.A. Orouji, "Nanoscale triple material double gate (TM-DG) MOSFET for improving short channel effects." Proceedings of Advances in Electronics and Microelectronics, 2008, pp. 11–14.
[52] T.K. Chiang, "A new two-dimensional analytical subthreshold behavior model for short-channel tri-material gate-stack SOI MOSFET's." Microelectronics Reliability, 49 (2009): 113–119.
[53] Santosh Kumar Gupta, Achinta Baidya and S. Baishya, "Simulation and analysis of gate engineered triple metal double gate (TM-DG) MOSFET for diminished short channel effects." International Journal of Advanced Science and Technology, 38 (2012): pp. 15–24.
[54] Jong-Ho Lee. "Bulk FinFETs: Fundamentals, modeling, and application." Semiconductor Materials and Device Laboratory.
[55] Min-hwa Chi, "Challenges in Manufacturing FinFET at 20 nm node and beyond." Technology Development, Global Foundries, Malta, NY 12020, USA.
[56] J.P. Colinge, M.H. Gao, A. Romano, H. Maes and C. Claeys, "Silicon-on-insulator gate-all-around device." Technical Digest of IEDM, 1990.
[57] Christopher Patrick Auth, "Physics and technology of vertical surrounding gate MOSFETs." ProQuest Dissertations and Theses; Thesis (PhD)–Stanford University, 1998, Publication Number: AAI9837171, Source: Dissertation Abstracts International, Vol.: 59–06, Section: B, p. 2923.
[58] Guang-Xi Hu and Ting-Ao Tang, "Some physical properties of a surrounding-gate MOSFET with undoped body." Journal of the Korean Physical Society, 49.2 (2006): 642–645.
[59] Rik Myslewski, "The 'pigs in a blanket' of process Silicon nanowires: The Next Big Thing™ in chip design technology." San Francisco, 16 March 2012.
[60] Yi Cui, Zhaohui Zhong, Deli Wang, Wayne U. Wang and Charles M. Lieber, "High performance silicon nanowire field effect transistors."
[61] C.N. Ram Rao and A. Govindaraj,"Nanotubes and nanowires." RSC Nanoscience & Nanotechnology, 2011, Vol. 2, pp. 1–530.
[62] Klaus D. Sattler, *Handbook of Nanophysics: Nanotubes and Nanowires*, CRC Press, UK, 2010.
[63] R. Shivastav and K. Fitzpatrick "A simple model for the overlap capacitance of VLSI MOS devices." IEEE Transactions on Electron Devices, ED-29 (1982): 1870–1875, Dec. 1982.
[64] H.J. Park and P.K. Ko, "An analytical model for intrinsic capacitances of short channel MOSFETs." IEDM Technical Digest 1984, pp. 301–303.
[65] Nihar R. Mohapatra, M.P. Desai, S.G. Narendra and V. Ramgopal Rao, "Modeling of parasitic capacitances in deep submicrometer conventional and high-k dielectric MOS transistors." IEEE Transactions on Electron Devices, 50.4 (2003): 959–966.
[66] Bawedin Maryline, "Transient floating body effects for memory applications in fully depleted SOI MOSFETs ." UCL Presses Universitaries De Louvain, September 2011.
[67] Debarati Das, Swapnadip De, Manash Chanda and C.K. Sarkar, "Modelling of sub threshold surface potential for short channel double gate dual material double halo MOSFET," The IUP Journal of Electrical & Electronics Engineering, ICFAI University Press, 7.4 (2014): 19–42.
[68] Swapnadip De, Angsuman Sarkar and C.K. Sarkar, "Effect of fringing field in modeling of subthreshold surface potential in dual material gate (DMG) MOSFETS." ICECE 2008 (Vol. 1), 20–22 December 2008, pp. 148–151.
[69] Angsuman Sarkar, Swapnadip De, Nagarajan, M., C.K.Sarkar and S.Baishya, "Effect of fringing fields on subthreshold surface potential of channel engineered short channel MOSFETs." IEEE Region 10 Conference Tencon 2008, 19–21 November, 2008, pp. 1–6.
[70] Swapnadip De, Angsuman Sarkar and C.K. Sarkar, "Fringing capacitance based surface potential model for pocket DMG n-MOSFETS." Journal of Electron Devices, 12 (2012): 704–712.
[71] Swapnadip De, M. Bhattacharya, A Kumari, P Dutta and I Gupta, " Comparative study of surface potential for non conventional double gate MOSFETs." International Journal of VLSI Design and Technology, Journalspub, 1.1 (2015): 1–19.

[72] T. Sakurai Perspectives of low power VLSIs. IEICE Transactions on Electron, E87-C (2004): 429–436.

[73] K. Bernstein, R.K. Cavin, W. Porod, A.C. Seabaugh and J. Welser Device and architectures outlook for beyond CMOS switches. Proceedings of IEEE, 98 (2010) (2010) 2169–2184.

[74] A.C. Seabaugh and Q. Zhang Low voltage tunnel transistors for beyond CMOS logic. Proceedings of IEEE, 98 (2010): 2095–2110.

[75] S.M. Sze, *Physics of Semiconductor Devices*, 1st edn. John Wiley United States, 1969.

[76] M.S. Lundstrom, The MOSFET revisited: device physics and modeling at the nanoscale. Proceedings of IEEE International SOI Conference 1–3 (2006).

[77] Daeyeon Kim, Yoonmyung Lee, Jin Cai, Isaac Lauer, Leland Chang , Steven J. Koester, Dennis Sylvester, David Blaauw, Heterojunction tunneling transistor (HETT)-based extremely low power applications. *Proc. Int. Symp. Low Power Electron. Design* 219–224 (IEEE/ACM, 2009).

[78] K. Bhuwalka, J. Schultze and I. Eisele A simulation approach to optimize the electrical parameters of a vertical tunnel FET. IEEE Transactions on Electron Devices 52 (2005): 1541–1547.

[79] Kailash Gopalakrishnan, Peter B. Griffin and James D. Plummer, "Impact ionization MOS (I-MOS)—Part I: Device and circuit simulations." IEEE Transactions on Electron Devices, 52.1 (2005): pp. 69–76.

[80] Caner Onal, Raymond Woo, H.-Y. Serene Koh, Peter B. Griffin and James D. Plummer, "A novel depletion-IMOS (DIMOS) device with improved reliability and reduced operating voltage." IEEE Electron Device Letters, 30.1 (2009): 64–67.

[81] Eng-Huat Toh, Grace Huiqi Wang, Lap Chan, Guo-Qiang Lo, Ganesh Samudra and Yee-Chia Yeo, "Strain and materials engineering for the I-MOS transistor with an elevated impact-ionization region." IEEE Transactions on Electron Devices, 54.10 (2007): pp. 2778–2785.

[82] W.Y. Choi, J.Y. Song, J.D. Lee, Y.J. Park, and B.-G. Park, "100-nm n-/p-channel I-MOS using a novel self-aligned structure." Electron Device Letters, 26.4 (2005): 261–263.

3 Advanced Energy-reduced CMOS Inverter Design

3.1 Introduction

In digital integrated circuit (IC), inverters, which can complement the logic level of input signal, are the fundamental logic gate doing Boolean operation also. Ideally as inverter has infinite input resistance and zero output resistance, it can be used for analog amplification also if we set the operating point in transition region. In this chapter we will concentrate on the basic digital operation of complementary metal oxide semiconductor (CMOS) inverter circuit and the different issues like power dissipation, delay, noise margin, area etc., in super-threshold region and then in subthreshold region [1–10].

3.1.1 Transfer Characteristics of Inverter

In this section first we analyze the ideal inverter and then the practical CMOS inverter circuit in detail in the next section.

In Figure 3.1, input and output waveform and voltage transfer characteristics of ideal inverter circuit are given. When we apply logic "0" (equivalent to zero volt) or logic "1" (equivalent to supply voltage V_{DD} volt) to an inverter, we will obtain the reverse logic in output node. It can be observed from the transfer characteristics of ideal inverter, if any input voltage, lying in the range of 0 to $V_{DD}/2$ volt, i.e. low logic; we get V_{DD} volt or high logic at output node. Similarly, $V_{DD}/2$ to V_{DD} volt range is considered as logic high level, which, when applied to the ideal inverter, produces 0 or logic "0" level at output. In case of ideal operation as delay is absent during switching, voltage transfer characteristics will rise vertically straight at $V_{DD}/2$, which shows infinite gain in transition region. This property is efficacious for analog amplification if we drive the inverter circuit in transition region. Ideal inverter is capable of restoring the logic level or suppressing the noise completely [1–4].

In practical operation, switching threshold voltage shifts from ideal position, i.e. $V_{DD}/2$, as given in Figure 3.2. Due to nonzero noise margin, transition region becomes less stepper or gain becomes finite. The characteristics parameters of VTC, given in Figure 3.3, can be defined as follows:

- V_{Oh}: High o/p voltage
- V_{Ol}: Low o/p voltage
- V_M: Switching threshold voltage
- V_{Ih}: Minimum input voltage for logic "1" range
- V_{Il}: Maximum input voltage for logic "0" range
- NM_H: Noise margin high ($V_{Oh} - V_{Ih}$)

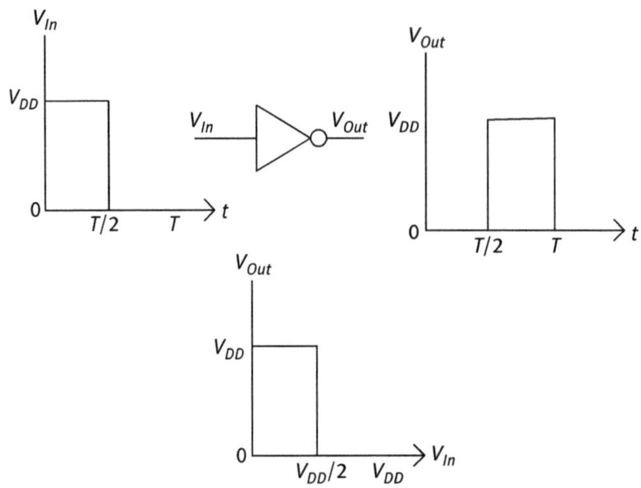

Figure 3.1: Input and output waveform and transfer characteristics of digital inverter.

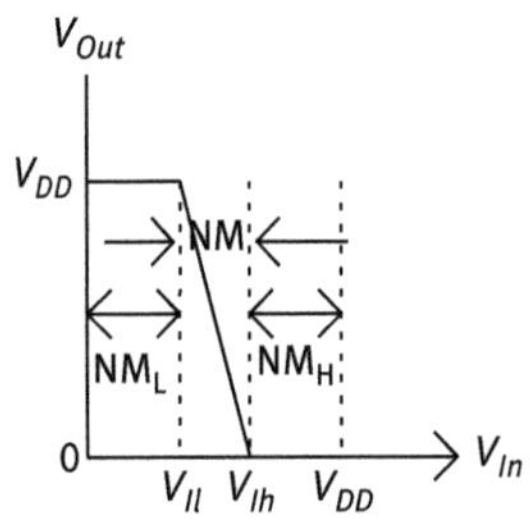

Figure 3.2: Voltage Transfer Characteristics (VTC) of practical inverter.

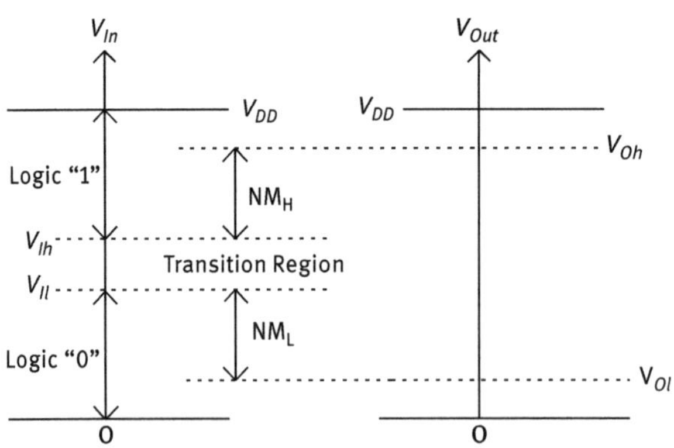

Figure 3.3: Voltage transfer characteristics and noise margin of inverter.

- $\mathrm{NM_L}$: Noise margin low ($V_{Il} - V_{Ol}$)
- NM: Noise margin ($\mathrm{NM_H} - \mathrm{NM_L}$)

In case of ideal inverter circuit, the above-mentioned transfer characteristic parameters will have the following values:

$$V_{Oh} = V_{DD} \text{ and } V_{Ol} = 0, \tag{3.1}$$

$$V_{Ih} = V_{Il} = V_{DD}/2, \tag{3.2}$$

$$\mathrm{NM_H} = \mathrm{NM_L} = V_{DD}/2. \tag{3.3}$$

So NM becomes zero in case of ideal inverter circuit. $\mathrm{NM_H}$ and $\mathrm{NM_L}$ also become less than $V_{DD}/2$, and noise margin becomes nonzero, as given in Figure 3.2. Also switching threshold voltage deviates from the ideal value, i.e. $V_{DD}/2$; by ΔV amount, where ΔV is small voltage drop.

3.1.2 Static CMOS Inverter in Super-threshold Regime

3.1.2.1 Structure and Operation

In 1963, Frank Wanlass at Fairchild described CMOS logic gate to us, but it was widely accepted after two decades of invention. CMOS [3–10] is considered as the most powerful transistor topology for minimum speed, cost per function and power dissipation per function. CMOS structure consists of pMOS and nMOS transistors, which are also complement to each other from the structural or operational point of view. In CMOS logic gate structure has two complementary network: pull-up network (PUN) and pull-down network (PDN), as given in Figure 3.4(a). Here we consider that logic "1" and logic "0" as supply voltage and ground. When we apply the low logic inputs (≈0), PUN turns on and PDN remains off. Under this situation, V_{DD} or logic "1" passes through the PUN to output load, and we obtain logic "1" as output. This situation is given in Figure 3.4(b). So PUN raises or pulls up the output voltage level to logic "1" under on condition. Similarly for high inputs (≈supply voltage) PDN turns on and PUN turns off. Then PDN passes the logic "0" from ground to output node. This situation is given in Figure 3.4(c). So PDN lowers or pulls down the output voltage level from logic "1" to logic "0" under on condition. As PUN and PDN are used to pass logic "1" and logic "0", respectively, we use pMOS and nMOS transistor, respectively, to construct them.

In a CMOS inverter single pMOS and nMOS transistors replace the PUN and PDN, as shown in Figure 3.5. pMOS (M_P) and nMOS (M_N) connect the output node with the supply voltage and ground, respectively. Hence V_{DD} and ground are taken as logic "1" and logic "0", respectively. The gate terminals of M_P and M_N are common, considered as the input of inverter, where we apply the input. Similarly the drain terminals of M_P and M_N are common, considered as the output, from where we measure the output. These charging and discharging operations are given in Figure 3.4 (b) and 3.4(c), respectively. The output and input waveforms of a CMOS inverter circuit are given

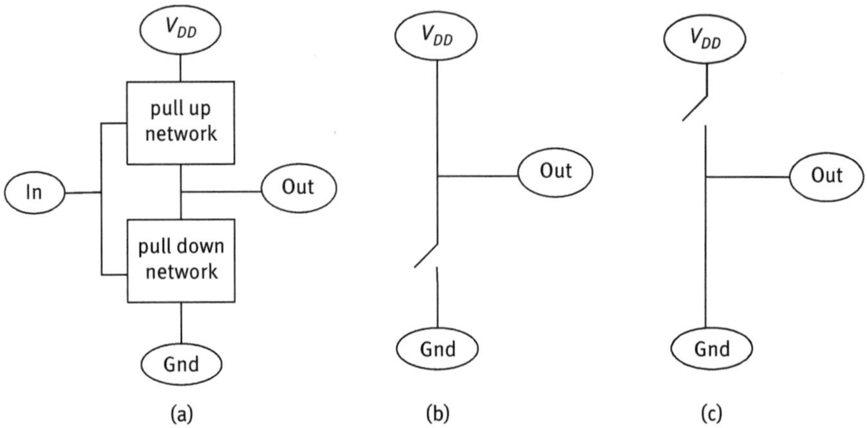

Figure 3.4: CMOS logic structure: (a) general structure, (b) PUN→ ON and PDN→ OFF, (c) PUN→OFF and PDN→ ON.

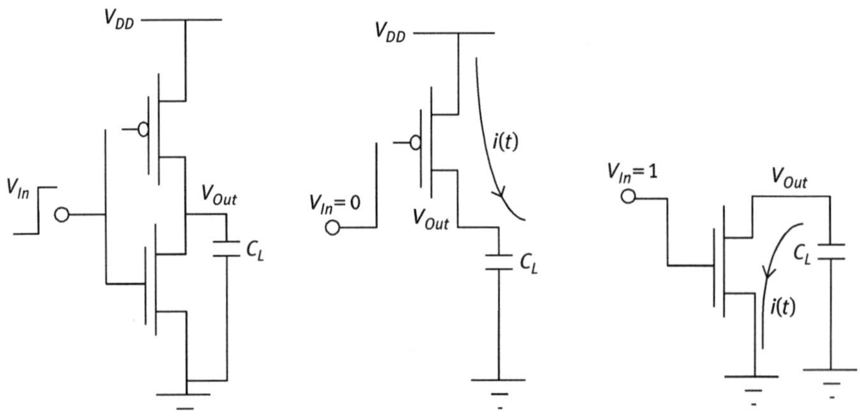

Figure 3.5: CMOS inverter: (a) inverter schematic; (b) charging phase; (c) discharging phase.

in Figure 3.6. Hence under 1 volt supply voltage, we are almost getting full swing in output node, so V_{Oh} = 1 volt and V_{Ol} = 0 volt.

When we apply low input ($V_I = 0$), M_P is turned on and M_N is turned off. The output node is charged up and the node voltage rises to supply voltage (logic "1") level through the pMOS transistor completely. As M_N is turned off, output remains at logic "1" state. Similarly, when we apply high input ($V_I = V_{DD}$) volt, input M_N is turned on and M_P is turned off. So the output node is discharged through the M_N transistor completely, produces logic "0" output. When M_P is on, M_N will be off. To understand

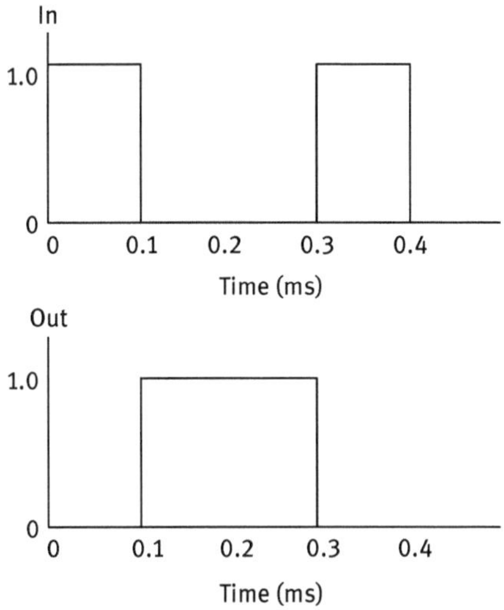

Figure 3.6: Input and output waveform of CMOS inverter in 22 nm technology with a 1 V supply voltage and 10 fF load.

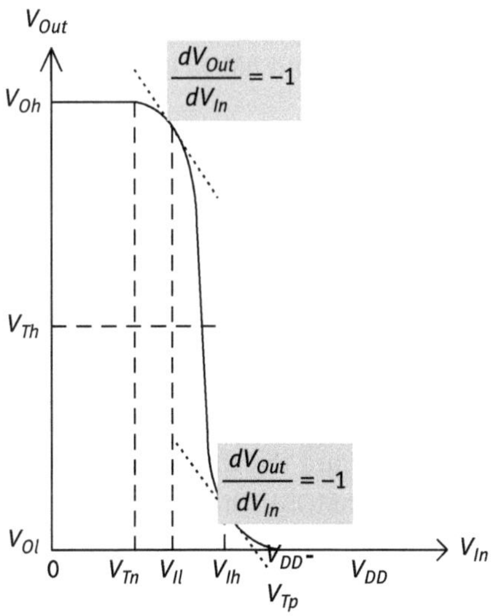

Figure 3.7: Transfer characteristics of CMOS inverter.

the VTC parameters of CMOS inverter, as given in Figure 3.7, we have to know the operating region of the pMOS and nMOS transistors properly first.

Hence from Figure 3.5 (a),

$$V_{GS,n} = V_{In} \text{ and } V_{SG,p} = V_{DD} - V_{In}, \tag{3.4}$$

$$V_{DS,n} = V_{Out} \text{ and } V_{Sd,p} = V_{DD} - V_{Out}. \tag{3.5}$$

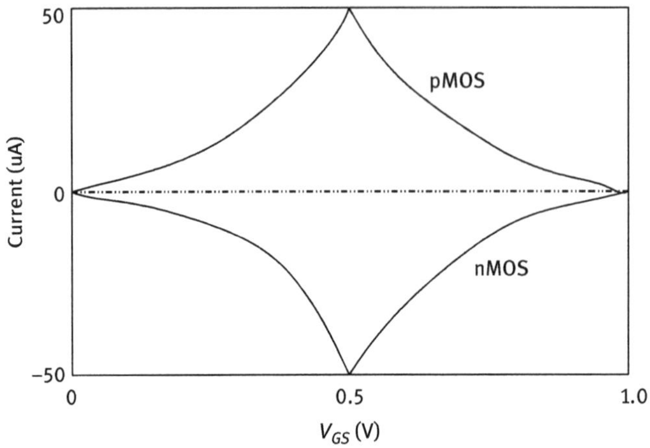

Figure 3.8: Current in pMOS and nMOS transistor in inverter (using CADENCE SPICE spectra).

To turn on nMOS or pMOS, the conditions are $V_{GS,n} \geq V_{Tn}$ or $V_{SG,p} \geq |V_{Tp}|$, respectively. When input switches from high to low, pMOS and nMOS transistors are turned off and on, respectively. Similarly the nMOS turns on and the pMOS turns off as the input level goes high. R_n and R_p, the turn on resistance of pMOS and nMOS transistor, can be expressed as

$$R_n = \frac{1}{K_n \left(\frac{W}{L}\right)_n (V_{DD} - V_{In})}; \tag{3.6}$$

$$R_p = \frac{1}{K_p \left(\frac{W}{L}\right)_p (V_{DD} - |V_{Tp}|)}. \tag{3.7}$$

Hence for pMOS and nMOS transistors, K_n and K_p are the transconductance parameters; $(\frac{W}{L})_n$ and $(\frac{W}{L})_p$ are the aspect ratios; V_{Tn} and $|V_{Tp}|$ are the threshold voltages, respectively; I_{Dx} is the drain current of X (X = pMOS/nMOS). Drain currents of X (X = pMOS/nMOS) transistor in a complete cycle is given in Figure 3.8 using CADENCE simulation, considering 1 volt supply voltage. pMOS and nMOS currents have almost same magnitude but they are opposite in direction. Peak amplitude of the drain current, for pMOS or nMOS, is approximately 48 μA for 1 volt supply voltage in 22 nm technology.

3.1.2.2 Noise Margin Calculation

The VTC of the CMOS inverter [11–16] can be divided into five different regions according to Figure 3.7. The regions are as follows:

- Region 1: $0 \leq V_{In} < V_{Tn}$
- Region 2: $V_{Tn} \leq V_{In} \leq V_{Il}$
- Region 3: $V_{Il} < V_{In} < V_{Ih}$

- Region 4: $V_{Ih} \leq V_{In} < V_{DD} - |V_{Tp}|$
- Region 5: $V_{DD} - |V_{Tp}| \leq V_{In} \leq V_{DD}$

Depending on the input conditions, the operating regions of the pMOS and nMOS transistors in these five regions are given in Table 3.1.

Region 1 In region 1, due to cutoff mode of nMOS, $I_{Dp} = I_{Dn} = 0$. Hence the output voltage is equivalent to supply voltage (V_{DD}) as the voltage drop across the pull-up path is negligibly small. So $V_{Oh} = V_{DD}$.

Region 2 When input voltage crosses the threshold of nMOS (V_{Tn}), current flow starts through the resistive path of PUN or PDN. Here pMOS remains in linear region but nMOS enters in the saturation. Due to nonzero current flow through the pull-up or pull-down path, voltage drop increases across the path; resultantly, output voltage decreases from maximum value. As $I_{Dp} = I_{Dn}$, we get the following:

$$\frac{K_n}{2}\left(\frac{W}{L}\right)_n (V_{GS,n} - V_{Tn})^2 = \frac{K_p}{2}\left(\frac{W}{L}\right)_p \left[2(V_{SG,p} - |V_{Tp}|)V_{Sd,p} - V_{Sd,p^2}\right]. \tag{3.8}$$

Now differentiating the above and by putting $\frac{dV_{Out}}{dV_{In}} = -1$, we can rewrite the above equation as

$$\frac{K_n}{2}\left(\frac{W}{L}\right)_n (V_{In}-V_{Tn})^2 = \frac{K_p}{2}\left(\frac{W}{L}\right)_p [2(V_{DD}-V_{In}-|V_{Tp}|)(V_{DD}-V_{Out})-(V_{DD}-V_{Out})^2]. \tag{3.9}$$

As $V_{In} = V_{Il}$,

$$V_{Il} = \frac{2V_{Out} + \dfrac{K_n\left(\frac{W}{L}\right)_n}{K_p\left(\frac{W}{L}\right)_p} V_{Tn} - V_{DD} - |V_{Tp}|}{1 + \dfrac{K_n\left(\frac{W}{L}\right)_n}{K_p\left(\frac{W}{L}\right)_p}}. \tag{3.10}$$

Table 3.1: Operating region of the pMOS and nMOS in an inverter circuit.

Region	pMOS	nMOS
Region 1	Linear region	Cutoff region
Region 2	Linear region	Saturation region
Region 3	Saturation region	Saturation region
Region 4	Saturation region	Linear region
Region 5	Cutoff region	Linear region

Assuming $|V_{Tp}| = V_{Tn} = V_{Th}$ and $r = \frac{K_p\left(\frac{W}{L}\right)_p}{K_n\left(\frac{W}{L}\right)_n}$, finally we get

$$V_{Il} = \frac{2V_{Out} + (\frac{1}{r} - 1)V_{Th} - V_{DD}}{(1 + \frac{1}{r})}. \tag{3.11}$$

Region 3 In region 3, pMOS and nMOS transistors both operate in saturation region. Output voltage drops rapidly as current reaches its maximum point in this transition region, providing very high gain. If we set the operating point in region 3, very high amplification or gain can be obtained. So inverter can be used as an analog amplifier if we set the operating point in the transition region. Equating the drain current of both transistors, we get

$$\frac{K_p}{2}\left(\frac{W}{L}\right)_p (V_{DD} - V_{In} - |V_{Tp}|)^2 = \frac{K_n}{2}\left(\frac{W}{L}\right)_n (V_{In} - V_{Tn})^2. \tag{3.12}$$

Hence we define the switching threshold voltage as $V_{In} = V_{Out} = V_{Th}$; So by putting $V_{In} = V_{Th}$ we get

$$V_{Th} = \frac{V_{Tn} + \sqrt{\frac{K_p\left(\frac{W}{L}\right)_p}{K_n\left(\frac{W}{L}\right)_n}}(V_{DD} - |V_{Tp}|)}{1 + \sqrt{\frac{K_p\left(\frac{W}{L}\right)_p}{K_n\left(\frac{W}{L}\right)_n}}}. \tag{3.13}$$

The above equation further can be simplified as

$$V_{Th} = \frac{V_{DD}\sqrt{r} + (1 - \sqrt{r})V_{Th}}{1 + \sqrt{r}}. \tag{3.14}$$

The V_{Th} of the CMOS inverter can be set to V_{DD} after proper sizing of the pMOS and nMOS transistors. Assuming $r = 1$, we get $V_{Th} = V_{DD}/2$. If the sizing parameter (r) is set to 1 we get

$$K_p\left(\frac{W}{L}\right)_p = K_n\left(\frac{W}{L}\right)_n,$$

assuming same C_{Ox} for both pMOS and nMOS transistors.

Or $\frac{\mu_p}{\mu_n} = \frac{\left(\frac{W}{L}\right)_n}{\left(\frac{W}{L}\right)_p}$,

where μ_p and μ_n are the mobility of pMOS and nMOS transistors.

Here assuming $\mu_p = n\mu_n$, we get $(\frac{W}{L})_p = n(\frac{W}{L})_n$.

So width of pMOS transistor is widened to match the resistance of nMOS by 3 to 3.5 times, or the effective resistances of PUN and PDN are matched completely, which provides symmetric propagation delays. It can be said clearly that as *r* varies, switching threshold voltage of the inverter also shifts. Wider pMOS shifts the switching threshold upward (more than $V_{DD}/2$), whereas if we increase the strength of the nMOS, switching threshold voltage moves closer to GND. For inverter circuit, layout and VTC using CADENCE SPICE spectra are shown in Figures 3.9 and 3.10. Layout of inverter circuit is useful to estimate the parasitic capacitances as well as intrinsic capacitances accurately.

Region 4 Here pMOS operates in saturation region but nMOS enters in linear region. We get the following equations after equating the drain currents:

$$\frac{K_p}{2}(V_{SG,p} - V_{Tp}|)^2 = \frac{K_n}{2}[2(V_{GS,n} - V_{Tn})V_{DS,n} - V_{DS,n^2}]. \tag{3.15}$$

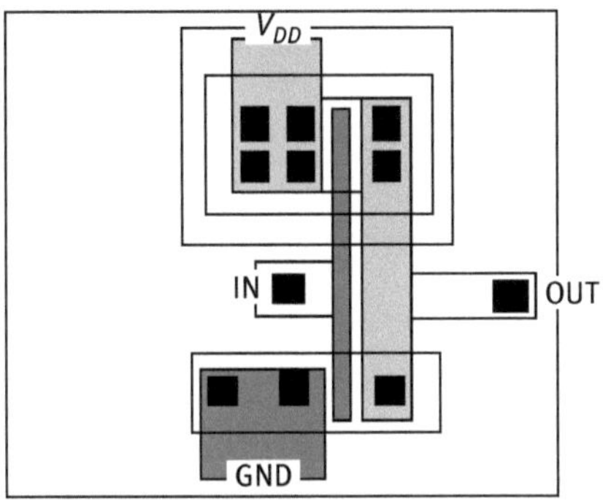

Figure 3.9: Layout of static CMOS inverter using CADENCE SPICE Spectra.

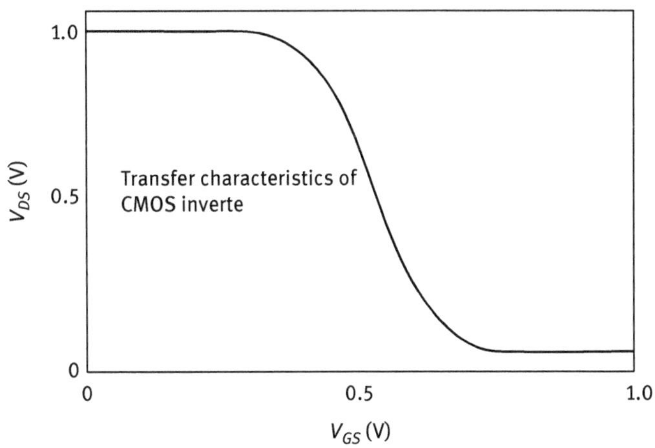

Figure 3.10: Transfer characteristics of inverter using CADENCE SPICE spectra.

Now differentiating the above and by putting $\frac{dV_{Out}}{dV_{In}} = -1$, we can rewrite the above equation as

$$\frac{K_p}{2}\left(\frac{W}{L}\right)_p (V_{SG,p} - V_{Tp}|)^2 = \frac{K_n}{2}\left(\frac{W}{L}\right)_n [2(V_{GS,n} - V_{Tn})V_{DS,n} - V_{DS,n}^2]. \tag{3.16}$$

As $V_{In} = V_{Ih}$,

$$V_{Ih}\left(1+\frac{K_p\left(\frac{W}{L}\right)_p}{K_n\left(\frac{W}{L}\right)_n}\right) = 2V_{Out} + V_{Tn} + \frac{K_p\left(\frac{W}{L}\right)_p}{K_n\left(\frac{W}{L}\right)_n}(V_{DD} - |V_{Tp}|); \tag{3.17}$$

$$\text{or } V_{Ih} = \frac{2V_{Out} + V_{Tn} + \frac{K_p\left(\frac{W}{L}\right)_p}{K_n\left(\frac{W}{L}\right)_n}(V_{DD} - |V_{Tp}|)}{\left(1+\frac{K_p\left(\frac{W}{L}\right)_p}{K_n\left(\frac{W}{L}\right)_n}\right)}. \tag{3.18}$$

Putting $|V_{Tp}| = V_{Tn} = V_{Th}$; and $r = \frac{K_p\left(\frac{W}{L}\right)_p}{K_n\left(\frac{W}{L}\right)_n}$, finally we get

$$V_{Ih} = \frac{2V_{Out} + (1-r)V_{Th} + rV_{DD}}{(1+r)}. \tag{3.19}$$

Region 5 When the input voltage crosses (V_{DD}-$|V_{Tp}|$) point, pMOS transistor is turned off, whereas the nMOS remains in previous active region and drags small amount of current from supply voltage. As input is set to high logic, pull-down path becomes active, setting the output node at ground potential. Output node discharges completely through the nMOS transistor. So $V_{Ol} = 0$. The sizing parameter plays the most important role in the operation of CMOS inverter. By setting $r = 1$, we almost get the ideal operation as V_{Th} becomes $V_{DD}/2$. From eqs. (3.11) and (3.18), the noise margin of the CMOS inverter can be derived easily. In case of symmetrical inverter NM_H and NM_L, both are almost equal to $(3V_{DD} + 2V_{Th})/8$, as V_{Ih} and V_{Il} can be approximated as $(5V_{DD} - 2V_{Th})/8$ and $(3V_{DD} + 2V_{Th})/8$, respectively.

However, noise margin [13–17] also depends on this sizing parameter strictly. If r increases NM_L increases and NM_H decreases. Similarly, if r decreases, then NM_H increases and NM_L decreases. Variation of transfer characteristics curve under

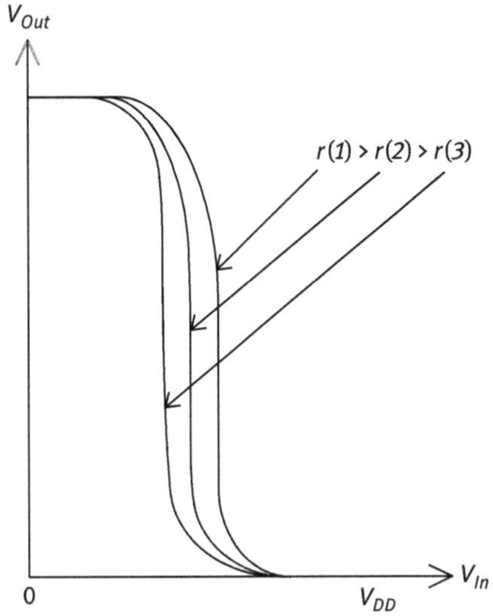

Figure 3.11: Effect of transistor's width on transfer characteristics.

different value of r is given in Figure 3.11. Here $r(2)$ is nominal, i.e. $(\frac{W}{L})_p = (\frac{W}{L})_n$. But we choose $r(1)$ and $r(3)$, where the relations between the aspect ratios are $(\frac{W}{L})_p > (\frac{W}{L})_n$ and $(\frac{W}{L})_p < (\frac{W}{L})_n$, respectively. Changing the aspect ratio from the nominal value can shift the transition region.

3.1.2.3 Delay of CMOS Inverter

Propagation delay [18–28] of the CMOS inverter is mainly due to the charging and discharging of the load capacitor. Estimation of load capacitance [19, 21] is most important in delay calculation, as delay depends on load directly. So here we first find out the effective load capacitance of an inverter and then will discuss the propagation delay in depth.

Figure 3.12 details the component of capacitive load. The components are $C_{Gd,p}$ (gate to drain capacitance of pMOS), $C_{Gd,n}$ (gate to drain capacitance of nMOS), $C_{Bd,p}$ (body to drain capacitance of pMOS) $C_{Bd,n}$ (body to drain capacitance of pMOS), C_W (wiring capacitance) and C_G (and interconnect input capacitance of the next stage) mainly. Here $C_{Gd,p}$ or $C_{Gd,n}$ are mainly overlap capacitances of pMOS or nMOS, respectively. Body to drain capacitances which is nonlinear in nature is due to reverse-biased *pn*-junction. Due to interconnects, different length and width of connecting wires also wiring capacitance grows and will be a component of output load capacitance. Gate capacitance of next stage is also taken into account to consider the number of fan-out gates. Above-mentioned effects can be combined to get the net output load capacitance as

$$C_L = C_{Gd,p} + C_{Gd,n} + C_{Bd,p} + C_{Bd,n} + C_W + C_G. \tag{3.20}$$

T_{Phl} or high-to-low propagation delay can be defined by the time gap between the 50% transition of the rising input voltage and the falling output voltage. Similarly T_{Plh} or low-to-high propagation delay is the time gap between the 50% transition of the falling input voltage and the rising output voltage. These propagation delays are given in Figure 3.13. Average propagation delay is the average of these T_{Phl} and T_{Plh}. So

$$T_{Pd} = \frac{T_{Phl} + T_{Plh}}{2}. \tag{3.21}$$

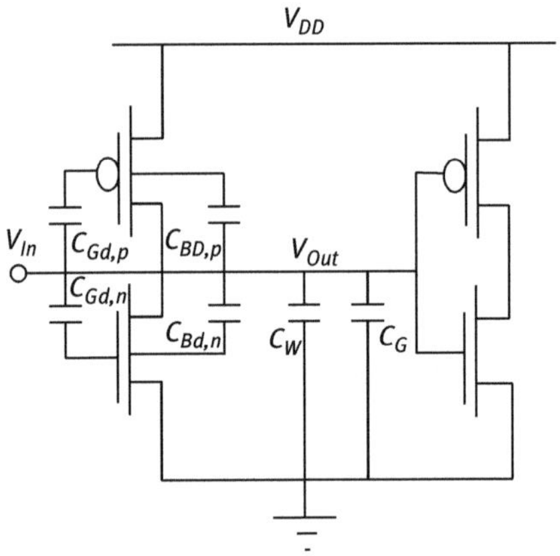

Figure 3.12: Components of load capacitance in CMOS inverter.

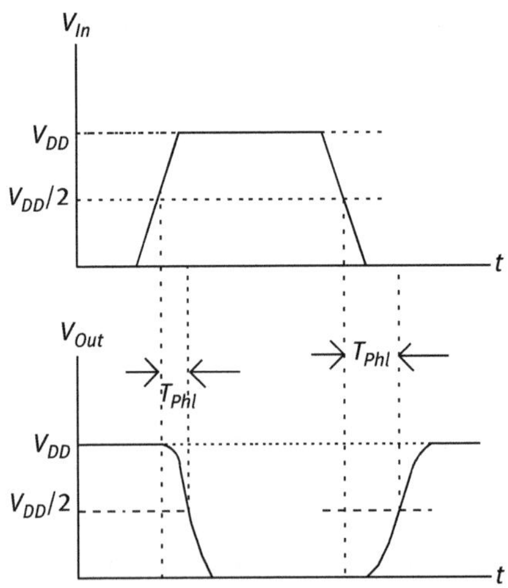

Figure 3.13: Graphical presentation of propagation delay.

SPICE simulation can be done to measure the delay accurately. If we want to find out the analytical expression of propagation delay, we need to integrate the discharging current going through the output node capacitance. To make the solution of differential equation little bit easier, we estimate the first-order delay first. When output node discharges to zero potential, nMOS lies in linear region. Assuming the turn-on resistance of pull-up path and the output load as R_N and C_L, respectively, low-to-high propagation delay can be calculated as

$$T_{Plh} = (\ln 2)R_N C_L. \tag{3.22}$$

From eq. (3.6), putting the value of R_N we get

$$T_{Phl} = \ln 2 \frac{C_L}{K_n\left(\frac{W}{L}\right)_n (V_{DD} - V_{In})}. \tag{3.23}$$

Similarly, for high-to-low transition, the T_{Phl} will be

$$T_{Plh} = \ln 2 \frac{C_L}{K_p\left(\frac{W}{L}\right)_p (V_{DD} - |V_{Tp}|)}. \tag{3.24}$$

So the average propagation delay can be approximated as

$$T_{Pd} = \frac{T_{Phl} + T_{Plh}}{2} = 0.345 C_L \left(\frac{1}{K_n\left(\frac{W}{L}\right)_n (V_{DD} - V_{In})} + \frac{1}{K_p\left(\frac{W}{L}\right)_p (V_{DD} - |V_{Tp}|)} \right). \tag{3.25}$$

In Figure 3.14, we consider the discharging operation through nMOS to estimate the more accurate value of T_{Plh}. From t_1 to t_2, nMOS will be in saturation and from t_2 to t_3

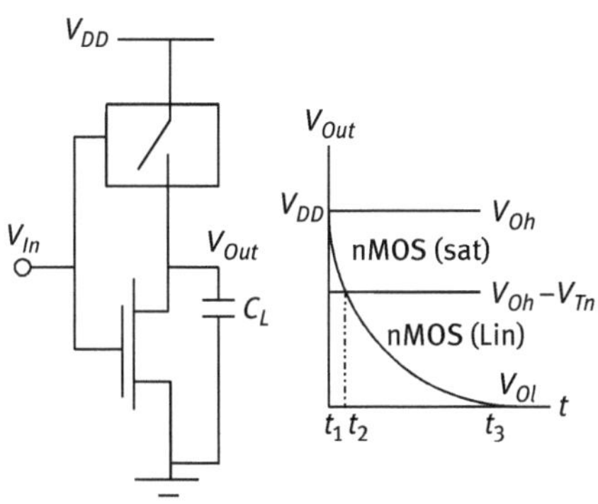

Figure 3.14: Delay time calculation.

nMOS will be in linear region. So high-to-low propagation delay can be obtained using the following steps:

$$T_{Phl} = -C_L \int_{V_{Oh}}^{V_{DD}/2} \frac{dV_{Out}}{I_{Dn}}; \tag{3.26}$$

$$= -C_L \int_{V_{Oh}}^{V_{Oh}-V_{Tn}} \frac{dV_{Out}}{I_{Dn}(sat)} - C_L \int_{V_{Oh}-V_{Tn}}^{V_{DD}/2} \frac{dV_{Out}}{I_{Dn}(lin)};$$

$$= -2C_L \int_{V_{Oh}}^{V_{Oh}-V_{Tn}} \frac{dV_{Out}}{K_n \left(\frac{W}{L}\right)_n (V_{Oh} - V_{Tn})^2} - 2C_L \int_{V_{Oh}-V_{Tn}}^{V_{DD}/2} \frac{dV_{Out}}{K_n \left(\frac{W}{L}\right)_n [2(V_{Oh} - V_{Tn})V_{Out} - V_{Out}^2]};$$

$$= \frac{C_L}{K_n \left(\frac{W}{L}\right)_n (V_{Oh} - V_{Tn})} \left\{ \frac{2V_{Tn}}{(V_{Oh} - V_{Tn})} + \ln\left[\frac{4(V_{Oh} - V_{Tn})}{(V_{Oh} + V_{Ol})} - 1\right] \right\}. \tag{3.27}$$

Putting the value of V_{Oh} and V_{Ol}, finally we get

$$T_{Phl} = \frac{C_L}{K_n \left(\frac{W}{L}\right)_n (V_{DD} - V_{Tn})} \left\{ \frac{2V_{Tn}}{(V_{DD} - V_{Tn})} + \ln\left[3 - 4\frac{V_{Tn}}{V_{DD}}\right] \right\}. \tag{3.28}$$

Similarly during charging operation, T_{Phl} can be expressed as

$$T_{Plh} = \frac{C_L}{K_p \left(\frac{W}{L}\right)_p (V_{Oh} - |V_{Tp}|)} \left\{ \frac{2|V_{Tp}|}{(V_{Oh} - |V_{Tp}|)} + \ln\left[\frac{4(V_{Oh} - |V_{Tp}|)}{(V_{Oh} + V_{Ol})} - 1\right] \right\} \tag{3.29}$$

or

$$T_{Plh} = \frac{C_L}{K_p \left(\frac{W}{L}\right)_p (V_{DD} - |V_{Tp}|)} \left\{ \frac{2|V_{Tp}|}{(V_{DD} - |V_{Tp}|)} + \ln\left[3 - \frac{4|V_{Tp}|}{V_{DD}}\right] \right\}. \tag{3.30}$$

Average propagation delay will be the average of T_{Phl} and T_{Plh}. From eqs. (3.28) and (3.30), it can be said that the propagation delay depends on load capacitance, supply voltage and the aspect ratios of the transistors mainly. Normalized delay under different supply voltage is given in Figure 3.15. It can be said from Figure 3.15 that to enhance the performances we need to increase the supply voltage to reduce the delay. Then other issues like oxide breakdowns, hot electron effects, power dissipations, etc.,

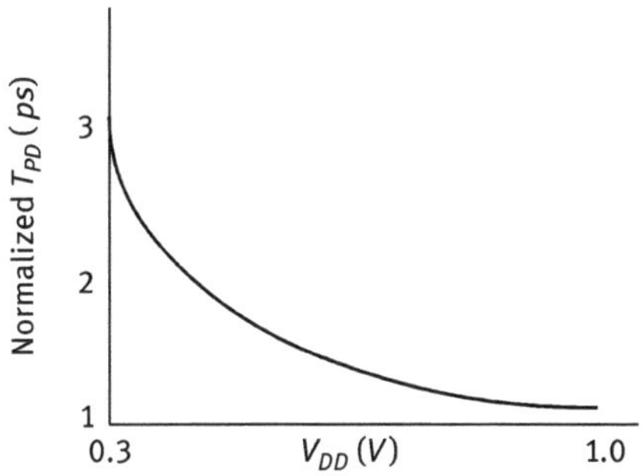

Figure 3.15: Normalized delay under different supply voltage for CMOS inverter.

will start to deteriorate the performance of the circuit. So we need to choose some optimum supply voltage for which the product of delay and power dissipation becomes minimum. This will be discussed in the next section in depth. From eqs. (3.28) and (3.30), we get

$$\left(\frac{W}{L}\right)_p = \frac{C_L}{T_{Plh}K_p(V_{Oh} - |V_{Tp}|)}\left\{\frac{2|V_{Tp}|}{(V_{Oh} - |V_{Tp}|)} + \ln\left[\frac{4(V_{Oh} - |V_{Tp}|)}{(V_{Oh} + V_{Ol})} - 1\right]\right\}; \tag{3.31}$$

$$\left(\frac{W}{L}\right)_n = \frac{C_L}{T_{Phl}K_n(V_{DD} - V_{Tn})}\left\{\frac{2V_{Tn}}{(V_{DD} - V_{Tn})} + \ln\left[3 - 4\frac{V_{Tn}}{V_{DD}}\right]\right\}. \tag{3.32}$$

Effect of the aspect ratios of pMOS and nMOS in CMOS inverter can be explained using the CADENCE simulation, as given in Figure 3.16. Larger transconductance parameter and aspect ratio of the MOSFETs increases the performances. Low-to-high propagation delay can be reduced by widening the pMOS transistor or increasing the charging current, but it may cause larger parasitic capacitance. Indeed larger width of the device increases the gate capacitances and increases fan-out, which can deteriorate the performances. So we need to reduce the load to achieve minimum delay [21, 22, 25, 29, 30] by proper sizing of the transistors [21, 26]. Choosing proper layout, we can reduce the effect of parasitic capacitances and resistances, which in turn reduces the load capacitances. Reduction of the load capacitance can enhance the performances of the inverter.

Here we discuss the effect of load capacitances on the circuit delay in detail. From eq. (3.20), we can represent the load capacitance (C_L) as a summation of internal capacitances (C_{In}) and external capacitance (C_{Ex}) along with the extra load. Internal capacitances are basically due to the internal capacitances of MOSFET which are $C_{Gd,p}$, $C_{Gd,n}$, $C_{Bd,p}$ and $C_{Bd,n}$; whereas the external capacitances are due to wiring and parasitic capacitances mainly. Under the assumption of symmetric delay (equal rise and fall time) and equal pull-up and pull-down current, first-order delay expression can

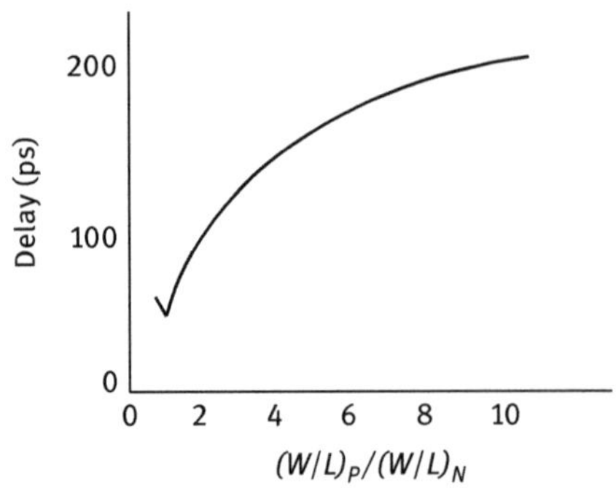

Figure 3.16: Variation of propagation delay for different aspect ratios of pMOS under 1 volt supply voltage.

be written as

$$T_{Pd} = (\ln 2)R_S(C_{In} + C_L). \tag{3.33}$$

Here R_S is the path resistance of pull-up and pull-down path under symmetric delay condition. C_L is substituted by $(C_{In} + C_L)$. So

$$T_{Pd} = \tfrac{1}{2}(\ln 2)R_S\, C_{Int}\left(1 + \frac{C_L}{C_{In}}\right); \tag{3.34}$$

$$= T_{Int,Delay}\left(1 + \frac{C_L}{C_{In}}\right). \tag{3.35}$$

Hence $T_{Int,Delay}$ (= $[\tfrac{1}{2}\ln 2]R_S\, C_{Int}$) is the intrinsic or unloaded (under-zero load) time delay of the inverter circuit, depending on the layout, process technology, etc. If we further assume that two identical CMOS inverters are connected in cascade and pMOS devices are made n times larger than nMOS, we get the following expression for the output load capacitance of the first stage:

$$C_L \approx C_{Gd,p} + C_{Gd,n} + C_{Bd,p} + C_{Bd,n} + C_{G,p} + C_{G,n} + C_W. \tag{3.36}$$

Hence neglecting extra load and the drain to body capacitances of the transistors, i.e. $C_{Bd,p}$ and $C_{Bd,n}$, we get

$$C_L \approx (1 + n)(C_{Gd,n} + C_{G,n}) + C_W. \tag{3.37}$$

Putting the value of C_L, we can rewrite eq. (3.25) as

$$T_{Pd} = \{½(\ln 2)(1+n)(C_{Gd,n} + C_{G,n}) + C_W\}\left(R_N + \frac{R_P}{n}\right);$$
$$= ½(\ln 2)\{(1+n)(C_{Gd,n} + C_{G,n}) + C_W\}R_N\left(1+\frac{\sigma}{n}\right), \quad (3.38)$$

where

$$\frac{R_P}{R_N} = \sigma.$$

Now the optimal value of n can be obtained by differentiating eq. (3.38) and putting it to equal to zero, i.e. $\frac{dT_{Pd}}{dn} = 0$
or

$$n = \sqrt{\sigma\left(1+\frac{C_W}{C_{Gd,n} + C_{G,n}}\right)}. \quad (3.39)$$

Finally we get

$$\left(\frac{W}{L}\right)_p = \left(\frac{W}{L}\right)_n\sqrt{\sigma\left(1+\frac{C_W}{C_{Gd,n} + C_{G,n}}\right)}. \quad (3.40)$$

Now here we are going to find the relationship between the input gate capacitance C_G and the output capacitance $(C_L + C_{Ex})$ to explain the loading effect. The inverter chain is given in Figure 3.17. Here $C_{G,i}$ represents the gate capacitance of the $(i+1)$ stage. To drive some extra load we insert a C_{Lf}. For a single stage inverter,

$$T_{Pd} = ½(\ln 2)R_S\, C_{Int}\left(1+\frac{C_L + C_{Ex}}{C_{In}}\right), \quad (3.41)$$

where, $C_{In} = \beta\, C_G$. β is a process-dependent parameter and is assumed to be 1 approximately. So

$$T_{Pd} = T_{Int,\ Delay}\left(1+\frac{C_L + C_{Ex}}{\beta C_G}\right);$$
$$= T_{Int,\ Delay}\left(1+\left(1+\frac{f_{Eff}}{\beta}\right).\right) \quad (3.42)$$

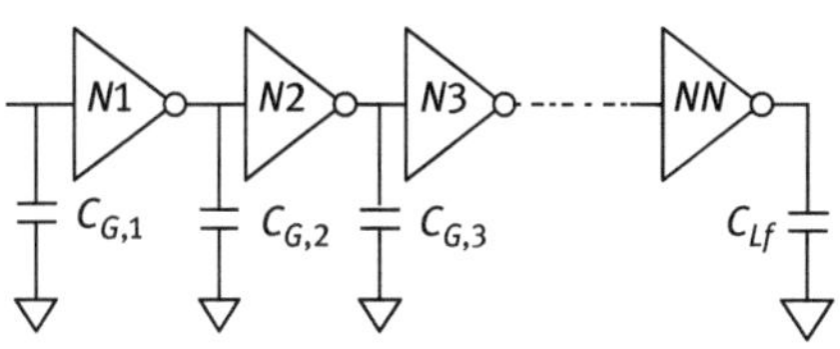

Figure 3.17: Inverter chain for optimum sizing.

Here f_{Eff} is the effective fan-out of a single inverter. In case of an N number of inverter's chain, the total delay can be estimated by adding up individual delay element. So total propagation delay becomes

$$\begin{aligned} T_{Pd,N} &= T_{Pd,1} + T_{Pd,2} + T_{Pd,3} + \cdots + T_{Pd,N} \\ &= \sum_{i=1}^{N} T_{Pd,i} = \sum_{i=1}^{N} \left(1 + \frac{C_{G,i+1}}{\beta C_{G,i}}\right). \end{aligned} \tag{3.43}$$

Solving the above equation we get approximately the optimal size of i-th inverter

$$C_{G,i} = \sqrt{C_{G,i-1} C_{G,i+1}}. \tag{3.44}$$

So each stage has same effective fan-out and same delay also. Total fan-out (F_{Eff}) becomes

$$F_{Eff} = \frac{C_{Lf}}{C_{G,1}} = f_{Eff}{}^{N}. \tag{3.45}$$

Thus minimum path delay becomes

$$T_{Pd,N}(\text{min}) = N T_{Int,\ Delay} \left(1 + \frac{\sqrt[N]{F_{Eff}}}{\beta}\right). \tag{3.46}$$

The above equation can be rearranged as

$$T_{Pd,N} = \frac{\ln F_{Eff}}{\ln f_{Eff}} T_{Int,\ Delay} \left(1 + \frac{\sqrt[N]{F_{Eff}}}{\beta}\right) \ as\ N = \frac{\ln F_{Eff}}{\ln f_{Eff}}; \tag{3.47}$$

$$= \frac{\ln F_{Eff}}{\beta} T_{Int,Delay} \left(\frac{\beta}{\ln f_{Eff}} + \frac{f_{Eff}}{\ln f_{Eff}}\right). \tag{3.48}$$

Hence to find out the optimal number of stages we can do the following:

$$\frac{dT_{Pd,N}}{df_{Eff}} = 0$$

or

$$T_{Int,\ Delay} \frac{\ln f_{Eff} - 1 - \frac{\beta}{f_{Eff}}}{(\ln f_{Eff})^2} = 0. \tag{3.49}$$

If we neglect the self-loading or if the load capacitor consists of the fan-out only then

$$N = \ln F_{Eff}. \tag{3.50}$$

3.1.2.4 Power Dissipation in CMOS Inverter

Power dissipation is a pivotal challenge in digital IC design. Though static CMOS logic structure dissipates almost zero power, yet in case of CMOS digital ICs having high performance and high packing density, power dissipation increases. The disadvantages of adding extra heat sinks is that it requires more silicon area, cost and weight [31–38]. Therefore proper analysis of power dissipation becomes essential for the efficacy of high performance VLSI circuits.

The main sources of power dissipation in a static inverter are mainly dynamic power dissipation (P_{Dyn}), static power dissipation (P_{Stat}) and finally short circuit power dissipation (P_S).

Dynamic Power dissipation

Dynamic or switching power dissipation is due to the charging and discharging of the output node, as given in Figure 3.5. In static CMOS when we charge a node having capacitance C_L, E_T ($=C_L V_{DD^2}$) energy is extracted from the supply source. Out of the total energy (E_T), E_C ($=\frac{1}{2} C_L V_{DD^2}$) is stored across the capacitor and remaining $\frac{1}{2} C_L V_{DD^2}$ energy is dissipated across the pull-up path. During the discharging phase, the charge, stored in capacitor, flows through the pull-down path to the ground, causing some energy dissipation across pull-down path. E_T and E_C can be determined on the basis of the following steps:

$$E_T = \int_0^{\infty} i(t) V_{DD} dt = C_L V_{DD} \int_0^{\infty} \frac{dV_{Out}}{dt} dt;$$

$$= C_L V_{DD} \int_0^{V_{DD}} dV_{Out} = C_L V_{DD}^2. \tag{3.51}$$

Similarly $E_C = \int_0^{\infty} i(t) V_{Out} dt = C_L V_{Out} \int_0^{\infty} \frac{dV_{Out}}{dt} dt;$

$$= C_L \int_0^{V_{DD}} V_{Out} dV_{Out} = \frac{C_L V_{DD}^2}{2}. \tag{3.52}$$

In short for every switching of output node $C_L V_{DD}^2$J of energy dissipates. In order to compute the dynamic power dissipations more accurately we need to introduce switching activity parameter (α). If input signal changes rapidly, switching of output node also becomes frequent and hence dissipation [35]. Overall network topology and the implemented function also influence the switching activity or the dynamic power dissipation. Assuming that the inverter circuit is switched on and off α time in 1 s, the expression of dynamic power dissipation will be

$$P_{Dyn} = \alpha C_L V_{DD}^2 f_{Op}. \tag{3.53}$$

Here α is the switching activity parameter. Some conclusions can be made in case of dynamic power dissipations. They are as follows:

- Dynamic power dissipation increases proportionately with the square of supply voltage. So if we decrease the supply voltage by two times, dynamic power will be reduced by four times. The drawback is reduction of supply voltage increases the delay or deteriorates the performances. Also as we cannot reduce the supply voltage below $V_{Tn} + |V_{Tp}|$ in static inverter, so aggressive scaling of supply voltage is not possible.
- In high performance designs, drain diffusion areas are kept as small as possible to reduce the parasitic capacitances which in turn reduce the effect of loading.
- Switching activity in CMOS digital ICs can be reduced by algorithmic optimization (based on characteristics and statistics of the data transmission), architecture optimization (balancing of delay path, ordering of input signals, etc.), logic topology and circuit optimization.

Short circuit Power Dissipation

When input switches from higher logic to lower or vice versa, both the transistors operate in saturation region, conduct for short time span and create a short circuit path between supply voltage and ground. Due to this direct current flow, short circuit power dissipation (P_S) [39–41] occurs, as shown in Figure 3.18. If we assume a zero rise or fall time then this short circuit power dissipation will be negligibly small. In realistic situations we have to consider some finite rise and fall time for every input pulse.

According to Figure 3.18, during finite rise (T_{Rise}) and fall (T_{Fall}) time, short circuit current of maximum value I_{Peak} is flowing through the transistors. Hence transistor sizing, capacitive load and transition time of the input signal influence this short circuit current. Short circuit current is also a strong function of the ratio between input and output slopes. Then the total power dissipation due to this event in a full cycle can be estimated roughly by the following expressions:

$$\begin{aligned} P_S &= V_{DD}\left(\frac{I_{Peak}T_{Rise}}{2} + \frac{I_{Peak}T_{Fall}}{2}\right)f_{Op}; \\ &= V_{DD}I_{Peak}\left(\frac{T_{Rise} + T_{Fall}}{2}\right)f_{Op}. \end{aligned} \tag{3.54}$$

It has been observed that this short circuit power dissipation dominates for shorter channel length.

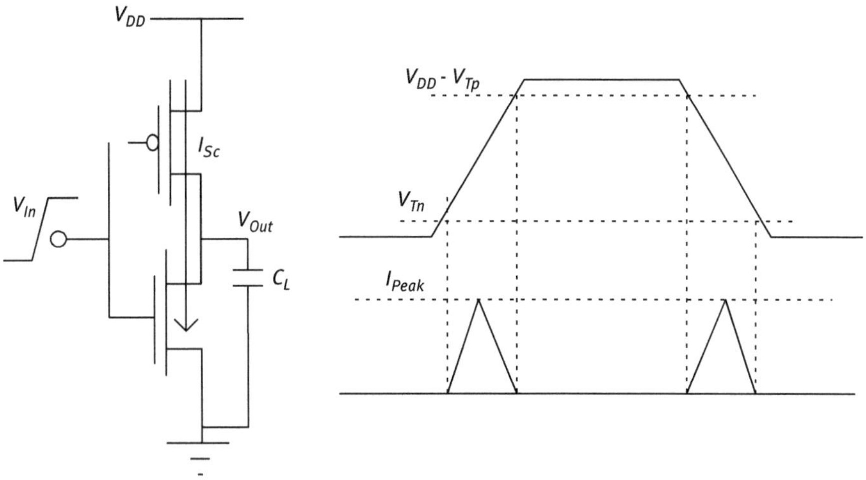

Figure 3.18: Generation and flow of short circuit current and power dissipation.

Static Power Dissipation

With downscaling of CMOS device architecture leakage current increases alarmingly and static power dissipation contributes [42–44] a significant portion of power. The leakage current components are given in Figure 3.19.

Main leakage components in static CMOS inverter are the following:

- Subthreshold leakage; when transistor is turned off, i.e. $V_{GS} < V_{Th}$
- Gate leakage through the ultra-thin oxide layers [44–46]
- Leakage current due to reverse-biased drain- and source-substrate junction
- Band-to-band tunneling (BTBT)

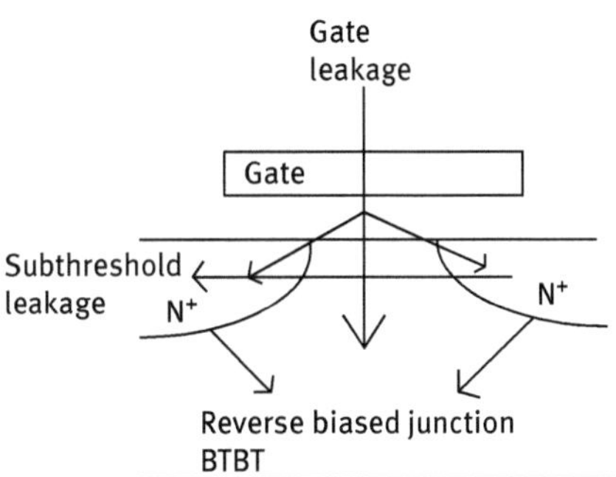

Figure 3.19: Leakage components of MOSFET in inverter.

Considering I_{Leak} accounts all the above-mentioned components, we get the static power dissipation as follows:

$$P_{Stat} = V_{DD} I_{Leak}. \tag{3.55}$$

Ignoring BTBT effect, we get

$$I_{Leak} = W I_S e^{\frac{V_{Th}}{\eta}}. \tag{3.56}$$

Here, W and V_{Th} are the transistor width and threshold voltage of the transistor, respectively; I_S is the leakage current at $V_{Th} = 0$, η is the subthreshold slope factor. So the static energy dissipation becomes

$$\begin{aligned} .E_S &= V_{DD} I_{Leak} T_{OP} \\ &= W I_S e^{\frac{V_{Th}}{\eta}} V_{DD} T_{Op}. \end{aligned} \tag{3.57}$$

T_{OP} is the circuit time. So the total power dissipation in a CMOS inverter becomes

$$\begin{aligned} P_{Total} &= P_{Dyn} + P_{Stat} + P_S \\ &= \alpha C_L V_{DD}^2 f_{Op} + V_{DD} I_{Peak} \left(\frac{T_{Rise} + T_{Fall}}{2}\right) f_{Op} + I_{Leak} V_{DD} f_{Op}. \end{aligned} \tag{3.58}$$

Total power dissipation of CMOS inverter circuit for 1 volt supply voltage in a complete cycle and for different supply voltages is given in Figures 3.20 and 3.21, respectively. Though for higher gate length switching energy dissipation dominates, yet in sub-90 nm technology static power dissipation starts to dominate over dynamic power due to aggressive scaling.

Here we are introducing a new metric, EDP (energy delay product), which is basically the product of total energy dissipation and gate delay. Using the EDP model, given in this section, one can find the optimal conditions.

Total energy dissipation in a CMOS inverter is given in eq. (3.58). Here we want to introduce a new parameter circuit time (T_{Op}), which can be written as $T_{Op} = L_D \tau$. Here L_D and τ are the logic depth and the gate delay of inverter. L_D represents the number of inverter in a ring oscillator. Hence we assume that the operating frequency of ring oscillator is same as the maximum operating frequency of the chip. Gate delay [47] can be represented as

$$\tau = K \frac{V_{DD}}{(V_{DD} - V_{Th})^{\alpha}}. \tag{3.59}$$

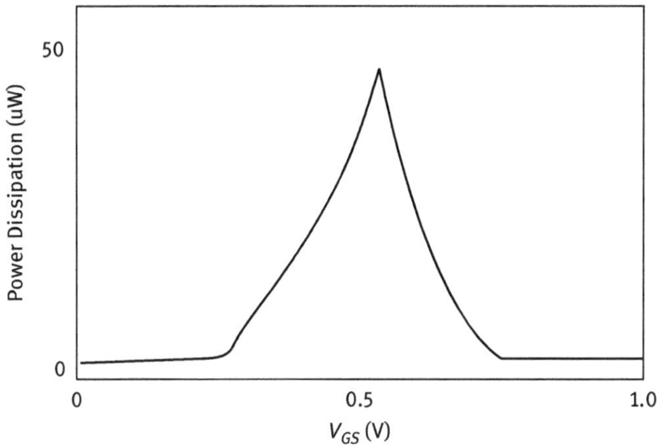

Figure 3.20: Power dissipation of CMOS inverter in a full cycle under 1 V supply.

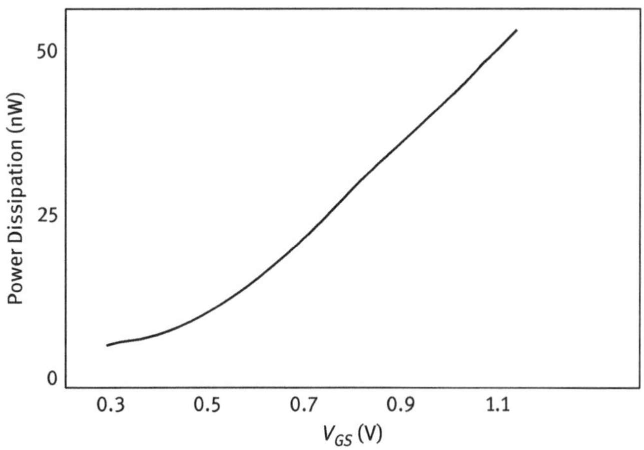

Figure 3.21: Power dissipation of CMOS inverter for different supply voltages in 90 nm technology.

Here K is a constant process parameter. So combining the above two equations we get the EDP as

$$EDP = \left(\alpha C_L V_{DD}^2 + WI_S e^{\frac{V_{Th}}{\eta}} V_{DD} T_{Op} \tau\right) K \frac{V_{DD}}{(V_{DD} - V_{Th})^{\alpha}}. \tag{3.60}$$

EDP depends on supply voltage strongly. To find out the optimum supply voltage for minimum EDP [47, 48] we have to differentiate the above equation and then set to zero. Finally we obtain

$$V_{DD,Optimum} = \frac{3V_{Th}}{3-\alpha} + \frac{3\alpha}{3-\alpha}\eta. \tag{3.61}$$

3.1.3 Introduction to Subthreshold Logic

To fulfill the demand of ultralow power dissipation in portable application, design of subthreshold CMOS ICs becomes attractive. The pivotal advantages of operating devices in the subthreshold region compared to super-threshold ($V_{DD} > V_{Th}$) counterpart are reduced power due to low supply voltage (V_{DD}) and small gate capacitance. Reduction of supply voltage decreases the power dissipation significantly by operating the circuit at a lower supply voltage. In subthreshold region, low gate capacitance produces more switching power reduction. Since the subthreshold leakage current is very small in magnitude and decreases at higher frequency range, these circuits cannot perform at very high frequencies. So this subthreshold logic family [49–54] is suitable for the application where ultralow power is pivotal issue instead of speed or performances [50, 52].

Subthreshold ICs depend on temperature and process parameters variations. The main effects of temperature fluctuation are the following:

- Due to temperature increment saturation of velocity degrades.
- Rising die temperature increases the gate overdrive voltage and the drain current.
- Drain current also increases due to the decrease in carrier mobility, as a result of an increase in die temperature.

Extra care should be taken on PVT analysis of subthreshold circuit family [49–51, 54], as circuits become less robust at ultralow voltages. In the next section we will discuss the inverter in subthreshold region. With that we also discuss the different issues like power dissipation, speed, noise margin, etc., in detail to provide a complete idea of these ultralow voltage logic operations [55, 56].

3.1.3.1 Subthreshold Inverter Circuit and Operation

In subthreshold voltage, the drain current of an n-channel MOSFET can be given by [50]

$$I_{Sub} = I_S e^{\frac{(V_{GS} - V_{Th})}{n_n u_T}} \left(1 - e^{\frac{-V_{DS}}{u_T}}\right). \tag{3.62}$$

Hence we have assumed body and drain terminals of MOSFET transistors are tied up, i.e. $V_{BS} = 0$; V_{GS} and V_{DS} are the gate-to-source and drain-to-source voltages, respectively; V_{Th} is the threshold voltage of nMOS transistor; u_T $(= \frac{KT}{q})$ is the thermal voltage; n is the process dependent parameter and can be termed as slope factor; I_S is the specific current which can be given by

$$I_S = 2n_n \mu C_{Ox} u_T^2 \left(\frac{W}{L}\right). \tag{3.63}$$

Here μ is mobility, W/L is the aspect ratio of nMOS and C_{Ox} is the oxide capacitances per unit gate area. Considering DIBL and body effects, threshold voltage can be represented as

$$V_{Th} = V_{Th0} - \lambda V_{DS} + yV_{BS}, \tag{3.64}$$

where λ and y are the DIBL and body effect parameter. In 45 nm technology, for NMSO (pMOS) the value of λ and y are 0.12 (0.06) and 0.20 (0.16), respectively. By substituting the value of V_{Th}, eq. (1) can be expressed as

$$I_{Sub} = 2n_n\mu C_{Ox}u_T^2\left(\frac{W}{L}\right)e^{\frac{(V_{GS}-V_{Th0}+\lambda V_{DS}-yV_{BS})}{n_nu_T}}\left(1-e^{\frac{-V_{DS}}{u_T}}\right)$$

$$= \alpha_n e^{\frac{V_{GS}+\lambda V_{DS}}{n_nu_T}}\left(1-e^{\frac{-V_{DS}}{u_T}}\right), \tag{3.65}$$

where $\alpha_n\left(=2n_n\mu C_{Ox}u_T^2\left(\frac{W}{L}\right)e^{\frac{(-V_{Th0}-yV_{BS})}{n_nu_T}}\right)$ can be termed as the strength of the nMOS. α_n depends on the aspect ratio of the transistor, operating temperature and also the C_{Ox}.

In subthreshold regime transistors cannot be considered as "ON" or "OFF" as $V_{GS} < V_{Th}$. On the basis of high and low drain-to-source voltage we can draw the equivalent model of a MOSFET in subthreshold regime. In case of a two-terminal device, if the i/p and o/p are almost at same potential, the device can be said to be in conducting mode, whereas if the i/p and o/p potential differs largely, the device can be said to be in nonconducting mode.

Hence, when $V_{DS} \gg \frac{KT}{q}$ (approximately $V_{DS} \gg 4\frac{KT}{q}$) then the MOSFET device can be considered to be in nonconducting mode. So subthreshold drain current becomes

$$I_{Sub} = \alpha_n e^{\frac{V_{GS}+\lambda V_{DS}}{n_nu_T}}\left(1-e^{\frac{-V_{DS}}{u_T}}\right)$$

$$\approx \alpha_n e^{\frac{V_{GS}+\lambda V_{DS}}{n_nu_T}} \quad as \left(1-e^{\frac{-V_{DS}}{u_T}}\right) \approx 1. \tag{3.66}$$

So the MOSFET will behave as voltage-controlled current source. Similarly for low drain-to-source voltage, or $V_{DS} \ll \frac{KT}{q}$ then the MOSFET device can be considered to be in conducting mode. Then subthreshold drain current can be expressed as

$$I_{Sub} = \alpha_n e^{\frac{V_{GS} + \lambda V_{DS}}{n_n u_T}} \left(1 - e^{\frac{-V_{DS}}{u_T}}\right). \tag{3.67}$$

By Taylor series expansion and neglecting the higher orders we can rewrite the equation as

$$I_{Sub} = \alpha_n e^{\frac{V_{GS}}{n_n u_T}} \left(\frac{V_{DS}}{u_T}\right).$$

So

$$R_{Ds} = \frac{u_T}{\alpha_n} e^{-\frac{V_{GS}}{n_n u_T}}. \tag{3.68}$$

In short, nMOS transistor will behave as a resistor under low V_{DS}. The equivalent model of MOSFET in subthreshold regime under low and high drain-to-source voltages is given in Figure 3.22. Table 3.2 gives the extracted parameter for 45 nm technology file with 120 nm/80 nm aspect ratio.

The inverter circuit structure in subthreshold regime as shown in Figure 3.23 (a) is similar to the super-threshold circuit structure. Here M_P and M_N are the pMOS and nMOS transistor in which body is connected with the source terminal. For low and high input signal the equivalent model of inverter is also given in Figure 3.21 (b) and 3.21 (c). Input and output waveforms of subthreshold inverter and the currents through

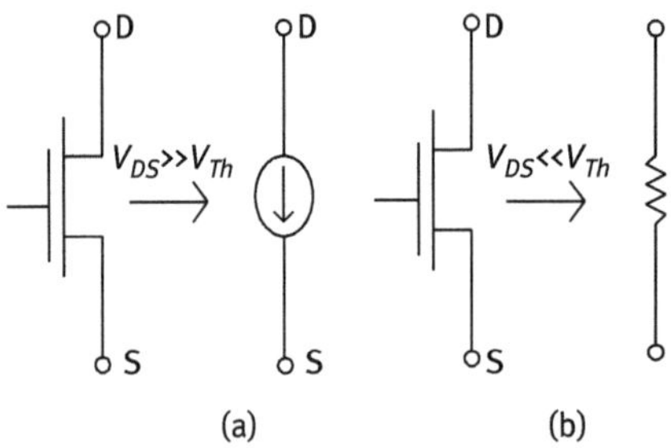

Figure 3.22: Equivalent model of MOSFET in subthreshold regime under high and low drain-to-source voltage.

Table 3.2: 45 nm (W/L = 120 nm/80 nm) CMOS process.

Device	λ_{Ds}	ψ_{Bs}	Section 1.01 n	I_0 (A)	V_{Tho} (V)
nMOS	0.12	0.20	1.9	2.89×10^{-7}	0.34
pMOS	0.05	0.16	1.9	7.63×10^{-8}	0.23

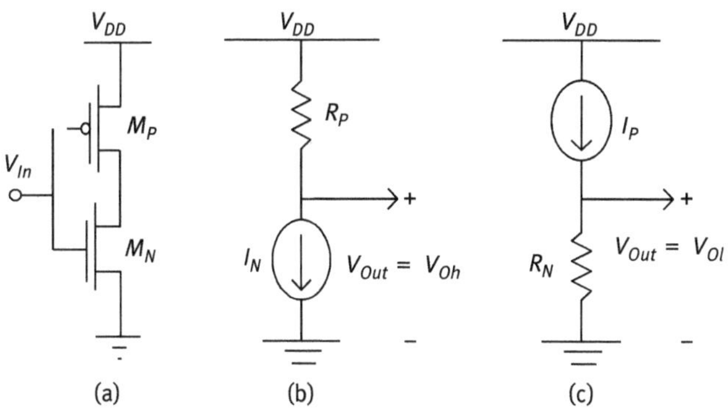

Figure 3.23: Subthreshold CMOS inverter circuit.

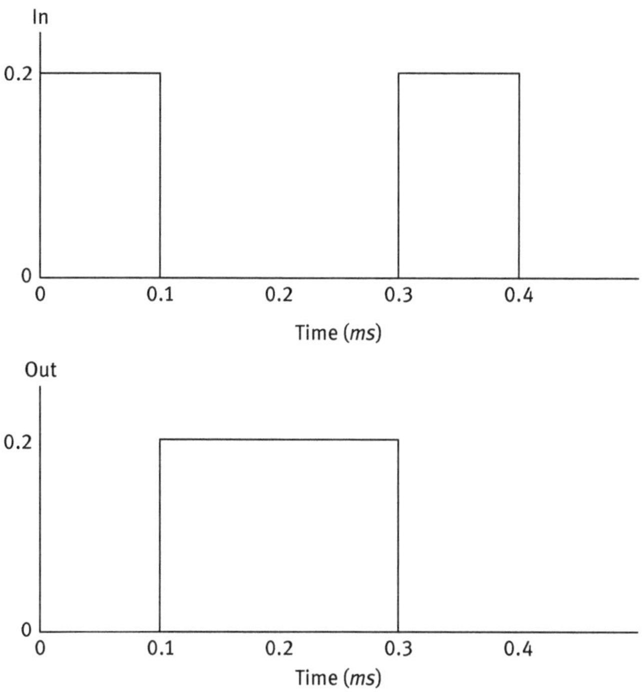

Figure 3.24: Output waveform of CMOS inverter obtained from CADENCE SPICE Spectra.

the transistors are given in Figures 3.24 and 3.25, respectively. For low input, M_P and M_N can be approximated as current source and resistor, having value I_P and R_N, respectively. So output node will be charged up through R_P. As a leakage current is flowing through nMOS, some charge will flow to the ground, so output high voltage (V_{Oh}) will be less than the supply voltage (V_{DD}). So the output high voltage (V_{Oh}) can

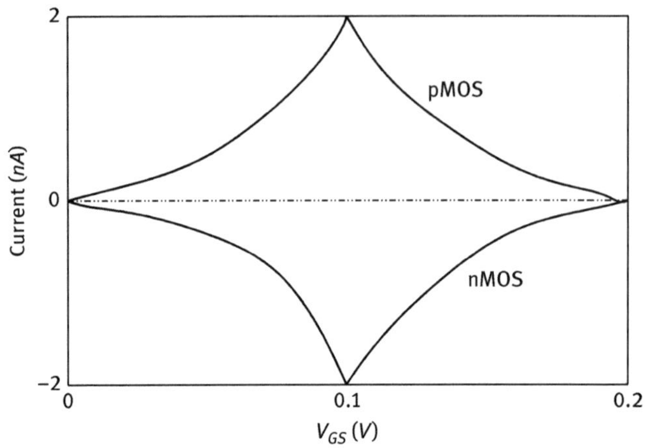

Figure 3.25: Current in pMOS and nMOS transistor in a subthreshold CMOS inverter.

be expressed as

$$\begin{aligned} V_{Oh} &= V_{DD} - I_N R_P \\ &= V_{DD} - \frac{u_t n_n}{\lambda_{Ds,n}} L\left(\frac{\lambda_{Ds,n}\alpha_n}{n_n\alpha_p} e^{\left(\frac{1}{n_p}-\frac{\lambda_{Ds,n}}{n_n}\right)\frac{V_{DD}}{u_t}}\right). \end{aligned} \tag{3.69}$$

In simplified form, the above equation can be represented as

$$V_{Oh} = V_{DD} - \frac{n_n n_p}{n_n + n_p} u_t. \tag{3.70}$$

Here $L(x)$ is the "Lambert function of x" given by $x = L(x)e^{L(x)}$. So output high voltage depends on the strength of the transistor strongly.

Similarly for high input M_P and M_N can be approximated as resistor and current source, having value R_P and I_N, respectively. Now the output node will be discharged through the R_N , but due to the presence of leakage current I_P full discharging is not possible. So output node voltage will be little bit higher than the ground potential. This output low voltage can be expressed as

$$\begin{aligned} V_{Ol} &= I_p R_n \\ &= \frac{u_t n_p}{\lambda_{Ds,p}} L\left(\frac{\lambda_{Ds,p}\alpha_p}{n_p\alpha_n} e^{\left(\frac{1}{n_n}-\frac{\lambda_{Ds,p}}{n_p}\right)\frac{V_{DD}}{u_t}}\right). \end{aligned} \tag{3.71}$$

The above expression can be simplified as

$$V_{Ol} = \frac{n_n n_p}{n_n + n_p} u_t. \tag{3.72}$$

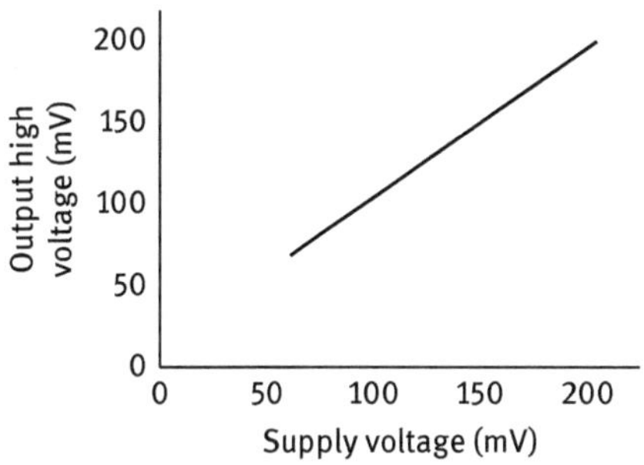

Figure 3.26: Output high voltage variation for different supply voltage in subthreshold inverter.

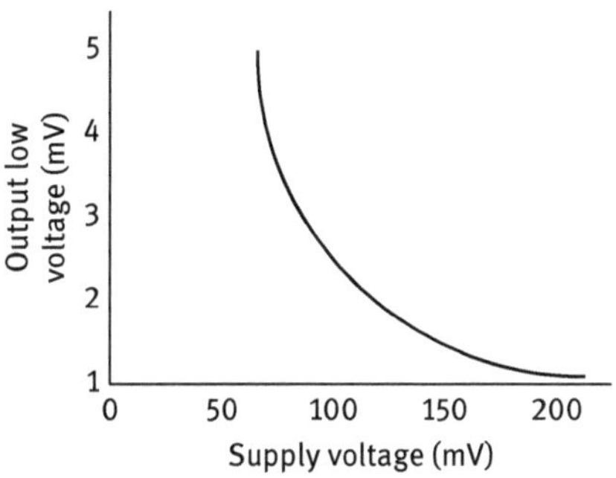

Figure 3.27: Output low voltage variation for different supply voltage in subthreshold inverter.

Variation of output high and low voltages with different supply in case of subthreshold inverter is given in Figures 3.26 and 3.27, respectively. So in case of subthreshold CMOS inverter the voltage swing becomes

$$V_{Sw} = V_{Oh} - V_{Ol}$$
$$= V_{DD} - \frac{u_t n_n}{\lambda_{Ds,n}} L\left(\frac{\lambda_{Ds,n}\alpha_n}{n_n\alpha_p} e^{\left(\frac{1}{n_p}-\frac{\lambda_{Ds,n}}{n_n}\right)\frac{V_{DD}}{u_t}}\right) - \frac{u_t n_p}{\lambda_{Ds,p}} L\left(\frac{\lambda_{Ds,p}\alpha_p}{n_p\alpha_n} e^{\left(\frac{1}{n_n}-\frac{\lambda_{Ds,p}}{n_p}\right)\frac{V_{DD}}{u_t}}\right). \quad (3.73)$$

After simplification we get

$$V_{Sw} = V_{DD} - 2\frac{n_n n_p}{n_n + n_p} u_t. \quad (3.74)$$

There must be a limit of supply voltage to ultralow voltage operation in subthreshold regime. For example, for the considered 22 nm technology with $\frac{\alpha_n}{\alpha_p} = 7.5$ roughly, the minimum voltage becomes 114 mV.

3.1.3.2 Analysis of Logic Threshold and Noise Margin

Transfer characteristic of CMOS inverter in subthreshold regime is given in Figure 3.28. The logic threshold V_{Th} of an inverter can be found by equating the current of the transistors, i.e. pMOS and nMOS, with the assumption $V_{Out} = V_{In} = V_{Lth}$.

Under this condition,

$$I_p = I_n \tag{3.75}$$

$$\text{i.e. } \alpha_{\mathrm{p}} e^{\left[\left(\frac{1+\lambda_{Dsp}}{n_{\mathrm{p}}}\right)\left(\frac{V_{DD}-V_{Lth}}{u_{\mathrm{t}}}\right)\right]} = \alpha_{\mathrm{n}} e^{\frac{V_{Lth}}{u_{\mathrm{t}}}\left(\frac{1+\lambda_{Dsn}}{n_{\mathrm{n}}}\right)}. \tag{3.76}$$

Solving eq. (3.76) for V_{Lth} we obtain the following:

$$\begin{aligned} V_{Lth} &= \frac{V_{DD}(\frac{1+\lambda_{Dsp}}{n_p}) + u_t \ln(\frac{\alpha_p}{\alpha_n})}{\left(\frac{1+\lambda_{Dsp}}{n_p}\right) + \left(\frac{1+\lambda_{Dsn}}{n_n}\right)}, \\ &= \frac{V_{DD}}{2} + \frac{n_n n_p}{n_n + n_p} u_t \ln\left|\frac{\alpha_p}{\alpha_n}\right|. \end{aligned} \tag{3.77}$$

This logic threshold voltage depends on supply voltage, ratios of transistor strength $(\frac{\alpha_{\mathrm{p}}}{\alpha_{\mathrm{n}}})$ and temperature mainly. Hence obtained logic threshold voltage is greater than $V_{DD}/2$, if the pull-up transistor is stronger than the nMOS transistor. Similarly logic threshold voltage falls below $V_{DD}/2$ if nMOS is made stronger than the pMOS. Variation of logic threshold voltage under different supply voltage is given in eq. (3.27). Fig. 3.29 and Fig. 3.30 denotes the variation of voltage swing and logic threshold voltage for different supply voltage in subthreshold inverter. For example, with 160 mV supply voltage, the theoretical V_{Lth} occurs in the range of 85 to 95 mV in 22 nm technology.

The noise margin characteristic [52, 59] for a CMOS inverter circuit has been discussed in the previous section. Hence V_{Oh}, V_{Ol}, V_{Ih} and V_{Il} are output high voltage, output low voltage, input high voltage and input low voltage, respectively, and have usual meaning.

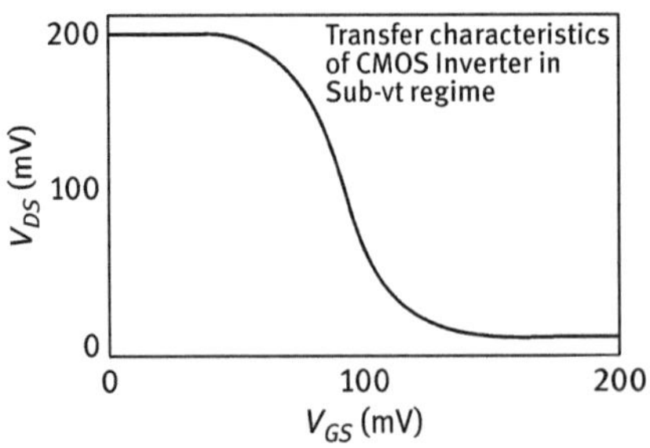

Figure 3.28: Transfer characteristic of subthreshold CMOS inverter.

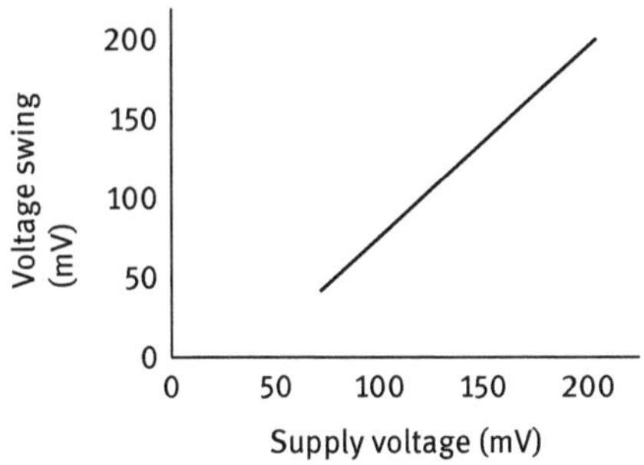

Figure 3.29: Voltage swing variation for different supply voltage in subthreshold inverter.

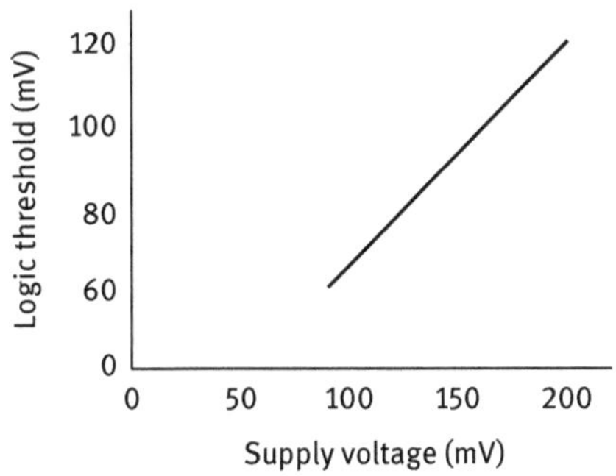

Figure 3.30: Logic threshold voltage variation for different supply voltage in subthreshold inverter.

Hence calculation of both V_{Ih} and V_{Il} follows the same procedures as given in the previous section. In simplified form V_{Ih} and V_{Il} can be expressed as

$$V_{Ih} = \frac{V_{DD} n_n}{n_n + n_p} + \frac{n_n n_p}{n_n + n_p} u_t \left(\ln \left| \frac{\alpha_p}{\alpha_n} \right| + \ln \left| \frac{n_n + n_p}{n_n n_p} \right| \right); \tag{3.78}$$

$$IL = V_{DD} \frac{n_n}{n_n + n_p} - \frac{n_n n_p}{n_n + n_p} u_t \left(\ln \left| \frac{\alpha_n}{\alpha_p} \right| + \ln \left| \frac{n_n + n_p}{n_n n_p} \right| \right). \tag{3.79}$$

V_{Ih} and V_{Il} variations under different supply voltage are given in Figure 3.31. The noise margin low becomes

$$\begin{aligned} NM_L =& V_{Il} - V_{Ol} \\ =& V_{DD}(1 - \frac{n_p}{n_n + n_p}) - \frac{n_n n_p}{n_n + n_p} u_t (\ln \left| \frac{\alpha_n}{\alpha_p} \right| + \ln \left| \frac{n_n + n_p}{n_n n_p} \right|) - \frac{n_n n_p}{n_n + n_p} u_t; \\ =& V_{DD} \frac{n_n}{n_n + n_p} - \frac{n_n n_p}{n_n + n_p} u_t (\ln \left| \frac{\alpha_n}{\alpha_p} \right| + \ln \left| \frac{n_n + n_p}{n_n n_p} \right| + 1). \end{aligned} \tag{3.80}$$

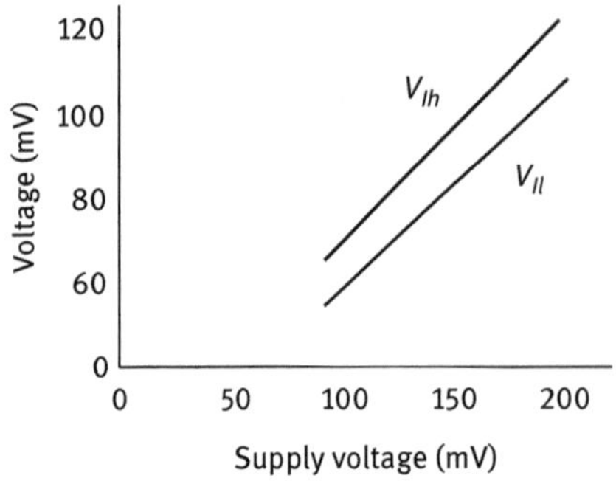

Figure 3.31: High and low input voltage variation for different supply voltages in subthreshold inverter.

Similarly noise margin high can be represented as

$$\begin{aligned}\mathrm{NM_H} &= V_{Oh} - V_{Ih};\\ &= V_{DD} - \frac{n_n n_p}{n_n + n_p} u_t - \frac{V_{DD} n_n}{n_n + n_p} - \frac{n_n n_p}{n_n + n_p} u_t \left(\ln \left| \frac{\alpha_p}{\alpha_n} \right| + \ln \left| \frac{n_n + n_p}{n_n n_p} \right| \right);\\ &= V_{DD} \frac{n_p}{n_n + n_p} - \frac{n_n n_p}{n_n + n_p} u_t \left\{ \left(\ln \left| \frac{\alpha_p}{\alpha_n} \right| + \ln \left| \frac{n_n + n_p}{n_n n_p} \right| \right) + 1 \right\}. \end{aligned} \tag{3.81}$$

$\mathrm{NM_H}$ and $\mathrm{NM_L}$ for different supply voltages are given in Figure 3.32. Finally noise margin can be calculated as

$$\mathrm{NM} = \mathrm{NM_H} - \mathrm{NM_L}.$$

Minimum possible noise margin in case of subthreshold inverter is

$$\mathrm{NM(min)} = \frac{V_{DD}}{2} - u_t - u_t \ln \eta, \tag{3.82}$$

where $\eta = \max(\frac{\alpha_n}{\alpha_p}, \frac{\alpha_p}{\alpha_n})$.

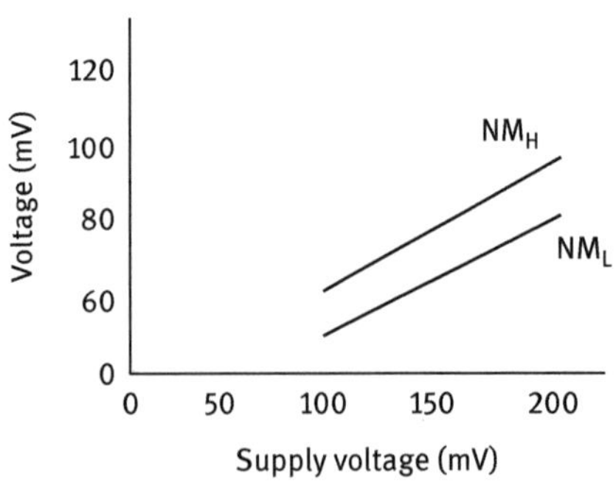

Figure 3.32: High and low noise margin variation for different supply voltages in subthreshold inverter.

Change in temperature may affect the noise margin. From a design point of view, n-lowering supply voltage may result in negative noise margin. Analytically the minimum supply voltage for positive noise margin can be expressed as

$$V_{DD}(\text{min}) = \frac{n_n n_p}{n_n + n_p} u_t \left[\ln \eta + \ln \frac{n_n + n_p}{n_n n_p} + 1 \right]. \tag{3.83}$$

3.1.3.3 Power Dissipation in Subthreshold Inverter

Before discussing power dissipation of CMOS inverter, let us take a view of the operation of the circuit in subthreshold regime. For low input, pMOS transistor conducts and a current flows from the supply voltage to charge up the output load capacitor. As the nMOS transistor remains in nonconducting mode for low input, a leakage current flows through it, causing leakage power dissipation across it. For high input, the nMOS transistor starts to conduct and output load capacitor discharges through the conducting path of nMOS. As pMOS transistors remain in nonconducting mode due to high input, some leakage current also flows through the supply voltage. In short, the total energy (E_{Tot}) of the inverter in subthreshold region is divided into dynamic energy (E_{Dyn}) and leakage energy (E_{Leak}). Total energy can be written as

$$E_{Tot} = E_{Dyn} + E_{Leak}. \tag{3.84}$$

Here dynamic energy can be modeled as

$$E_{Dyn} = \xi N_S C_L V_{DD}^2, \tag{3.85}$$

where ξ = Switching activity parameter;
- N_S = Number of switched nodes to perform the operation;
- C_L = Output load capacitor and
- V_{DD} = DC supply voltage.

Switching power dissipation depends on supply voltages mainly. Reduction of supply voltage can downscale the dynamic power dissipation significantly. For simplicity, in case of an inverter we can set the switching activity parameter to ½.

Leakage energy dissipation occurs due to the subthreshold leakage current ($I_{Sub-Leak}$) and gate tunneling current (I_{Gate}) mainly. With gate tunneling current through the insulating layer underneath the gate also increases due to the direct tunneling of electrons and holes. The leakage energy (E_{Leak}) can be expressed as

$$E_{Leak} = V_{DD} I_{Leak} \mathrm{T} = \mathrm{T} V_{DD} \alpha_n e^{\frac{V_{GS} + \lambda V_{DS}}{n_n u_T}} \left(1 - e^{\frac{-V_{DS}}{u_T}} \right). \tag{3.86}$$

Leakage energy decreases exponentially when increasing because of shorter delay, provided that the devices operate in subthreshold regime. Leakage energy is directly proportional to $W_{\min} C_L 10^{\frac{-V_{DD}}{n}}$. So C_L, W_{Min} or subthreshold slope play pivotal role to

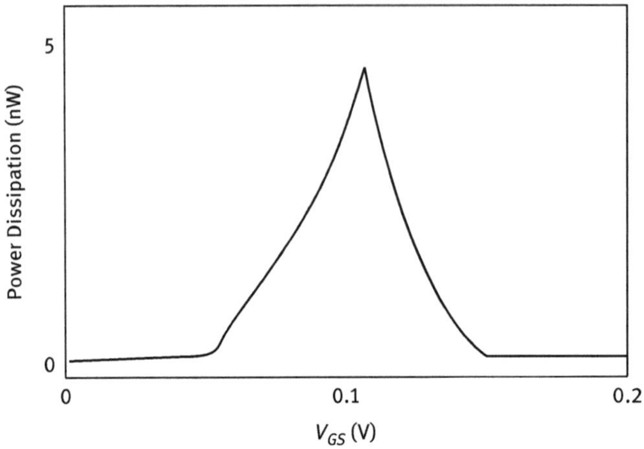

Figure 3.33: Power dissipation of subthreshold inverter in a full cycle under 250 mV.

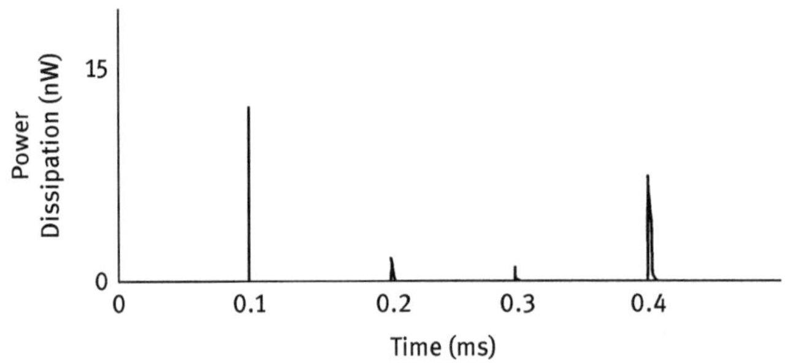

Figure 3.34: Power dissipation graph of inverter in subthreshold regime for input stream.

determine the leakage energy. For example, by downscaling W (width of the MOSFET), we can reduce the leakage energy significantly under 250 mV supply voltage.

Power dissipation of a CMOS inverter under 250 mV of a full cycle is given in Figure 3.33. For an input stream the power dissipation of same inverter is given in Figure 3.34.

To calculate minimum energy [51, 53, 57–64], it can be assumed that sizing of transistors, supply voltage, etc., can be reduced to a moderate value. Conventionally, the pMOS transistor is sized properly so that it can pull the output node up to a great extent.

In this section we observe two important things. First, dynamic power increases quadratically with the increase of supply voltage and second, leakage power reduces exponentially with the increase of supply voltage. So there should be some optimum value of the supply voltage for which energy dissipation will be minimum. Selection of optimum supply voltage or minimum energy point is the most important design issue

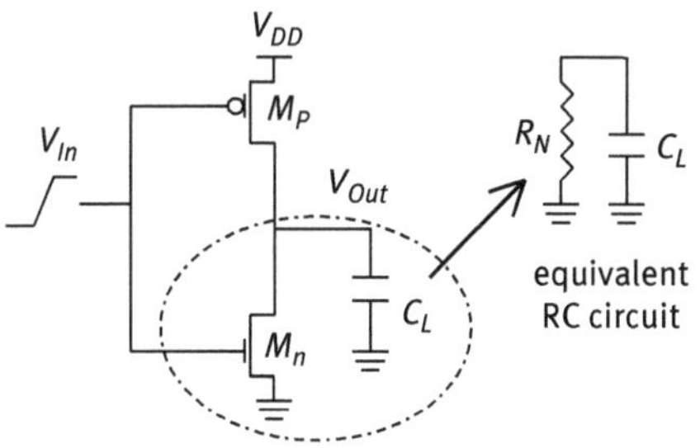

Figure 3.35: Equivalent circuit of inverter for delay modeling.

in the subthreshold regime. In general, optimum V_{DD} depends on load capacitance, minimum width of the transistor, subthreshold slope and circuit parameter like the switching activity too.

3.1.3.4 Propagation Delay of Subthreshold Inverter

In this section we will estimate the propagation delay in terms of T_{Phl} and T_{Plh} of an inverter in subthreshold regime [49–52, 57, 58]. The average propagation delay $\tau_P((\tau_{Phl} + \tau_{Plh})/2)$ of the inverter characterizes the average time required for the input signal to propagate through the inverter. Circuit delay of the inverter is related to the gate delay $C_L V_{DD}/I$, where V_{DD} and I are the supply voltage and on-current, respectively. C_L is the output load capacitance. To simplify the analysis, we assume that the input voltage has zero rise time. Recalling the drain current of an nMOS in subthreshold regime, we get

$$\begin{aligned} I_{Sub} &= 2n_n\mu C_{Ox}u_T^2\left(\frac{W}{L}\right)e^{\frac{(V_{GS}-V_{Th0}+\lambda V_{DS}-\gamma V_{BS})}{n_n u_T}}\left(1-e^{\frac{-V_{DS}}{u_T}}\right); \\ &= \alpha_n e^{\frac{V_{GS}+\lambda V_{DS}}{n_n u_T}}\left(1-e^{\frac{-V_{DS}}{u_T}}\right). \end{aligned} \tag{3.87}$$

Hence in $0 \to 1$ step input transition, the inverter can be simplified through the equivalent circuit depicted in Figure 3.35. Assuming the resistance of the discharging path be R_N and C_L as output load capacitances we can find first-order delay expression for this RC equivalent circuit. Hence during discharging event output node voltage can be expressed as

$$V_{Out}(t) = V_{DD}\exp(-t/\tau), \tag{3.88}$$

where τ is the time constant of the equivalent RC circuit. Again, according to the definition of gate propagation delay $V_{Out}(t = \tau_{Phl}) = V_{DD}/2$. Therefore, the above equation reduces to

$$\tau_{Phl} = \ln 2\,\tau, \tag{3.89}$$

where

$$\tau_{Phl} = R_N C_L = C_L \frac{u_T}{\alpha_n} e^{-\frac{V_{GS}}{n_n u_T}} \tag{3.90}$$

This model is efficacious for hand calculation. To estimate more accurate delay we are considering different input signals for fast and slow ramp input.

3.1.3.4.1 Fast Ramp Input

First we assume fast input. During discharging event, τ_{Phl} can be expressed as

$$\tau_{Phl} = -C_L \int_{V_{DD}}^{V_{DD/2}} \frac{dV_{Out}}{I_{Sub}}. \tag{3.91}$$

Hence gate-to-source voltage can be treated as constant parameter and equivalent to V_{DD}. Substituting the value of I_{Sub} from eq. (3.87) on eq.(3.91), we get

$$\begin{aligned}\tau_{Phl} &= -C_L \int_{V_{DD}}^{V_{DD/2}} \frac{dV_{Out}}{\alpha_n e^{\frac{V_{GS}+\lambda V_{DS}}{n_n u_T}} \left(1 - e^{\frac{-V_{DS}}{u_T}}\right)}; \\ &\approx -C_L \int_{V_{DD}}^{V_{DD/2}} \frac{dV_{Out}}{\alpha_n e^{\frac{V_{GS}}{n_n u_T}} \left(1 - e^{\frac{-V_{DS}}{u_T}}\right)}.\end{aligned} \tag{3.92}$$

Finally we get

$$\tau_{Phl} = \frac{C_L u_t}{\alpha_n e^{\frac{V_{DD}}{n_n V_{Th}}}} \left[\log\left(\frac{1 - e^{\frac{V_{DD}}{u_T}}}{1 - e^{\frac{V_{DD}}{2u_T}}}\right)\right]. \tag{3.93}$$

Similarly, if we calculate for τ_{Plh} we get

$$\tau_{Plh} = \frac{C_L u_t}{\alpha_p e^{\frac{V_{DD}}{n_p u_t}}} \left[\log\left(\frac{1 - e^{-\frac{V_{DD}}{2u_t}}}{1 - e^{-\frac{V_{DD}}{u_t}}}\right)\right]. \tag{3.94}$$

3.1.3.4.2 Slow Ramp Input

We assume the input and the output vary in the same time interval. Under this assumption, during the discharging of the output capacitance, nMOS current now depends on V_{DS} and V_{GS}. The delay expression can be expressed as

$$I_{Dn}(t, V_{Out}) = C_L \frac{dV_{Out}}{dt}. \tag{3.95}$$

Considering input as a slow ramp of slope K, replacing $V_{GS} = Kt$ (K is the slope of the ramp), we get

$$\int_0^{\tau_{Phl}} dt = -C_L \int_{V_{DD}}^{V_{DD/2}} \frac{dV_{Out}}{\alpha_n e^{\frac{Kt}{n_n u_T}} \left(1 - e^{\frac{-V_{DS}}{u_T}}\right)}. \quad (3.96)$$

After performing the integration, we get

$$\tau_{Phl} = \frac{n_n u_t}{K} \ln \left| \left(1 + \frac{K C_L u_t}{n_n u_t \alpha_n} \log \left(\frac{1 - e^{\frac{V_{DD}}{2u_t}}}{1 - e^{\frac{V_{DD}}{u_t}}}\right)\right| . \quad (3.97)$$

The time t represents the time from a zero point of the input to a half of V_{DD} point of the output. Hence τ_{Phl} becomes

$$\tau_{Phl} = \frac{n_p u_t}{K} \ln \left| \left(1 + \frac{K C_L u_t}{n_p u_t \alpha_n} \log \left(\frac{1 - e^{\frac{V_{DD}}{2u_t}}}{1 - e^{\frac{V_{DD}}{u_t}}}\right)\right)\right| - \frac{V_{DD}}{2K}. \quad (3.98)$$

Figure 3.36 represents the delay of inverter circuits for different supply voltage. From Figure 3.36, it is evident that as V_{DD} increases delay decreases significantly like the super-threshold circuit behavior. Effect of load capacitances on the delay in inverter is also shown in Figure 3.37.

3.1.4 Summary

This chapter deals with the design of CMOS inverter in super-threshold as well as sub-threshold regime. Design issues like power dissipation, noise margin and delay have

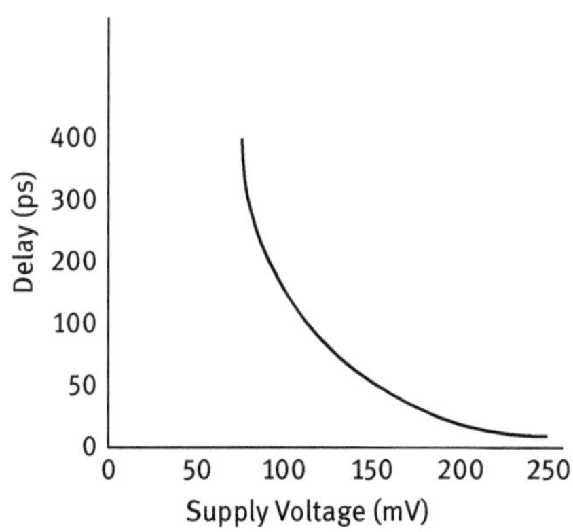

Figure 3.36: Delay for variable supply voltage.

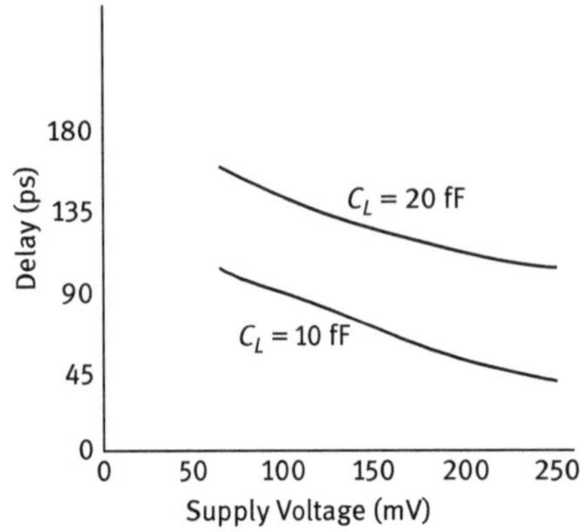

Figure 3.37: Delay of CMOS inverter for different load in subthreshold regime.

been expressed analytically to make it more reliable. Layout of CMOS inverter is also given in this chapter to make the analysis more compact. All the simulation results have been done using CADENCE SPICE spectra in sub-90 nm technology. Hopefully this chapter will be efficacious for the researchers who are interested in conventional to low power high speed VLSI design.

References

[1] P. Larsson and C. Svensson, "Noise in digital dynamic CMOS circuits." IEEE Journal of Solid-State Circuits, 29 (1994): 655–662.

[2] K.L. Shepard and V. Narayanan, "Noise in deep submicron digital design." Proceedings of International Conference on Computer Aided Design, 1996, pp. 524–531.

[3] R. Zimmermann and W. Fichtner, "Low-power logic styles: CMOS versus pass-transistor logic." IEEE JSSC, 1997, pp. 1079–1090.

[4] R.H. Krambeck, "High-speed compact circuits with CMOS." IEEE JSSC, 1982, pp. 614–619.

[5] M. Bohr, "A high-performance 0.25-Î¼ m logic technology optimized for 1.8 V operation." IEDM, 1996, pp.847–850.

[6] C. Piguet, J.-M. Masgonty, S. Cserveny and E. Dijkstra, "Low-power low-voltage digital CMOS cell design." Proceedings of PATMOS,'94, 1994, pp. 132–139.

[7] P. Ng, P.T. Balsara and D. Steiss, "Performance of CMOS differential circuits." IEEE Journal of Solid-State Circuits, 31 (1996): 841–846.

[8] K. Chu and D. Pulfrey, "A comparison of CMOS circuit techniques: Differential cascode voltage switch logic versus conventional logic." IEEE Journal of Solid-State Circuits, 22 (1987): 528–532.

[9] I.S. Abu-Khater, A. Bellaouar and M.I. Elmasry, "Circuit techniques for CMOS low-power high-performance multipliers." IEEE Journal of Solid-State Circuits, 31.10 (1996): 1535–1546.

[10] V. Navarro-Botello, J.A. Montiel-Nelson and S. Nooshabadi, "Analysis of high-performance fast feedthrough logic families in CMOS." IEEE Transactions on Circuits System II, Express Briefs, 54.6 (2007): 489–493.

[11] Y. Tsividis, *Operation and Modelling of the MOS Transistor*, 1st edn. Singapore: McGraw-Hill, 1988.

[12] Abdellatif Bellamour and Mohamed I. Elmasary, *Low Power Digital VLSI Design, VLSI Circuit and Systems*. Kluwer Academic, Singapore, 1995.

[13] C.F. Hill, "Noise margin and noise immunity in logic circuits." Microelectronics, 1 (1968): 16–21.

[14] F.J. List, "The Static Noise Margin of SRAM Cells." Digital Technical Papers, 1986.

[15] Mano, M. Morris and R. Charles Kime. *Logic and Computer Design Fundamentals*, Third Edition. Prentice Hall, 2004. p. 73. Upper Saddle River, USA.

[16] J. Lohstroh, E. Seevinck and J. de Groot, "Worst-case static noise margin criteria for logic circuits and their mathematical equivalence." IEEE Journal of Solid-State Circuits, SC-18.6 (1983): 803–807.

[17] R. Chau, S. Datta, M. Doczy, B. Doyle, B. Jin, J. Kavalieros, A. Majumdar, M. Metz and M. Radosavljevic, "Benchmarking nanotechnology for high-performance and low-power logic transistor applications." IEEE Transactions on Nanotechnology, 4.2 (2005): 153–158.

[18] A.A. Hamoui and N.C. Rumin, "An analytical model for current, delay, and power analysis of submicron CMOS logic circuits." IEEE Transactions on Circuits and Systems-II, 47.10 (2000): 999–1007.

[19] A.C. Deng and Y.C. Shiaw, "Generic linear RC delay modeling for digital CMOS circuits." IEEE Transactions on Computer-Aided Design, 9 (1990): 367–376.

[20] L. Brocco, S. Mccormic and J. Allen, "Macromodeling CMOS circuits for timing simulation." IEEE Transactions on Computer-Aided Design, 7 (1988): 1237–1249.

[21] Î'.S. Carlson and C.Y.R. Chen, "Performance enhancement of CMOS VLSI circuits by transistor reordering." Proceedings of DAC, 1993.

[22] K.A. Bowman, S.G. Duvall and J.D. Meindl, "Impact of die-to-die and within-die parameter fluctuations on the maximum clock frequency distribution for gigascale integration. IEEE Journal of Solid-State Circuits, 37.2 (2002): 183–190.

[23] A.I. Kayssi, K.A. Sakallah and T.M. Burks, "Analytical transient response of CMOS inverters." IEEE Transactions Circuits Systems I, 39 (1992): 42–45.

[24] B. Cline, K. Chopra, D. Blaauw, A. Torres and S. Sundareswaran, "Transistor-specific delay modelling for SSTA." Design, Automation and Test in Europe, March 2008, pp. 592–597.

[25] L.Z. Zhang, W. Chen, Y. Hu and C.C. Chen, "Statistical static timing analysis with conditional linear MAX/MIN approximation and extended canonical timing model." IEEE Transactions on Computer-Aided Design, 25.6 (2006): 1183–1191.

[26] A. Wroblewski, O. Schumacher, C.V. Schimpfte and J.A. Nassek, "Minimizing gate capacitances with transistor sizing." Proceedings of IEEE International Symposium on Circuits and Systems, May 2001, pp. 186–189.

[27] P. Liu and Y.J. Lee, "An accurate analytical propagation delay model of nano CMOS circuits." IEEE International SoC Design Conference (ISOCC), Oct. 2007, pp. 200–203.

[28] Z.C. Huang, H. Yu, A. Kurokawa and Y. Inoue, "Modeling the overshooting effect for CMOS inverter in nanometer technologies." Asia and South Pacific Design Automation Conference, 2007, pp. 565–570, 2007.

[29] P.S. Zuchowski, P.A. Habitz, J.D. Hayes and J.H. Oppold, "Process and environmental variation impacts on ASIC timing." IEEE/ACM International Conference on Computer Aided Design, November 2004, pp. 336–342.

[30] J. Tschanz, K. Bowman and V. De, "Variation-tolerant circuits: circuit solutions and techniques." Design Automation Conference, 2005, pp. 762–763.

[31] N. Ekekwe and R. Etienne-Cum-, "Power dissipation and possible control techniques in the ultra-deep sub-micron technologies." Microelectronics Journal, 37.9 (2006): 851–860.

[32] F. Payet, F. Boeuf, C. Ortolland and T. Skotnicki, "Nonuniform mobility-enhancement techniques and their impact on device performance." IEEE Transactions on Electron Devices, 55.4 (2008): 1050–1057.

[33] F. Boeuf, M. Sellier, F. Payet, B. Borot and T. Skotnicki, "Using model for assessment of technology and roadmaps (MASTAR) as a pre-simulation program with integrated circuit emphasis (SPICE) model generator for early technology and circuit simulation." Japanese Journal Applied Physics, 2008, Vol. 47, No. 5, pp. 3384–3389.
[34] F. Arnaud, B. Duriez, B. Tavel, L. Pain, J. Todeschini, M. Jurdit, Y. Laplanche, F. Boeuf, F. Salvetti, D. Lenoble, J.P. Reynard, F. Wacquant, P. Morin, N. Emonet, D. Barge, M. Bidaud, D. Ceccarelli, P. Vannier, Y. Loquet, H. Leninger, F. Judong, C. Perrot, I. Guilmeau, R. Palla, A. Beverina, V. DeJonghe, M. Broekaart, V. Vachellerie, R.A. Bianchi, B. Borot, T. Devoivre, N. Bicais, D. Roy, M. Denais, K. Rochereau, R. Difrenza, N. Planes, H. Brut, L. Vishnobulta, D. Reber, P. Stolk and M. Woo, "Low Cost 65 nm CMOS platform for low power & general purpose applications." VLSI Symposium Technical Digest, 2004, pp. 10–11.
[35] M. Xakellis and F. Najm, "Statistical estimation of the switching activity in digital circuits." 31st ACM/IEEE Design Automation Conference, 1994, pp. 728–733, 1994.
[36] A.P. Chandrakasan, S. Sheng and R.W. Brodersen, "Low-power CMOS digital design." IEEE Journal of Solid-State Circuits, 27.4 (1992): 473–483.
[37] S.C. Prasad and K. Roy, "Circuit optimization for minimization of power consumption under delay constraint." Proceedings of International Workshop on Low-power Design, 1994, pp. 15–20.
[38] C.H. Tan and J. Allen, "Minimization of power in VLSI circuits using transistor sizing, input ordering and statistical power estimation." Proceedings of 1994 International Workshop on Low Power Design, 1994.
[39] H.J.M. Veendrick, "Short-circuit dissipation of static CMOS circuitry and its impact on the design of buffer circuits." IEEE J Solid-State Circuits, SC-19 (1984): 468–473.
[40] Q. Wang and S.B.K. Vrudhula, "On short circuit power estimation of CMOS inverters." Proceedings of IEEE International Conference on Computer Design, 1998, 62–67.
[41] L. Bisdounis, O. Koufopavlou and S. Nikolaidis, "Accurate evaluation of the CMOS short-circuit power dissipation for short-channel devices." Proceedings of IEEE International Symposium Low-power Electron. Design, 1996, pp. 189–192.
[42] S.G. Narendra and A.P. Chandrakasan, *Leakage in Nanometer CMOS Technologies*. Springer, Berlin, Germany, 2006.
[43] W.M. Elgharbawy and M.A. Bayuomi, "Leakage sources and possible solutions in nanometer CMOS technologies." IEEE Circuits System Magazine, 5.4 (2005): 6–17.
[44] M.A. Elgamel and M.A. Bayoumi, "Interconnect noise analysis and optimization in deep submicron technology." IEEE Circuits System Magazine, 3.4 (2003): 6–17.
[45] J.W. Tschanz, S.G. Narendra, Y. Ye, B.A. Bloechel, S. Borkar and V. De, "Dynamic sleep transistor and body bias for active leakage power control of microprocessors. IEEE Journal of Solid-State Circuits, 38.11 (2003): 1838–1845.
[46] D. Lee, D. Blaauw and D. Sylvester, "Gate oxide leakage current analysis and reduction for VLSI circuits." IEEE Transactions on Very Large Scale Integration (VLSI) Systems, 12.2 (2004): 155–166.
[47] T. Sakurai and A.R. Newton, "Alpha-power law MOSFET model and its applications to CMOS inverter delay and other formulas." IEEE Journal of Solid-State Circuits, 25 (1990): 584–594.
[48] K. Chen and C. Hu, "Device and technology optimizations for low power design in deep sub-micron regime." International Symposium on Low Power Electronics and Design, 1997, pp. 312–316.
[49] A. Wang, B. Calhoun and A.P. Chandrakasan, *Sub-threshold Design for Ultra Low-power Systems* (Series on Integrated Circuits and Systems). Springer-Verlag New York, Inc., Secaucus, NJ, USA, 2006.
[50] C.C. Enz, F. Krummenacher and E.A. Vittoz, "An analytical MOS transistor model valid in all regions of operation and dedicated to low-voltage and low-current applications." Special Issue

Analog Integrated Circuits and Signal Processing J. Low-voltage and Low-power Design, 8 (1995): 83–114.

[51] S. Sapatnekar, C. Kim, J. Keane, H. Eom and T.-H. Kim, "Subthreshold logical effort: a systematic framework for optimal subthreshold device sizing." Proceedings of 43rd ACM/IEEE Design Automation Conference, 2006, pp. 425–428.

[52] M. Alioto, "Closed-form analysis of DC noise immunity in subthreshold CMOS logic circuits." ISCAS, June 2010, pp. 1468–1471.

[53] H. Veendrick, *Deep-Submicron CMOS ICs*. Kluwer, Norwell, U.S.A, pp.75–75, 1998.

[54] J. Rabaey, J. Ammer, B. Otis, F. Burghardt, Y.H. Chee, N. Pletcher, M. Sheets and H. Qin, "Ultra-low-power design–the roadmap to disappearing electronics and ambient intelligence." IEEE Circuits and Devices Magazine (2006): Vol. 22, No. 4, 23–29.

[55] A. Raychowdhury, B.C. Paul, S. Bhunia and K. Roy, "Computing with subthreshold leakage: device/circuit/architecture co-design for ultralow-power subthreshold operation." IEEE Transaction on VLSI systems, 13.11 (2005): 1213–1224.

[56] H. Soeleman, K. Roy and B.C. Paul, "Robust subthreshold logic for ultralow power operation." IEEE Transactions on Very Large Scale Integration (VLSI) System, 9.1 (2001): 90–99.

[57] A. Tajalli, M. Alioto and Y. Leblebici, "Improving power-delay performance of ultralow-power subthreshold SCL circuits." IEEE Transactions on Circuits System II, Express Briefs, 56.2 (2009):127–131.

[58] A. Tajalli, F.K. Gurkaynak, Y. Leblebici, M. Alioto and E.J. Brauer, "Improving the power-delay product in SCL circuits using source follower output stage." Proceedings of ISCAS, Seattle, WA, May 2008, pp. 145–148.

[59] B. Calhoun and A. Chandrakasan, "Static noise margin variation for sub-threshold SRAM in 65-nm CMOS." IEEE Journal of Solid-State Circuits, 41.7 (2006): 1673–1679.

[60] G. Ono and M. Miyazaki, "Threshold-voltage balance for minimum supply operation." IEEE Journal of Solid-State Circuits, 38.5 (2003): 830–833.

[61] R. Keyes and T.Watson, "On power dissipation in semiconductor computing elements." Proceedings of IRE, 50.12 (1962): 2485.

[62] J. Ammer, F. Burghardt, E. Y. Lin, B. Otis, R. Shah, M. Sheets and J. M. Rabaey, (2005). Ultra-Low Power Integrated Wireless Nodes for Sensor and Actuator Networks. In Ambient Intelligence (pp. 301–325). Springer Berlin Heidelberg.

[63] M. Alioto and Y. Leblebici, "Circuit techniques to reduce the supply voltage limit of subthreshold MCML circuits." Proceedings of VLSI-SoC, Rhodes Island, Greece, October 2008, pp. 239–244.

[64] M. Alioto and Y. Leblebici, "Analysis and design of ultra-low power subthreshold MCML gates." Proceedings of ISCAS, Taipei, Taiwan, May 2009, pp. 2557–2560.

4 Advanced Combinational Circuit Design

Introduction

A static logic gate is one that has a well-defined output once the inputs are stabilized and the switching transients have decayed away. In other words, in a static gate, we are not considering the transient response; we have to consider steady-state response. If the inputs are well defined, the outputs of a static logic are also well defined for all times. In a static logic circuit output node has a low resistance path to V_{DD} or ground. The output voltage is not changing with time. Periodic signals or clocks are not required for refreshing the voltage of the nodes, although it may be present for other purposes [1].

4.1 Static CMOS Logic Gate Design

In a CMOS logic gate, there will be logic array of pull-down transistors and logic array of pull-up transistors. The pull-up and pull-down networks (PUN and PDN) are operated by input voltages in a complementary fashion. When PDN conducts for an input combination, the output node establishes a low resistance path with ground (PDN transistors are ON, and PUN transistors are OFF), resulting in logic "0" output. When PUN conducts, the output node establishes a low resistance path with V_{DD} (PUN transistors are ON, PDN transistors are OFF), resulting a logic "1" output.

Thus PDN conduction results in logic "0" output and PUN conduction results in logic "1" output.

There is never a path through "ON" transistors from the "1" to the "0" supplies for any combination of inputs which is the basis for the low static power dissipation in CMOS logic.

The pull-down array consists of nMOS transistors as nMOS is good for passing logic "0" (without signal degradation) (refer pass transistor in earlier chapter). Similarly pMOS is used for pull-up array as pMOS is good for passing logic "1".

4.2 Complementary Properties of CMOS Logic

- Under static condition, PUN and PDN never turn on simultaneously. This is one complementary action property of CMOS logic.
- PDN comprises nMOSFETs, whereas PUN comprises pMOSFETs. This is another complementary property of CMOS logic.

- That PUN and PDN are dual of each other (series connection of pMOS transistors in PDN is equal to parallel connection of pMOS transistors in PUN) is another complementary property.
- AND function is implemented by the series connection of nMOSFETs and the parallel connection of pMOSFETs; in contrast OR function is implemented by the parallel connection of nMOSFETs and the series connection of pMOSFETs.

4.2.1 CMOS NAND Gate

To construct a CMOS circuit that provides this function we will use two complementary pairs, one for each of the inputs *A* and *B* and create the nMOS and pMOS arrays according to the needed outputs as shown in Fig. 4.1–4.2. First, note that there is only a single case where the output is a 0. This occurs when both inputs are at logic 1 values. Translating this observation into voltages then says that the output voltage $V_{Out} = 0$ V if and only if both of the two input voltages are high, i.e. $A = V_{DD} = B$. Since the nMOSFETs connect the output node to ground, this requires that the two nMOSFETs be connected in the series. If either input voltage is low, then $V_{Out} = V_{DD}$, indicating that the output node must be connected to the power supply. To accommodate these cases, we will wire the two pMOSFETs in parallel. Combining the requirements for the MOSFETs results in the circuit as shown in Figure 4.3.

4.2.2 CMOS NOR Gate

The NOR gate operation can be explained as follows: when either one or both inputs are high i.e. when the nMOS network creates a conducting path between the

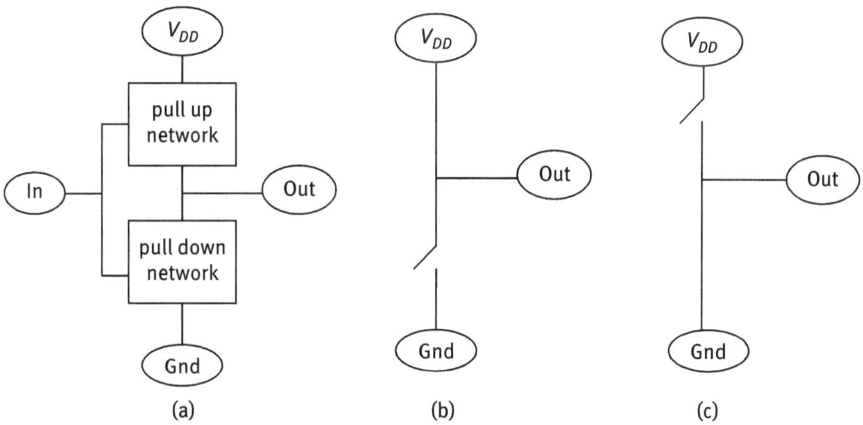

Figure 4.1: General structure of a CMOS logic gate.

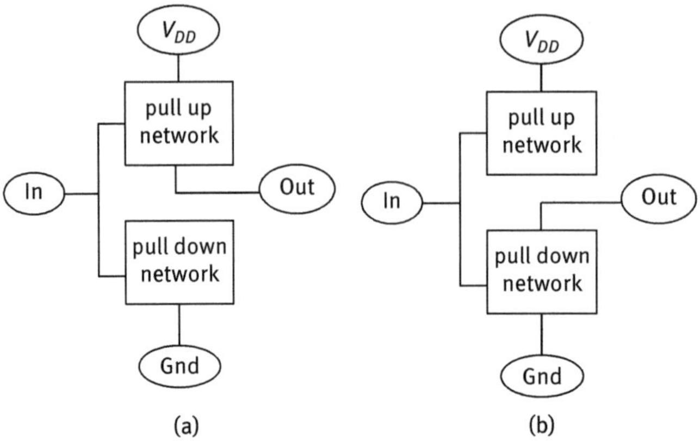

Figure 4.2: General structure and working of a CMOS logic gate.

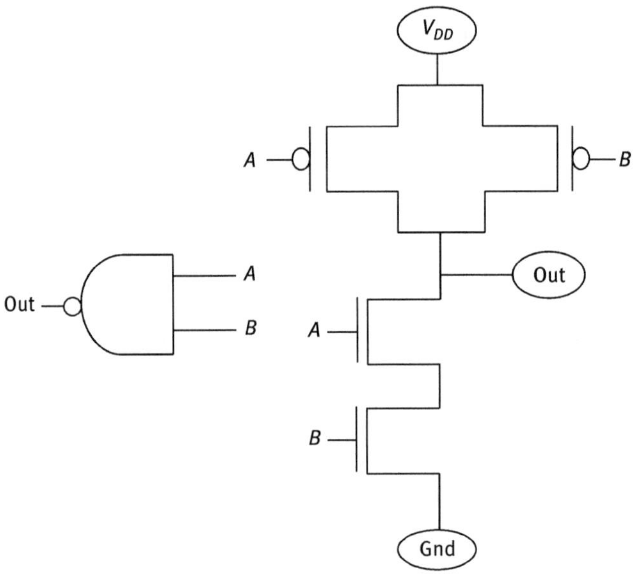

Figure 4.3: The function, truth table, symbol and structure of NAND gate using CMOS logic design style.

output node and ground, the pMOS network is cut off. On the other hand if both input voltages are low, i.e. the nMOS network is cut off, the pMOS network creates a conducting path between the output node and power supply voltage V_{DD}. Thus the dual or complementary circuit structure allows that, for any given input combination, the output is connected to V_{DD} or to ground via a low resistance path. Fig. 4.4–4.6 provides some example CMOS logic gates/circuits.

4.2.3 Some More Examples of CMOS Logic

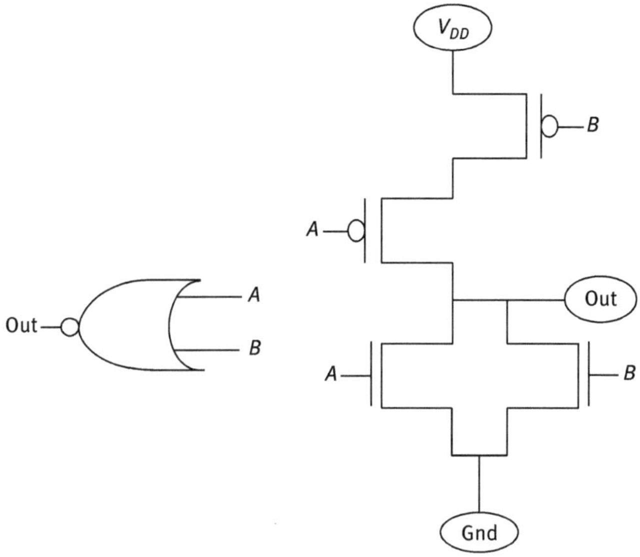

Figure 4.4: The function, truth table, symbol and structure of NOR gate using CMOS logic design style.

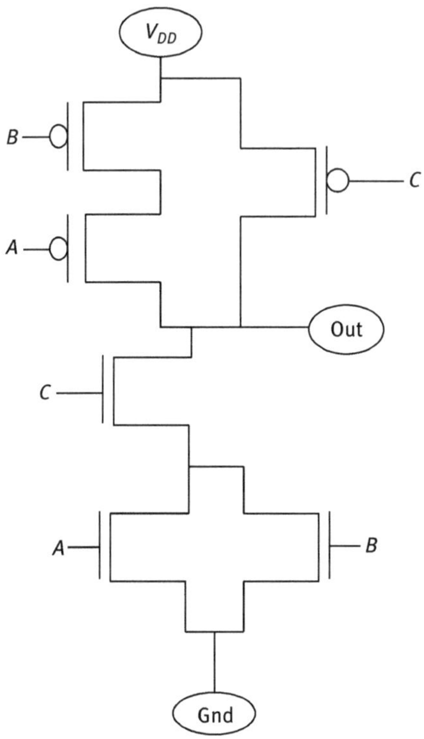

Figure 4.5: CMOS logic gate-1.

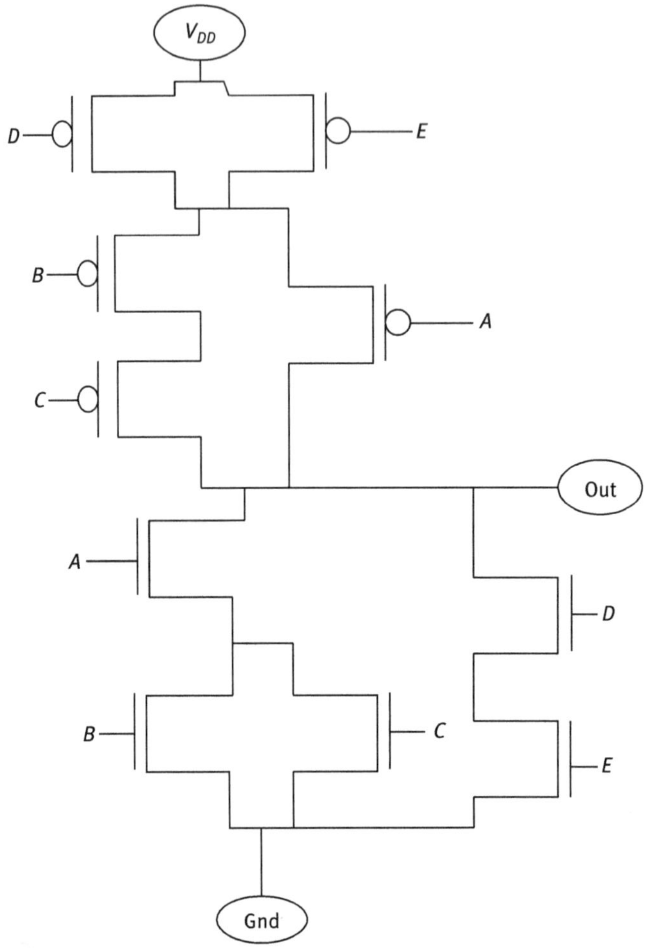

Figure 4.6: CMOS logic gate-2.

4.2.4 XOR or Nonequivalence Gate Using CMOS Logic

The reason why XOR gate is known as nonequivalence checker is because when the inputs *A* and *B* are not equivalent or different then the output is high as indicated from the truth table as shown in Figure 4.7. To realize XOR gate in CMOS we have to realize the function $f = \overline{A \oplus B} = AB + \overline{AB}$ as CMOS logic always generates inverted logic.

4.2.5 XOR–XNOR or Equivalence Gate Using CMOS Logic

Note that full CMOS implementation of XOR function requires 12 transistors including four additional transistors required to obtain inverse of *A* and *B* as shown in Fig. 4.8.

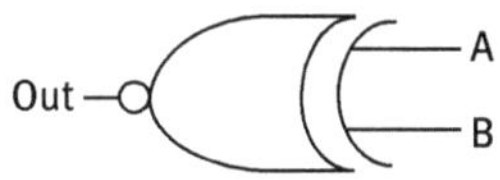

Truth table		
A	B	OUT
0	0	0
0	1	1
1	0	1
1	1	0

Figure 4.7: Function, truth table and symbol of XOR gate.

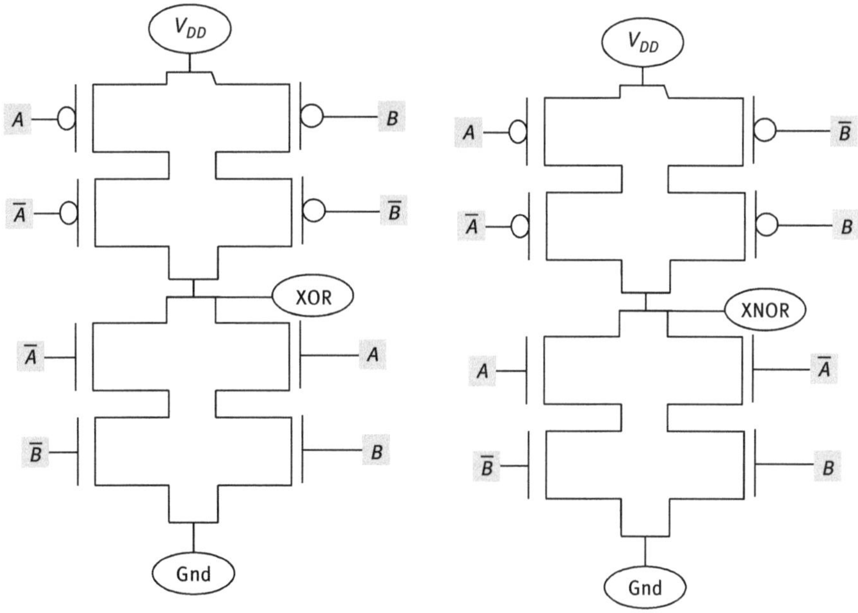

Figure 4.8: Structure of XOR and XNOR gates.

4.2.6 And-Or-Invert and Or-And-Invert Gates

There are two main important categories of complex CMOS gate. They are And-Or-Invert (AOI) and Or-And-Invert (OAI) gates [2]. The AOI gates enable sum-of-products (SOP) realization consisting of parallel braches of the series of connected nMOS pull-down transistors. The OAI enables products-of-sums (POS) realization consisting of series of braches of parallel-connected nMOS pull-down transistors. In both cases, PUN will be dual of the PDN as shown in Fig. 4.10–4.12.

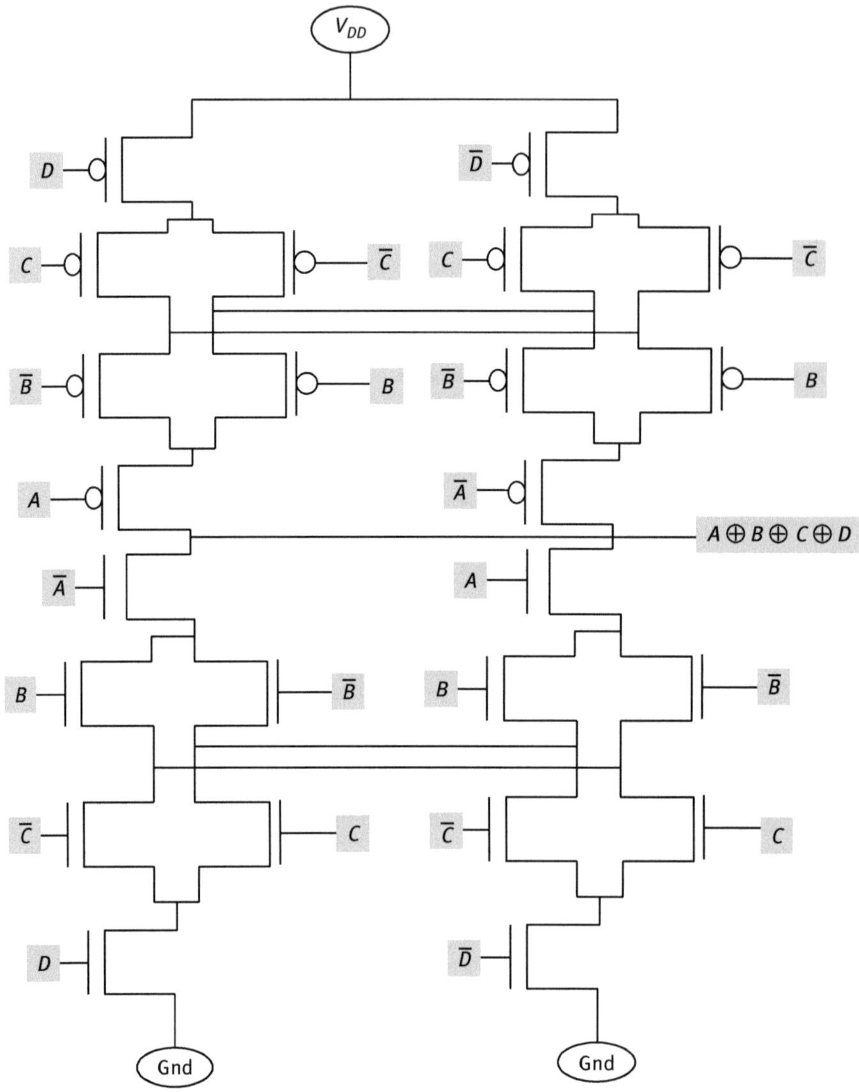

Figure 4.9: Four-input XOR gate.

4.2.7 Full Adder Circuits Using CMOS Logic

$$\text{sum\_out} = A \oplus B \oplus C = ABC + A\bar{B}\bar{C} + \bar{A}\bar{B}C + \bar{A}B\bar{C}$$
$$\text{carry\_out} = AB + BC + CA$$

To implement full adder, carry_out signal is used to generate sum output instead of realizing sum and carry separately. This implementation will reduce the circuit complexity and also save the chip area as shown in Fig. 4.13–4.15. Table 4.1 shows the truth table of the full adder. In a similar manner, a 4-bit ripple carry adder chain consisting of 4 full adders can also be constructed as shown in Fig. 4.16.

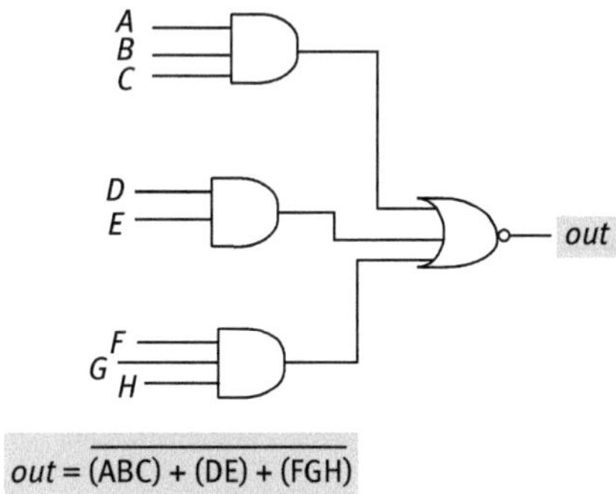

Figure 4.10: Gate level representation of AOI logic gate network.

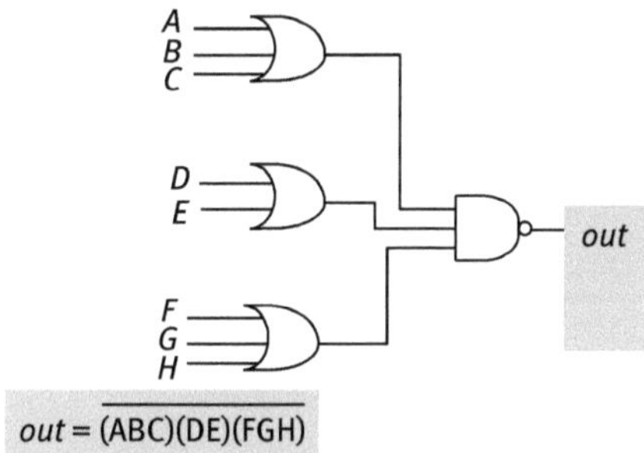

Figure 4.11: Gate level representation of OAI logic gate network.

Thus sum_out or output sum can be written as

$$\begin{aligned}
sum_out &= (A + B + C)(\overline{carry_out}) + ABC \\
(A + B + C)(\overline{AB + BC + CA}) + ABC &= (A + B + C)(\overline{AB}.\overline{BC}.\overline{CA}) + ABC \\
&= (A + B + C)[(\overline{A} + \overline{B})(\overline{B} + \overline{C})(\overline{C} + \overline{A})] + ABC \\
&= (A + B + C)[(\overline{A}.\overline{B} + \overline{A}.\overline{C} + \overline{B}.\overline{B} + \overline{B}.\overline{C})(\overline{C} + \overline{A})] + ABC \\
&= (A + B + C)[\overline{A}.\overline{B}.\overline{C} + \overline{A}.\overline{C}.\overline{C} + \overline{B}.\overline{B}.\overline{C} + \overline{B}.\overline{C}.\overline{C} + \overline{A}.\overline{B}.\overline{A} \\
&\quad + \overline{A}.\overline{C}.\overline{A} + \overline{B}.\overline{B}.\overline{A} + \overline{B}.\overline{C}.\overline{A}] + ABC \\
&= (A + B + C)(\overline{A}.\overline{B}.\overline{C} + \overline{A}.\overline{C} + \overline{B}.\overline{C} + \overline{A}.\overline{B}) + ABC \\
&= ABC + A\bar{B}\bar{C} + \bar{A}\bar{B}C + \bar{A}B\bar{C}.
\end{aligned}$$

Thus sum_out is same as before.

Thus in general to implement full adder, we can write

$$\begin{aligned}
Carry_out &= C_{n+1} = a_n.b_n + c_n\,(a_n + b_n) \\
Sum_out &= S_n = a_n \rho A b_n \rho A c_n = a_n.b_n.c_n + (a_n + b_n + c_n)\,.\overline{C_{n+1}}.
\end{aligned}$$

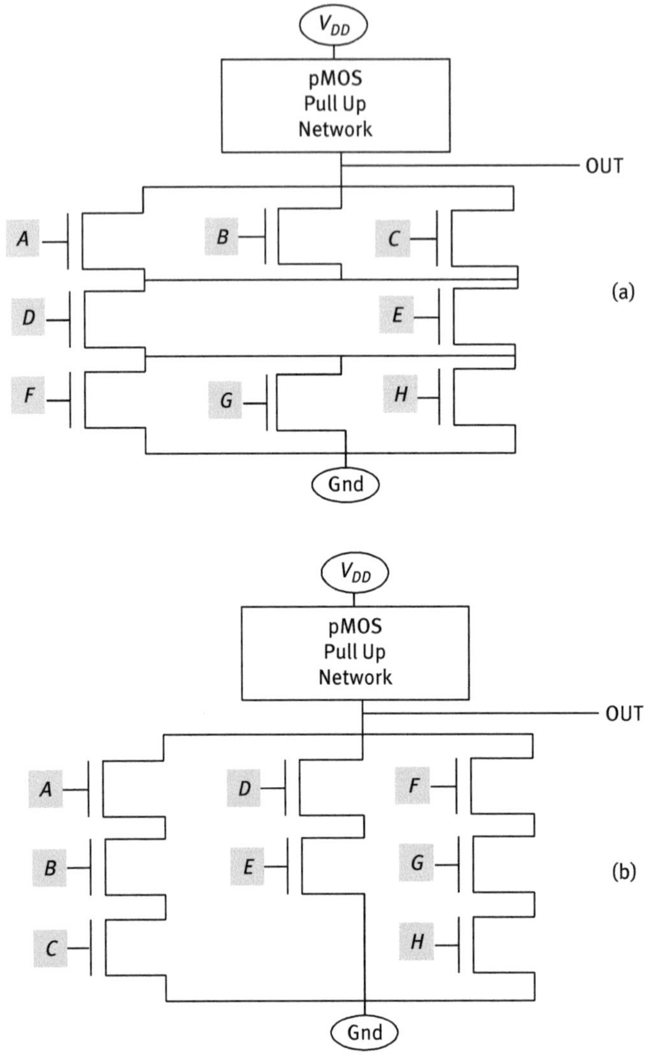

Figure 4.12: Structure of AOI and OAI logic gate network.

Note that the circuit contains 14 nMOS and 14 pMOS transistors, together with the two CMOS inverters required to generate the output.

4.3 Pseudo-nMOS Gates

In CMOS logic design, for *n*-input logic gate, total 2*n* number of transistors are needed. pMOS network pull-up structure requires *n* number of pMOS, and nMOS network

Table 4.1: Truth table of full adder.

a) a_n	*b)* b_n	*c)* c_n	*d)* s_n	*e)* C_{n+1}
0	0	0	0	0
0	0	1	1	0
0	1	0	1	0
0	1	1	0	1
1	0	0	1	0
1	0	1	0	1
1	1	0	0	1
1	1	1	1	1

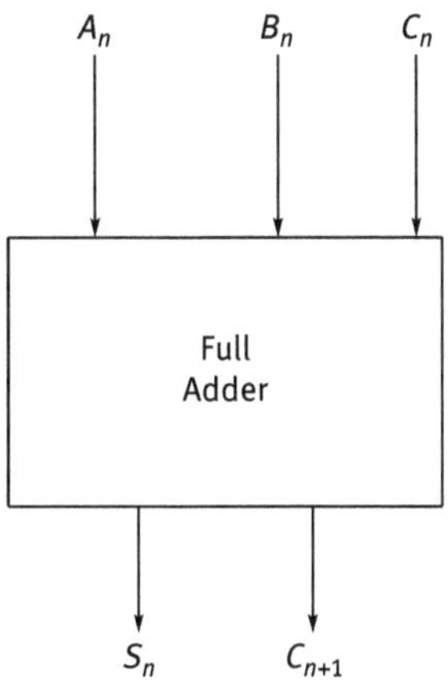

Figure 4.13: Block diagram of 1-bit full adder.

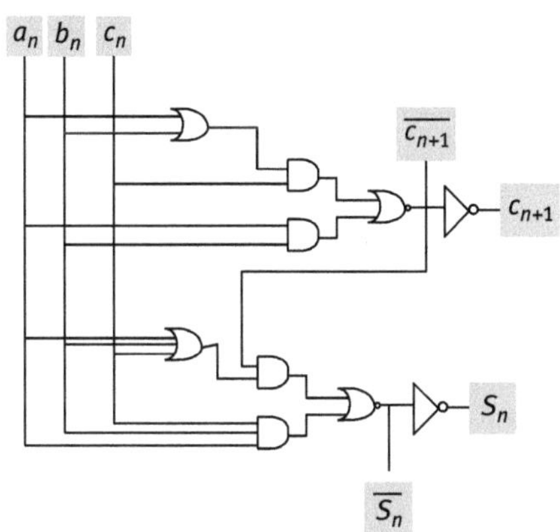

Figure 4.14: Logic gate diagram of 1-bit full adder.

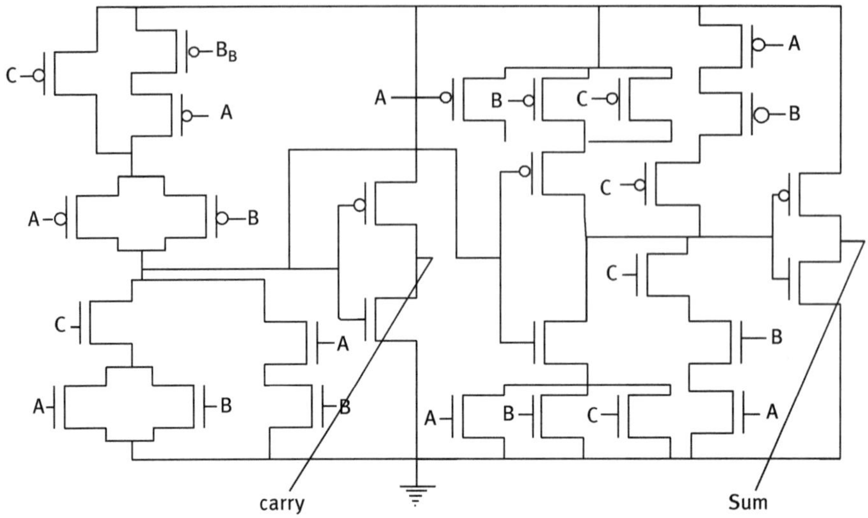

Figure 4.15: Transistor level schematic of 1-bit full adder circuit.

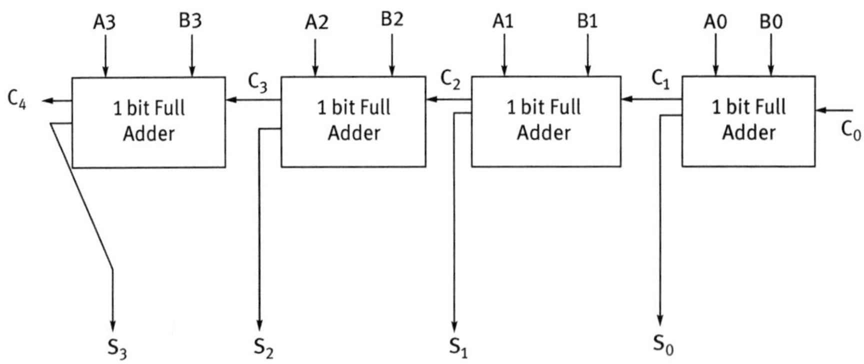

Figure 4.16: Block diagram of a 4-bit ripple carry adder chain consisting of 4 full adders.

pull-down structure requires *n* number of nMOS transistors [3]. This requires large area for the implementation of complex CMOS logic within a high density deigned IC [4].

Pseudo-nMOS logic design is one way to reduce the transistor count and also area requirement than CMOS logic design. In pseudo-nMOS logic design, n+1 number of transistors are required for implementing n input logic gate. Out of this n+1 transistors, one is pMOS transistor functioning as pull-up device and n number of nMOS transistor wired for implementation of PDN, which is same as CMOS logic design, which was discussed earlier in this chapter. The primary advantage to this type of circuit is simplified interconnect wiring due to the absence of a pMOS logic array.

Instead of using *n* number of pMOS transistor, only one pMOS transistor is used in this logic design. The purpose of the PUN in complementary CMOS is to provide a conditional path between V_{DD} and the output when the PDN is turned off. In ratioed logic, the entire PUN is replaced with a single always-ON unconditional load device that pulls up the output for a high output. Instead of a combination of active PDN and PUN, such a gate consists of an nMOS PDN that realizes the logic function, and a simple load device. The gate of the pMOS transistor is grounded. Thus the pMOS transistor remains always-ON, reminding the always-ON depletion mode inveter.

4.3.1 Why the Name Is Pseudo-nMOS?

The depletion type always-ON nMOS device used as pull-up device in depletion load inverter is replaced by always-ON pMOS device in pseudo-nMOS technology. This is the reason it is known as pseudo-nMOS technology as it reminds of the older nMOS technology. Historically, nMOS preceded CMOS as the dominant technology, but it is now obsolete. Pseudo-nMOS logic is a CMOS technique where the circuits resemble the older nMOS-only networks.

4.3.2 Ratioed Logic

Pseudo-nMOS logic gates are known as ratioed logic because the transfer function depends on the ratio of the strength of the pull-down transistor to the pull-up device. This type of logic gates exhibit static power dissipation in certain states (when the output is low), which is a major disadvantage of this type of technology. Therefore, ratioed circuits have limited use in certain limited circumstances where they offer critical benefits such as smaller area/low transistor count per logic gate [5].

When the output voltage is lower than V_{DD} then the pMOS device conducts a steady-state current. Thus under steady-state condition logic gate exhibits power. As it is a ratioed logic, the value of the V_{Ol} and the noise margin are also determined by the ratio of the strength of pMOS load transistor to the pull-down or driver transistors.

The value of V_{Ol} and noise margin depends upon the driver-to-load ratio

$$\frac{\beta_n}{\beta_p} = \frac{k_n'\left(\frac{W}{L}\right)_n}{k_p'\left(\frac{W}{L}\right)_p}.$$

Figure 4.17 shows an inverter, its VTC and general structure for pseudo-nMOS logic gates.

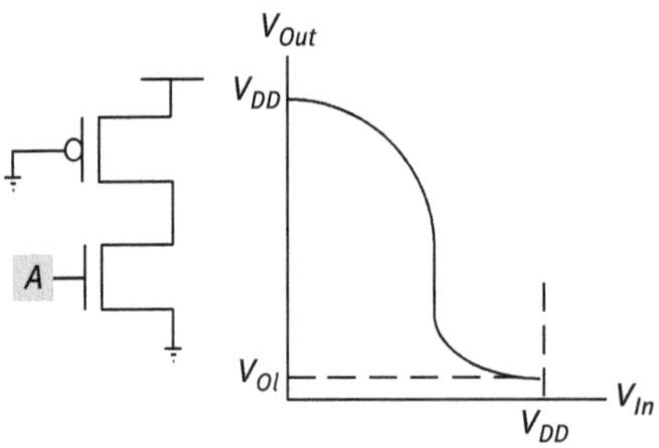

Figure 4.17: Pseudo-nMOS inveter and its VTC.

4.3.3 Operation of Pseudo-nMOS Inverter

From Figure 4.18 the pMOS voltages of the inverter are

$$V_{Sdp} = V_{DD} - V_{Out} \text{ and } V_{SG} = V_{DD}.$$

The value of the source gate voltage indicates that the pMOS is always biased into the active region and cannot be turned off; however, the actual value of the current is controlled by the nMOS driver transistor.

When input voltage is less than threshold voltage of nMOS device, i.e. $V_{In} < V_{Tn}$, the nMOS transistor is OFF and $V_{Oh} = V_{DD}$ as pMOS connects the output node to power supply V_{DD}. If one increases V_{In}, nMOS turns ON and starts conducting to pull down the output voltage. By setting $V_{In} = V_{Oh}$ we can calculate the output low voltage V_{Ol} circuit. Let us assume that V_{Ol} is small, so that nMOS is non-saturated; if $V_{Ol} < |V_{Tp}|$ also holds, then pMOS will be saturated. Equating the currents through the transistors gives

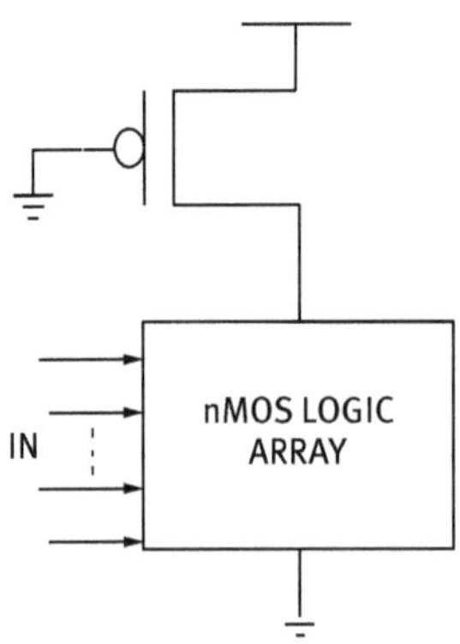

Figure 4.18: Structure of pseudo-nMOS logic gate and NAND gate using pseudo-nMOS logic.

$$\frac{\beta_n}{2}\left[2\left(V_{DD}-V_{Tn}\right)V_{Ol}-V_{Ol}{}^2\right]=\frac{\beta_p}{2}\left[V_{DD}-\left|V_{Tp}\right|\right]^2. \tag{4.1}$$

This is a quadratic equation with the solution

$$V_{Ol}=\left(V_{DD}-V_{Tn}\right)-\sqrt{\left(V_{DD}-V_{Tn}\right)^2-\frac{\beta_p}{\beta_n}\left(V_{DD}-\left|V_{Tp}\right|\right)^2}. \tag{4.2}$$

Figure 4.18 shows the general shape of the voltage transfer curve for this inverter circuit.

First, it is not possible to achieve a value of V_{Ol} = 0 V since the square root term can never equal $(V_{DD} - V_{Tn})$. Second, the value of V_{Ol} depends upon the *driver-to-load* ratio

$$\frac{\beta_n}{\beta_p}.$$

A small V_{Ol} requires a large driver-to-load ratio. Mathematically, this can be seen by noting that increasing the driver-to-load ratio moves the square root term closer to $(V_{DD} - V_{Tn})$. This corresponds to the physical viewpoint that we must make the nMOS more conductive to pull it closer to ground voltage. Finally, note that if input is at logic high, both transistors are ON and conducting, establishing a DC current path between supply and ground; this implies that static DC power dissipation occurs when the input is at a stable logic high level.

4.4 Pass-transistor Logic

This is a popular alternative to widely accepted CMOS logic design style where the main intension is to reduce the number of transistors. In the pass transistor logic the primary inputs drive gate terminals as well as the drain and/or source terminals. In contrast to the conventional CMOS logic design where primary inputs drive the gate terminals only [6].

The essential requirement in the design of the pass transistor logic is to ensure that every circuit node has at all times a low resistance path to ground or V_{DD}.

Figure 4.19 shows an AND gate realized by pass transistor logic using only nMOS transistors. If the input B is high, the input A is copied to the output Y, and if B is low, then bottom pass transistor turns on and passes logic “0” at the output. Initially at first glance the bottom switch looks redundant. But the bottom pass transistor is essential to ensure the low resistance path for the output node under the circumstance when B is low. If this pass transistor is omitted then output will be connected to high impedance when B is low.

This AND gate requires only 4 transistors (2 pass transistors, another 2 required to invert B), in contrast to CMOS logic AND gate which requires 6 transistors. In general we can conclude, pass transistor logic offers advantage of saving in transistor count as shown in Fig. 4.20.

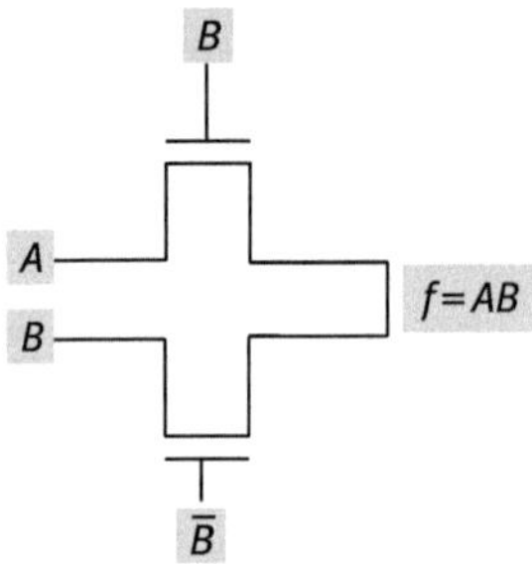

Figure 4.19: Pass transistor implementation of an AND gate.

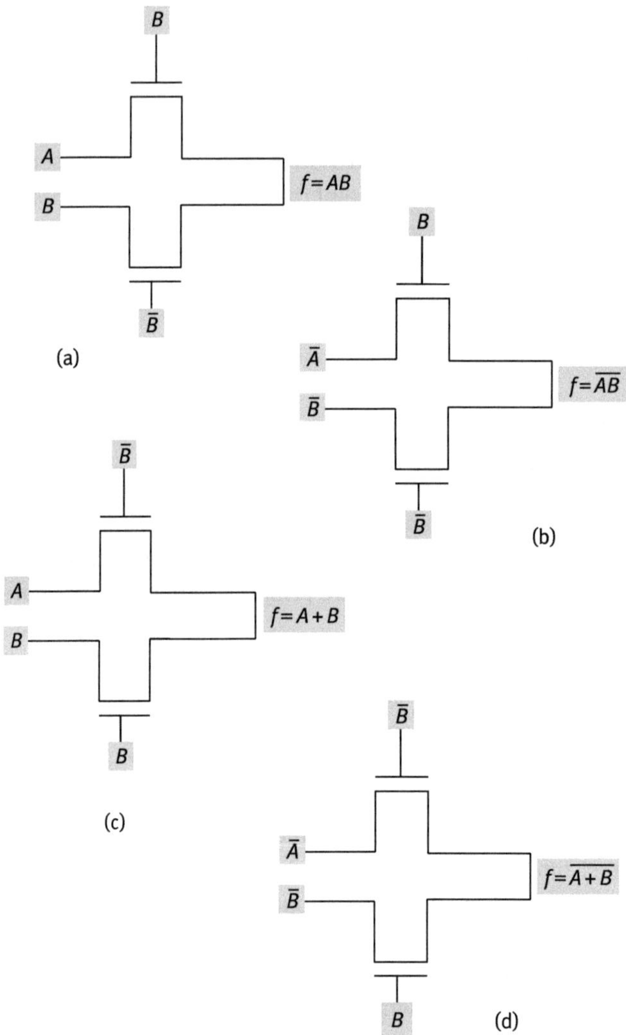

Figure 4.20: Pass transistor implementation of (a) AND (b) NAND (c) OR (d) NOR (e) XOR (f) XNOR gates.

But pass transistor suffers from limited output swing, in contrast to full rail to rail output swing in case of CMOS logic. We know that an nMOS device is effective at passing a 0 but is poor at pulling a node to V_{DD}. When the pass transistor pulls a node high, the output only charges up to $V_{DD} - V_{Tn}$. The situation will get worsen if the transistor suffers from substrate bias effect. It will increase the threshold voltage. Considering substrate bias effect, threshold voltage can be written as $V_{Tn}(V_{BS})$. Now as $V_{Tn}(V_{BS}) > V_{Tn}$, output swing limit is more for a pass transistor experiencing body effect.

4.5 Complementary Pass Transistor Logic

For high performance design, a differential pass transistor logic family called complementary pass transistor (CPL) is commonly used [6]. The main features of CPL are listed below:

- The basic idea is to accept true and complementary inputs and produce the true and complementary outputs.
- CPL circuit essentially consists of complementary inputs, nMOS pass transistor logic network to generate complementary outputs and CMOS inverters to restore output signals from pass transistors suffering with limited output swing.

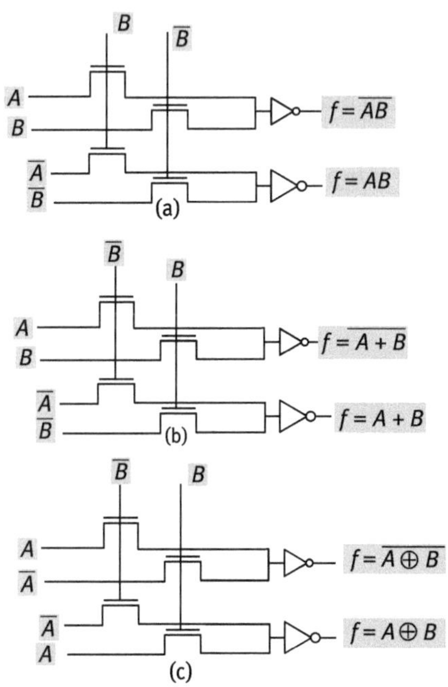

Figure 4.21: Circuit diagram of CPL implementation of (a) NAND/AND (b) NOR/OR and (c) XOR/XNOR gates.

- Complementary data inputs and outputs are available as these circuits are members of differential family. Although generating the differential signals requires extra circuitry, the differential style has the advantage that some complex gates such as XORs and adders can be realized efficiently with a small number of transistors.
- Furthermore, the availability of both polarities of every signal eliminates the need for extra inverters, as is often the case in static CMOS or pseudo-nMOS.
- CPL belongs to the class of static gates, because the output-defining nodes are always connected to either V_{DD} or GND through a low resistance path. This is helpful for the noise suppression.

The CPL realization always does not offer advantage of reduced transistor count as evident from Figure 4.21, where NAND/NOR gates require more transistor than their conventional CMOS counterpart, whereas XOR/XNOR functions (6 transistors) have a reduced complexity than their CMOS realizations (8 transistors).

4.6 Signal Restoring Pass Transistor Logic Design

Pass transistor logic and CPL suffer from static power dissipation and reduced noise margins, since the high input to the signal-restoring inverter only charges up to $V_{DD} - V_{Tn}$.

Level restorer circuit can be used to solve this problem as shown in Figure 4.22.

The gate of the pMOS device is connected to the output of the inverter, its drain connected to the input of the inverter and the source to V_{DD}. Assume node X is at 0 V (out is at V_{DD}, M_r is off, $B = V_{DD}$ and $A = 0$). Now if input A makes a transition from $0 \rightarrow V_{DD}$, pass transistor M_p charges node X to $V_{DD} - V_{Tn}$. This is, however, enough to switch the output of the inverter low, turning on the feedback device M_r and pulling node X all the way to V_{DD}. This eliminates any static power dissipation in the inverter.

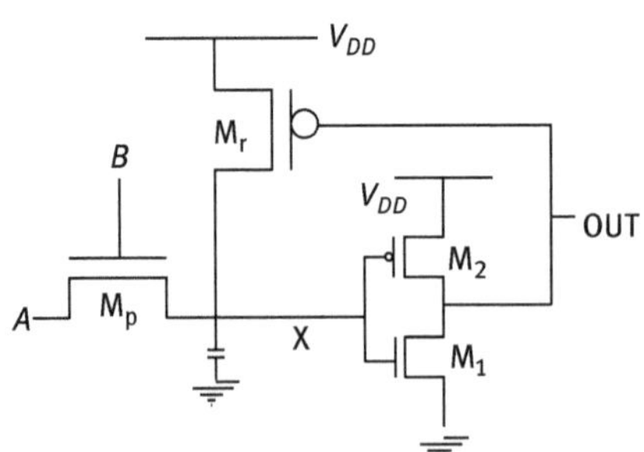

Figure 4.22: Use of logic restorer circuit to solve limited swing of output signal in pass transistor logic.

In summary, this circuit has the advantage that all voltage levels are either at GND or V_{DD}, and no static power is consumed.

The circuit should be carefully designed and must be ratioed to function properly. Consider node X make a transition from high to low. The pass transistor network attempts to pull-down node X while the level restorer pulls now X to V_{DD}. Therefore, the pull-down device must be stronger than the pull-up device in order to switch node X and the output.

Another solution to solve the limited output signal swing problem is to use multiple threshold devices as noted below.

Using zero threshold devices for the nMOS pass transistors eliminates most of the threshold drop and passes a signal close to V_{DD}. All devices other than the pass transistors (i.e. the inverters) are implemented using standard high threshold devices. Notice that even if the devices threshold was implanted to be exactly equal to zero, the body effect of the device prevents a swing to V_{DD}.

The problems are extra processing required to obtain ion implanted nMOS transistors. The zero threshold nMOS devices are prone to subthreshold conduction, which can generate some sneak path, through which significant energy leakage can occur [7].

4.7 Sizing of Transistor in CMOS Design Style

The sizing of the transistors in the CMOS design style is determined after designing the logic gate. Let us denote $(W/L)_n = n$ and $(W/L)_p = p$ for an inverter. For a CMOS inverter with symmetric VTC $p = \left(\frac{\mu_n}{\mu_p}\right) n$ [8].

For any logic gate, the W/L ratios for the MOS devices are chosen in such a way that the current driving capability of the gate in both directions (charging and discharging output capacitance through PUN and PDN) is equal to that of the basic inverter. For example, the W/L ratios for all the transistors in the PDN should be assigned a value so that the PDN should be able to provide a capacitor discharge current that at least is equal to that of an nMOS transistor with $W/L = n$, and the PUN should be able to provide a charging current that at least is equal to that of pMOS transistor with $W/L = p$. This will make sure that the worst-case gate delay will be equal to that of the basic inverter.

Worst-case means when we have to make decision about device size, we should consider the input combinations that result in the lowest output current and then choose sizes that will make this current equal to that of the basic inverter.

In CMOS logic gates the PUN and PDN consist of parallel- or series-connected MOSFETs. Thus we will first find the equivalent W/L ratio for parallel or series-connected MOSFETs.

We know that for a MOSFET larger W/L means more current driving capability and thus less on resistance. Thus we can conclude that the resistance (r_{DS}) is inversely proportional to (W/L) ratio. Thus if a number of MOSFETs are connected in series having ratios $(W/L)_1$, $(W/L)_{2,\ldots,}$ the equivalent series resistance can be obtained by adding the on resistances as

$$\begin{aligned} R_{\text{series}} &= r_{Ds1} + r_{Ds2} + \cdots \\ &= \frac{\text{const}}{(W/L)_1} + \frac{\text{const}}{(W/L)_2} + \cdots \\ &= \text{const}\left(\frac{1}{(W/L)_1} + \frac{1}{(W/L)_2} + \cdots\right) = \frac{\text{const}}{(W/L)_{\text{eq}}} \\ (W/L)_{\text{eq in series}} &= \frac{1}{\left(\frac{1}{(W/L)_1} + \frac{1}{(W/L)_2} + \cdots\right)}. \end{aligned}$$

Similarly one can show that parallel connection of transistors results in an equivalent W/L ratio of

$$(W/L)_{\text{eq in parallel}} = (W/L)_1 + (W/L)_2 + \cdots$$

Two identical MOS with $W/L = 2$ result in an equivalent W/L of 4 when connected in parallel and of 1 when connected in series as shown in Figure 4.23. It means that when transistors are connected in series, their current driving capability increases and it decreases when connected in series.

Now after determining the equivalent W/L for series and parallel-connected MOSFETs, let us consider 4 input NOR gates as shown in Figure 4.24. The worst-case or lowest current for PDN is obtained when only one of the nMOS transistors is conducting. We therefore select the W/L of each nMOS transistor that is equal to n, which is the W/L ratio of nMOS transistor in a CMOS inverter. For the PUN the only case and indeed the worst case is when all inputs are low and four series transistors are conducing. Since the W/L ratio of equivalent pMOS transistor will be 1/4th of each pMOS device, we therefore select the W/L ratio of each pMOS transistor to be four times that of pMOS transistor in a basic CMOS inverter, which is equal to 4p.

The proper sizing of a 4 input NAND gate is shown in Figure 4.25.

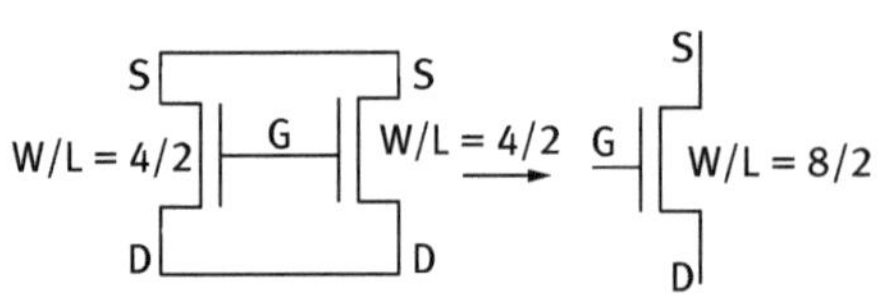

Figure 4.23: Parallel-connected identical MOSFETs result in equivalent MOSFETs with double current driving capability (double W/L).

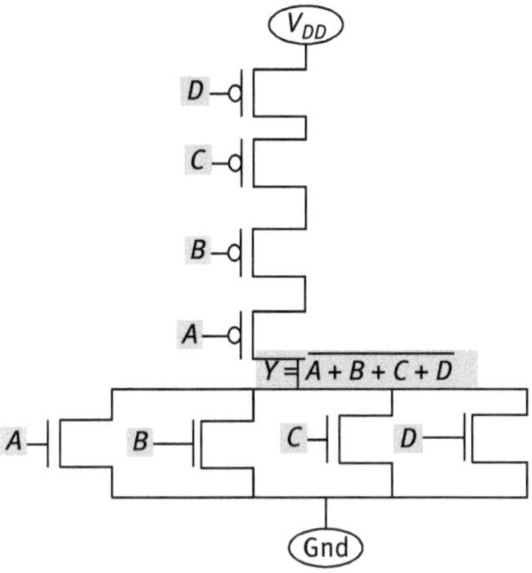

Figure 4.24: Proper transistor sizing of a 4 input NOR gate so that the worst-case gate delay equals to that of basic CMOS inverter, p denotes W/L ratio of pMOS and n denotes W/L ratio of nMOS present in the basic CMOS inverter.

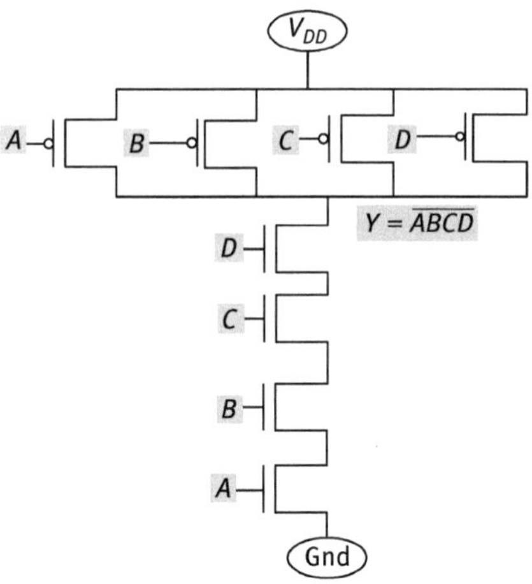

Figure 4.25: Proper transistor sizing of a 4 input NOR gate so that the worst-case gate delay equals to that of basic CMOS inverter; p denotes W/L ratio of pMOS and n denotes W/L ratio of nMOS present in the basic CMOS inverter.

If we compare Figures 4.24 and 4.25, we can observe that the NOR gate requires larger area than NAND gate as pMOS occupies much greater (almost twice) area than nMOS as $p = \left(\frac{\mu_n}{\mu_p}\right)n$.

For this reason NAND gate is preferred to NOR gate for implementation of CMOS logic.

Figure 4.26(a) and (b) shows the NOR and NAND gates in pseudo-nMOS logic design style.

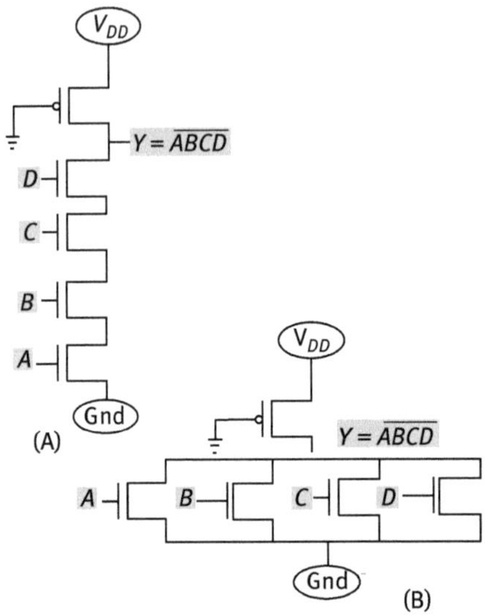

Figure 4.26: Proper transistor sizing of a 4 input NOR gate and 4 input NAND gate in pseudo-nMOS logic design style so that the worst-case gate delay equals to that of basic pseudo-nMOS inverter; p denotes W/L ratio of pMOS and n denotes W/L ratio of nMOS present in the basic pseudo-nMOS inverter.

NOR gates are preferred over NAND gates in pseudo-nMOS logic design. NOR gates are implemented by connecting nMOS transistor in parallel, which can be designed with minimum size nMOS transistor, requiring smaller area.

4.8 Introduction to Logical Effort

4.8.1 Definitions of Logical Effort

Definition 1: The logical effort of a logic gate is defined as the number of times worse it is at delivering output current than would be an inverter with identical input capacitance.

All logic gates including inverters deliver output currents. If all of them have identical input capacitance, then some of them will deliver more current while some of them will deliver less current. A CMOS inverter consists of a PUN and PDN with only one transistor. Thus to maintain same input capacitance, transistors will be maximum wider and thus will produce highest output current. A logic gate must have more transistors than an inverter, and so to maintain equal input capacitance, its transistors must be narrower (less width) on average and thus less able to conduct current than those of an inverter with identical input capacitance. Now in PDN or in PUN, the transistors may be connected in series or in parallel. If series topology is used, the logic gate cannot deliver output current same as the inverter with identical input capacitance. If parallel topology is used, we will consider only the worst case where only one of the transistors connected in parallel is conducting and will deliver less current than an inverter with identical input capacitance.

In general logic gate will deliver less current than an inverter irrespective of it being in series or in parallel. Logical effort is a measure of how much worse current it is than the inverter.

Definition 2: The logical effort of a logic gate is defined as the ratio of its input capacitance to that of an inverter that delivers equal output current.

This alternative definition is useful for computing the logical effort of a particular topology. To compute logical effort of a logic gate, adjust transistor sizes to obtain output current same as inverter and then compare input capacitance of each input signal. The ratio of this input capacitance to that of the standard inverter is the logical effort of that input to the logic gate.

Logical effort values for a few CMOS logic gates are shown in Table 4.2. Logical effort is defined so that an inverter has a logical effort of one. These unitless form means that all delays are measured relative to the delay of a simple inverter.

The logical effort of a logic gate tells how much worse it is at producing output current than is an inverter, given that each of its inputs may contain only the same input capacitance as the inverter [9]. Reduced output current means slower operation, and thus the logical effort number for a logic gate tells how much more slowly it will drive a load than an inverter would. Equivalently, logical effort is how much more input capacitance a gate presents to deliver the same output current as an inverter.

Assume the (W/L) ratio of a pMOS pull-up transistor is twice than (W/L) ratio of an nMOS pull-down transistor in an static CMOS inverter.

$p = \left(\frac{\mu_n}{\mu_p}\right)^n$ i.e. $p = 2n$ as mobility of electron is twice the mobility of holes.

If $n = (W/L)_n$ is selected as 1 then $p = (W/L)_p$ is equal to 2.

Table 4.2: Logical effort for inputs of static CMOS gates, assuming aspect ratio (W/L) of inverter's pull-up pMOS to pull-down nMOS is equal to 2.

	Number of inputs				
Gate type	**1**	**2**	**3**	**4**	***f*) *n***
Inverter	1				
NAND		4/3	5/3	6/3	$(n+2)/3$
NOR		5/3	7/3	9/3	$(2n+1)/3$
Multiplexer		2	2	2	2
XOR		4	12	32	

The logical effort of inverter is equal to 1. In the inverter the input "a" faces a gate capacitance of 2 due to pMOS and 1 due to nMOS; total equals $(2+1) = 3$.

There are two pMOS transistors in a 2 input NAND gate which are in parallel; thus in worst case any one of them will conduct. To make the $(W/L)_{\text{equivalent}}$ of the pMOS of the NAND gate same as the (W/L) of the pMOS of the inverter we must choose $p = (W/L)_p = 2$.

There are two nMOS transistors in a 2 input NAND gate which are in series; thus in worst case and only one case indeed, both of them will conduct. To make the $(W/L)_{\text{equivalent}}$ of the nMOS of the NAND gate same as the (W/L) of the nMOS of the inverter that equals 1, we must choose $n = (W/L)_n = 2$. As we know

$$(W/L)_{\text{eq in series}} = \frac{1}{\left(\frac{1}{(W/L)_1} + \frac{1}{(W/L)_2} + \cdots\right)}.$$

Now any input "a" or "b" in the NAND gate faces an input capacitance of 2 due to pMOS and another 2 due to nMOS; total equals to $2 + 2 = 4$. The input of the inverter faces an input capacitance that equals 3. Thus the logical effort is 4/3. Similarly from Figure 4.27(c) one can calculate the logical effort of 2 input NOR gate that equals 5/3.

As 4/3 is closer to 5/3, a NAND gate's delay characteristic is closer to the characteristic of an inverter than a 2 input NOR gate. Thus NAND gate is preferred to NOR gate for implementation of CMOS logic.

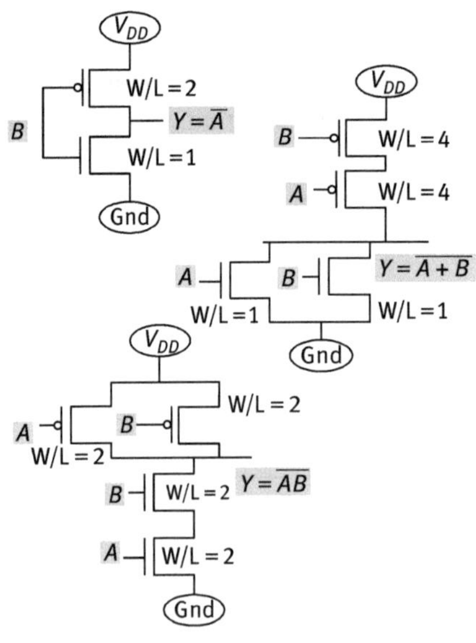

Figure 4.27: (a) Static inverter (b) 2 input NAND gate and (c) 2 input NOR gate. Numbers indicate relative size (aspect ratio) of the transistors.

It is interesting but not surprising to note from Table 4.2 that more complex logic functions have larger logical effort. Moreover, the logical effort of most logic gates grows with the number of inputs to the gate. Larger or more complex logic gates will thus exhibit greater delay. Logical effort makes the comparison and choice easy among different logical structures. Designs that minimize the number of stages of logic will require more inputs for each logic gate and thus have larger logical effort. Designs with fewer inputs and thus less logical effort per stage may require more stages of logic.

4.9 Delay Estimation by Logical Effort

The delay in an MOS circuit can be estimated by an easy method known as logical effort. This method also specifies the number of logic stages and optimized transistor sizes. This method is suitable for comparison of delay among various logic structures and optimization of logic design [8, 9].

We first define that total delay suffered by a logic gate consists of two parts: a fixed part known as parasitic delay, and a part proportional to the load on the gate's output called the effort delay. The parasitic delay of a logic gate is fixed, independent of the size of the logic gate and of the load capacitance it drives.

$$d = p + f$$

d denotes total delay, p denotes parasitic delay and f denotes effort delay.

The effort delay depends on the load and also on the properties of the logic gate. To incorporate the first factor, electrical effort denoted as h is introduced, which characterizes the load. On the other hand logical effort denoted as g is introduced to capture the properties of the logic gate. The effort delay of the logic gate is the product of these two factors given by

$$f = g * h$$

The logical effort g reflects the logic gate's topology and its ability to produce output current. The *logical effort* represents the fact that, for a given load, complex gates have to work harder than an inverter to produce a similar response. In other words, the logical effort of a logic gate tells how much worse it is at producing output current than an inverter, given that each of its inputs may contain only the same input capacitance as the inverter. Equivalently, logical effort is how much more input capacitance a gate presents to deliver the same output current as an inverter. Logical effort is a useful parameter, because it depends only on circuit topology.

The electrical effort h describes how the electrical environment of the logic gate affects performance and how the size of the transistors in the gate determines its load driving capability. The electrical effort is defined by

$$h = C_{Out}/C_{In},$$

where C_{Out} is the capacitance that loads the logic gate and C_{In} is the capacitance presented by the logic gate at one of its input terminals. Electrical effort is also called fan-out by many CMOS designers. The load driven by a logic gate is generally capacitive (C_{Out}) in nature and such load will slow down the circuit. The input capacitance of the circuit is a measure of the size of its transistors. The input capacitance term appears in the denominator of equation $h = C_{Out}/C_{In}$ because bigger transistors in a logic gate will drive a given load faster. Electrical effort is usually expressed as a ratio of transistor widths rather than actual capacitances. We know that the capacitance of a transistor gate is proportional to its area; if we assume that all transistors have the same minimum length, then the capacitance of a transistor gate is proportional to its width. Because most logic gates drive other logic gates, we can express both C_{In}, and C_{Out} in terms of transistor widths.

Combining equations above, we obtain the basic equation that models the delay through a single logic gate.

$$d = gh + p.$$

4.10 Introduction to Transmission Gate

The most widely used solution to deal with the voltage-drop problem is the use of transmission gates [11–13].

When an nMOS or pMOS transistor is used as a pass transistor, it acts as an imperfect switch, causing logic degradation. If pMOS and nMOS are connected in parallel, a switch is obtained where both logic 0 and logic 1 are passed without degradation and in an acceptable fashion. This switch is called transmission gate (TG) as shown in Figure 4.28. It acts as an excellent switch, providing bidirectional current flow, and it exhibits an on resistance that remains almost constant for wide range of input voltage.

These characteristics make TG not only an excellent switch in digital application but also excellent choice as analog switch.

Figure 4.28 (a–e) below shows five alternate symbols or representations of CMOS TG. nMOS transistor passes logic "0" and pMOS transistor passes logic "1" in a CMOS TG provided appropriate control signals are applied. So the output is always strongly driven and the levels are never degraded. This is called a fully restored logic gate and simplifies circuit design considerably. A CMOS TG requires both the control input and its complement to work properly. Thus CMOS TG is categorized in doubled rail logic.

TG consists of one nMOS and one pMOS in parallel. The gate voltages applied to these two transistors are also complementary signals. As such CMOS TG operates as a bidirectional switch between the node *A* and *B* controlled by signal *C*.

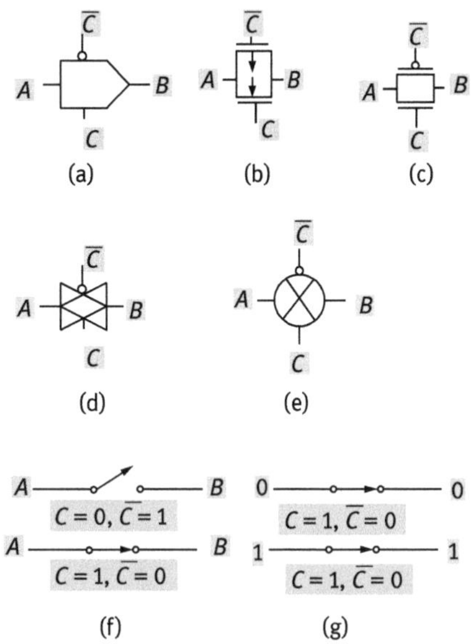

Figure 4.28: Five (a–e) different representations of the CMOS transmission gate (TG) (f) TG acting as an open gate or closed gate depending upon applied control signal (g) both high and low signals are passed through TG without degradation.

If the control signal C is logic high, i.e. equal to V_{DD}, then both transistors are turned on and provide a low resistance path between the nodes A and B. If, on the other hand, the control signal C is low, then both transistors will be off, and the path between the nodes A and B will be an open circuit. This condition is also called the high impedance state.

TG can be used as a basic building block to represent a class of logic families called CMOS transmission gate or pass gate logic. Pass transistor logic utilizes MOSFETs in series path from the input to the output to pass or block the signal transmission as discussed earlier. In pass transistor logic a single MOSFET is used as the switch. CMOS TG can be used to implement the switches, giving alternative name, transmission gate logic. The terms (*pass transistor logic* or *transmission gate logic*) can be interchanged without considering the actual implementation of the switches. In this section we will investigate how TG finds its application to be used in logic circuit design, sometimes resulting in compact circuit structures which may need smaller number of transistors than their standard CMOS counterparts.

We will examine here a simple switch circuit using TGs as their basic building blocks. The transmission gate is a fundamental and ubiquitous component in MOS logic. It is used as a multiplexing element, a logic structure, a latch element, and an analog switch.

4.10.1 Use of CMOS TG as Switch

Figure 4.29 shows a TG with input connected to an ideal step input signal at $t = 0$. Assume initial output voltage across the output capacitor is zero. The output node is connected to subsequent logic stages, which represent capacitive loads. The substrate terminal of nMOS is connected to ground and substrate terminal of pMOS is connected to V_{DD}. The control signals are applied in such a way that both the transistors will conduct.

4.10.1.1 Logic "1" Transfer

For an nMOS,

$$V_{DSn} = V_{In} - V_{Out}(t) = V_{DD} - V_{Out}(t)$$
$$V_{GSn} = V_{DD} - V_{Out}(t).$$

Assume the output voltage $V_{Out}(t)$ is zero. Therefore $V_{DSn} > V_{GSn} - V_{Tn}$. Thus initially nMOS is operating at saturation. The nMOS transistor will be turned off when $V_{GSn} < V_{Tn}$, i.e. $V_{DD} - V_{Out}(t) < V_{Tn}$ or in other words $V_{Out}(t) > V_{DD} - V_{Tn}$. Thus the output voltage cannot exceed $V_{DD} - V_{Tn}$. As long as $V_{Out}(t)$ is less than $V_{DD} - V_{Tn}$ the nMOS is operating in saturation mode. Once the $V_{Out}(t)$ equals $V_{DD} - V_{Tn}$, the nMOS is turned off.

Thus nMOS in the TG for rising input always works in saturation mode. The nMOS in saturation mode conducts a charging current I_{Dsn} which is given as

$$I_{Dsn} = \frac{\mu_n C_{Ox}}{2} \frac{W}{L} (V_{GS} - V_{Tn})^2$$
$$I_{Dsn} = \frac{\mu_n C_{Ox}}{2} \frac{W}{L} (V_{DD} - V_{Out} - V_{Tn})^2 .$$

This diminishing current through nMOS reduces to zero when $V_{Out}(t) = V_{DD} - V_{Tn}$.

The voltage at the source terminal of the nMOS equals $V_{Out}(t)$, which is not equal to zero. The substrate of the nMOS is connected to ground. Thus there is a voltage

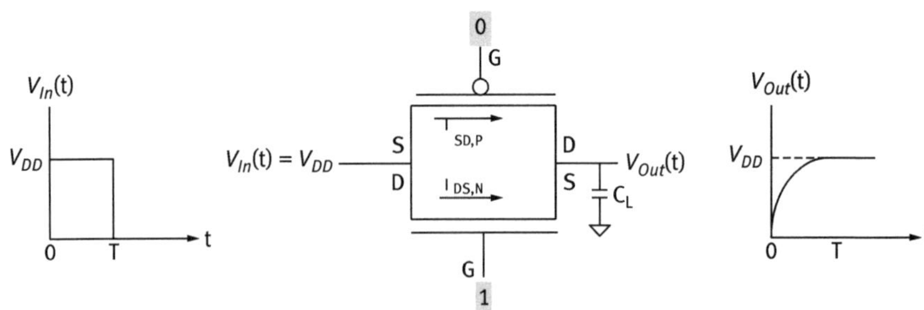

Figure 4.29: Logic "1" transfer by CMOS transmission gate.

difference V_{Sb} between the source and the substrate terminal of the nMOS. Thus the threshold voltage V_{Tn} will be determined by the body effect and is given by

$$V_{Tn} = V_{t0} + y\left(\sqrt{V_{Out}(t) + 2f_f} - \sqrt{2f_f}\right).$$

For pMOS,

$$\begin{aligned} V_{DSp} &= V_{Out}(t) - V_{DD} \\ V_{Sdp} &= V_{DD} - V_{Out}(t) \\ V_{GSp} &= -V_{DD}. \end{aligned}$$

The condition for pMOS transistor to work in the saturation is $V_{DSp} \le V_{GSp} - V_{Tp}$.
Substituting values, we obtain

$$V_{Out}(t) - V_{DD} \le -V_{DD} - V_{Tp}.$$

Thus as long as $V_{Out}(t) \le |V_{Tp}|$ the pMOS transistor works in saturation region, when $V_{Out}(t)$ crosses $|V_{Tp}|$ or when $V_{Out}(t) > |V_{Tp}|$ the pMOS operates in linear region. pMOS transistor remains always on for a logic “1” transfer, regardless of the output voltage level $V_{Out}(t)$.

Thus pMOS transistor operates initially at the saturation with $V_{SG} = V_{DD}$ with a current passing through it.

$$I_{Dsp} = \frac{\mu_p C_{Ox}}{2}\frac{W}{L}\left(V_{DD} - |V_{Tp}|\right)^2$$

Initially $V_{Out}(t) = 0$. However, $V_{Out}(t)$ increases with time, which decreases V_{Sdp} or V_{DSp} increases. As long as $V_{DSp} \le V_{GSp} - V_{Tp}$ or $V_{Out}(t) \le |V_{Tp}|$ the pMOS works in saturation and then it changes mode from saturation to linear after $V_{Out}(t)$ crosses $|V_{Tp}|$. After this occurs

$$I_{Dsp} = \frac{\mu_p C_{Ox}}{2}\frac{W}{L}\left[2\left(V_{DD} - |V_{Tp}|\right)\left(V_{DD} - V_{Out}(t)\right) - \left(V_{DD} - V_{Out}(t)\right)^2\right]$$

until $V_{Out}(t)$ reaches V_{DD}.

The substrate and the source terminal of the pMOS are connected to V_{DD}. Thus there is no body effect in the pMOS for logic “1” transfer.

Initially the total charging current equals to the sum of I_{Dsn} and I_{Dsp}. The pMOS will enter into the linear region when $V_{Out}(t) = |V_{Tp}|$ and will continue to conduct until capacitor C_{Out} is fully charged and $V_{Out}(t) = V_{Oh} = V_{DD}$. Thus pMOS will provide the gate with a good “1”. The MOSFET regions of operation are summarized by the plot in Figure 4.30.

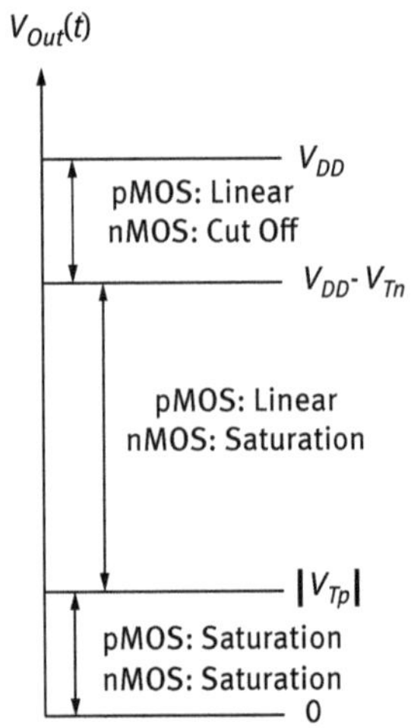

Figure 4.30: Various modes of operation for nMOS and pMOS present in a TG for logic "1" transfer.

4.10.1.2 Logic "0" Transfer

Figure 4.31 shows a TG transferring logic "0". Assume $V_{Out}(t) = 0$ initially before $t = 0$. If V_{In} goes low at $t = 0$, the nMOS and pMOS interchange roles as compared to logic "1" transfer.

For pMOS,

$$V_{Sdp} = V_{Out}(t)$$
$$V_{SGp} = V_{Out}(t).$$

So pMOS is initially saturated with current given by

$$I_{Dsp} = \frac{\mu_p C_{Ox}}{2} \frac{W}{L} \left(V_{Out}(t) - |V_{Tp}|\right)^2,$$

which is valid so long as $V_{SG,p} = V_{Out}(t) \geq |V_{Tp}|$.

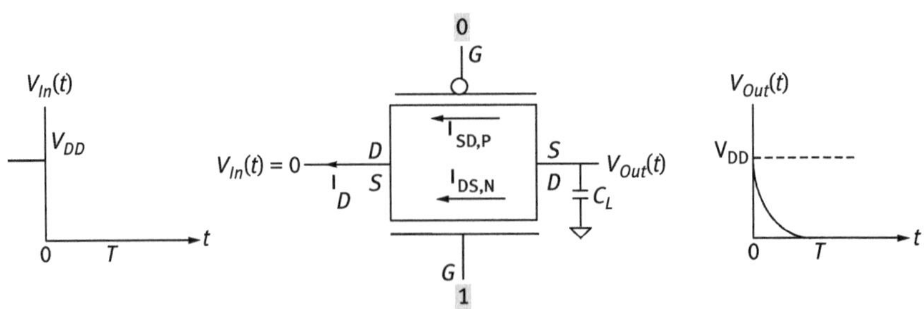

Figure 4.31: Logic "0" transfer by CMOS transmission gate.

Since the nMOS allows for a complete discharge of C_{Out} pMOS will go into cutoff when $V_{Out}(t)$ falls below $|V_{Tp}|$ and $I_{Dp} = 0$.

Thus pMOS will move into cutoff mode when $V_{Out}(t)$ falls below $|V_{Tp}|$, where $|V_{Tp}|$ is affected by body effect and is given by

$$V_{Tp} = V_{t0} + y\left(\sqrt{V_{DD} - V_{Out}(t) + 2f_f} - \sqrt{2f_f}\right).$$

For nMOS

$$V_{DSn} = V_{Out}\,(t)$$
$$V_{GSn} = V_{DD}.$$

So nMOS transistor was initially at saturation with current given by

$$I_{Dsn} = \frac{\mu_n C_{Ox}}{2}\frac{W}{L}\left(V_{Out}(t) - V_{Tn}\right)^2.$$

The output capacitor C_{Out} is discharging. nMOS changes mode from saturation to linear when $V_{Out}(t)$ falls below $V_{DD} - V_{Tn}$.

It continues to conduct in linear mode until C_{Out} is fully discharged with the current given by

$$I_{Dsn} = \frac{\mu_n C_{Ox}}{2}\frac{W}{L}\left[2\left(V_{DD} - V_{Tn}\right)V_{Out}(t) - V_{Out}(t)^2\right]$$

The remaining discharge continues until $V_{Out}(t) = V_{Ol} = 0$ V, a strong "0". The operational modes of the transistors for this case are summarized in Figure 4.32.

Thus TG provides no logic degradation – superior performance than single nMOS switches. The price paid is increased size, area and complexity.

4.10.2 2:1 Multiplexer Using TG

Figure 4.33 shows a two-input MUX circuit consisting of two TGs with its truth table shown in Table 4.3. If the control input *s* is logic high, then the bottom TG will conduct and output will be equal to B. If the control signal is low, the bottom TG will turn off and the top TG will connect the output to input A.

4.10.3 XOR Gate Using TG

The XOR gate using 8 MOS transistors using 2 TG and 2 CMOS static inverters is shown in Figure 4.34 with its truth table shown in Table 4.4. Full CMOS representation of XOR gate requires 12 transistors as shown in Figure 4.9. Thus TG logic reduces

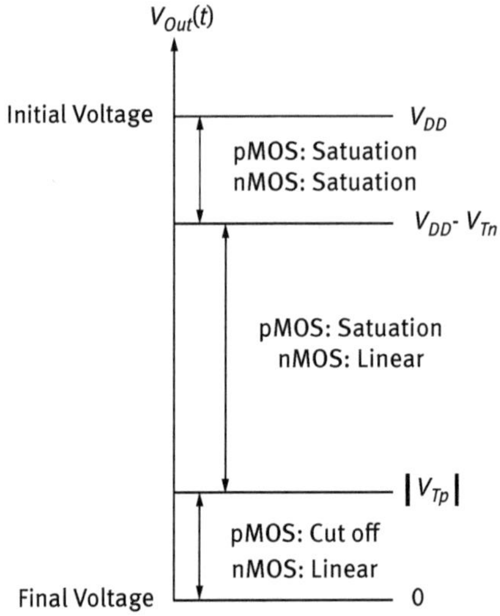

Figure 4.32: Various modes of operation for nMOS and pMOS present in a TG for logic "0" transfer.

Table 4.3: Multiplexer truth table.

$S/\overline{S}$	D_1	D_0	Y
0/1	X	0	0
0/1	X	1	1
1/0	0	X	0
1/0	1	X	1

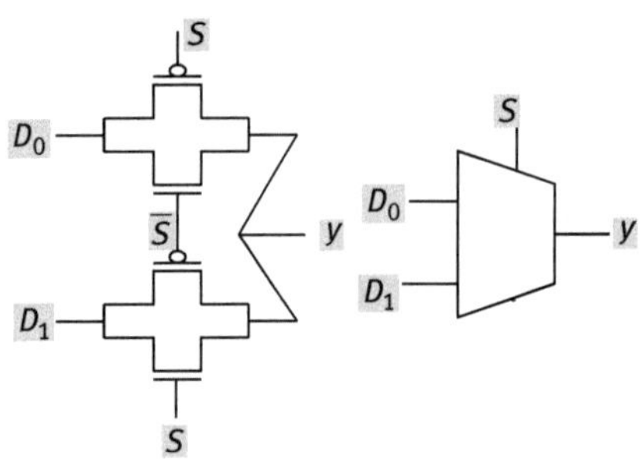

Figure 4.33: 2:1 Multiplexer implemented using two CMOS TGs.

transistor count than standard CMOS logic in this particular case. A better XOR gate, implemented using only 6 transistors, is shown in Figure 4.35.

If B is low then o/p = A.

Otherwise if B is high o/p = not(A).

Table 4.4: Truth table of XOR/nonequivalence function.

g) *A*	h) *B*	*Y*(XOR)
0	0	0
0	1	1
1	0	1
1	1	0

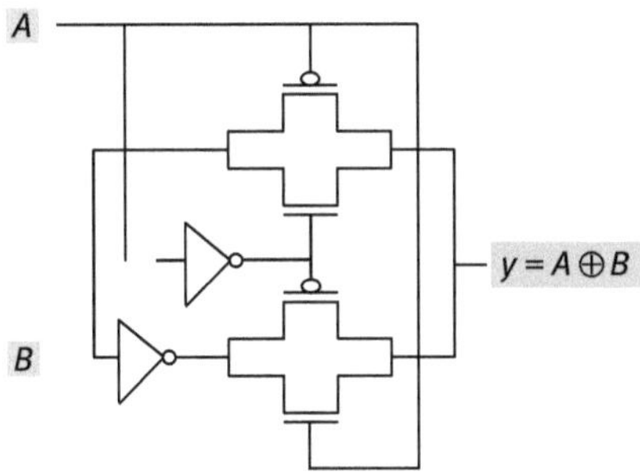

Figure 4.34: Eight-transistor CMOS TG implementation of the XOR function.

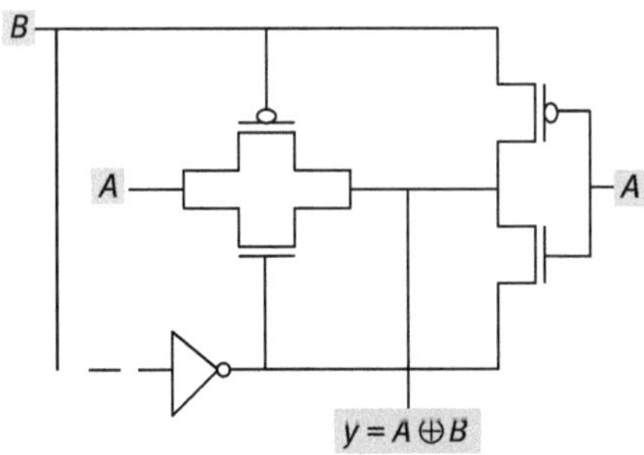

Figure 4.35: Six-transistor XOR CMOS TG implementation.

4.10.4 XNOR Gate Using TG

XNOR gates are known as equivalence function gate because when both inputs *A* and *B* are equal, the output turns high; otherwise it remains low as shown in its truth table in Table 4.5. Figure 4.36 shows a XNOR using TG with only 6 transistors.

Table 4.5: Truth table of XNOR/equivalence function.

i) *A*	*j*) *B*	*Y*(XNOR)
0	0	1
0	1	0
1	0	0
1	1	1

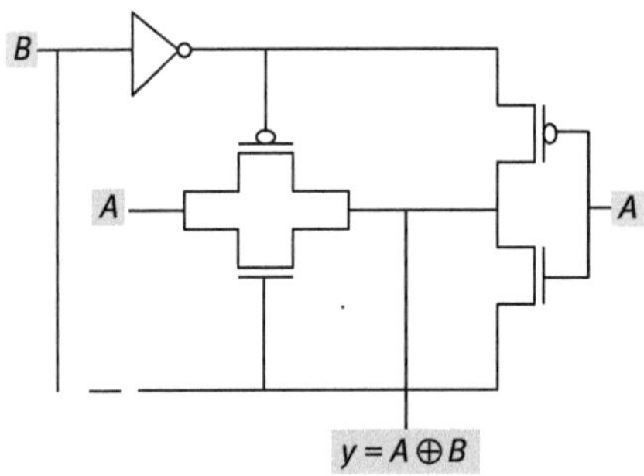

Figure 4.36: XNOR gate using 6 transistors.

4.10.5 Transmission Gate Adders

The full adder output, sum and carry can be written as

$$sum\_out = A\rho AB\rho AC = (A\rho AB).\overline{C} + \overline{(A\rho AB)}.C$$

$$carry\_out = AB + BC + CA = AB(C + \overline{C}) + BC(A + \overline{A}) + CA(B + \overline{B}) = \overline{A}BC + A\overline{B}C$$
$$+ ABC + AB\overline{C}$$
$$= C(\overline{A}B + A\overline{B}) + AB = C(\overline{A}B + A\overline{B}) + (AB + \overline{A}.\overline{B})A = (A\rho AB).C + \overline{(A\rho AB)}.A$$

The full adder is implemented using TG as shown in Figure 4.37.

4.10.6 More Examples of TG Logic

(1) OR gate (Fig. 4.38)

$$f = A + \overline{A}.B = A\,(1 + B) + \overline{A}.B = A + AB + \overline{A}.B = A + B(A + \overline{A}) = A + B$$

(2) Three-variable Boolean function (Fig. 4.39)

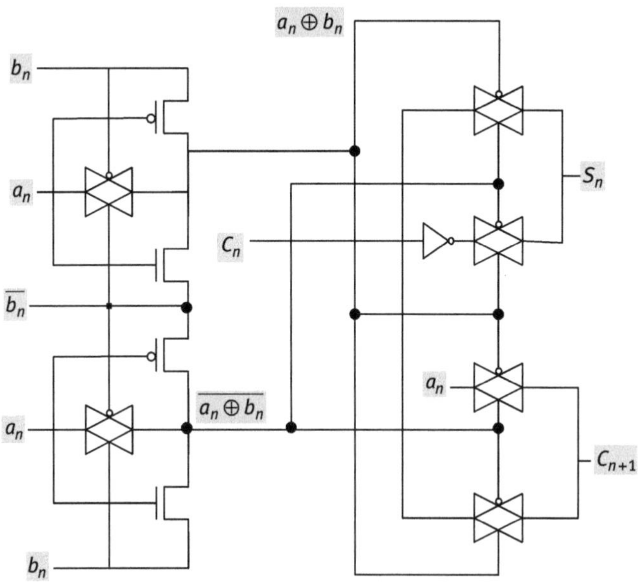

Figure 4.37: Full adder using TG logic.

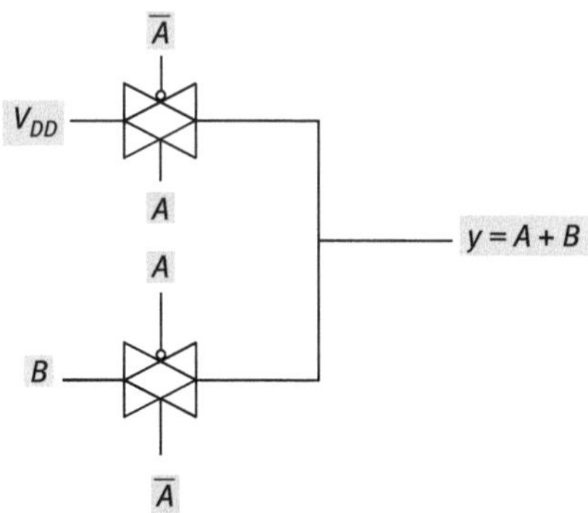

Figure 4.38: OR gate using TG.

(3) 4:1 Multiplexer (Fig. 4.40)

$$f = P_0.\left(\overline{S_1 S_0}\right) + P_1.\left(\overline{S_1} S_0\right) + P_2.\left(S_1 \overline{S_0}\right) + P_2.\left(S_1 S_0\right)$$

4.11 Tristate Buffer

We know that complementary CMOS gates are always inverting in nature. When inputs are logic "1", nMOSFETs turn on , leading to a "0" at the output. To make a non-inverting buffer we might be tempted to place the pMOS in place of the nMOS and nMOS in place of the pMOS. It will be a non-inverting gate as shown in Figure 4.41 but both the pMOS and nMOS produce degraded output. Thus this type of design is

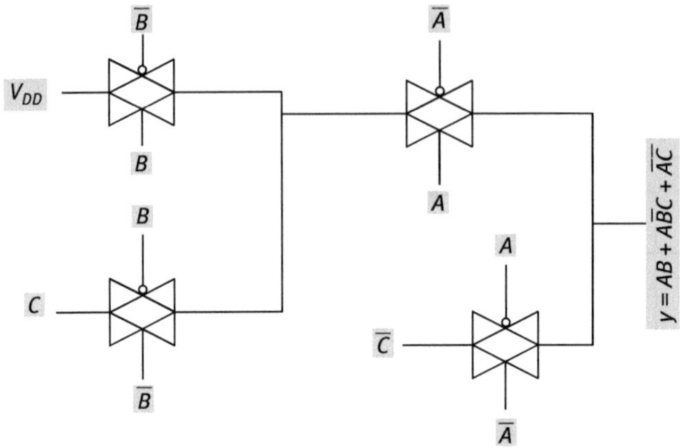

Figure 4.39: CMOS TG realization of a three-variable Boolean function.

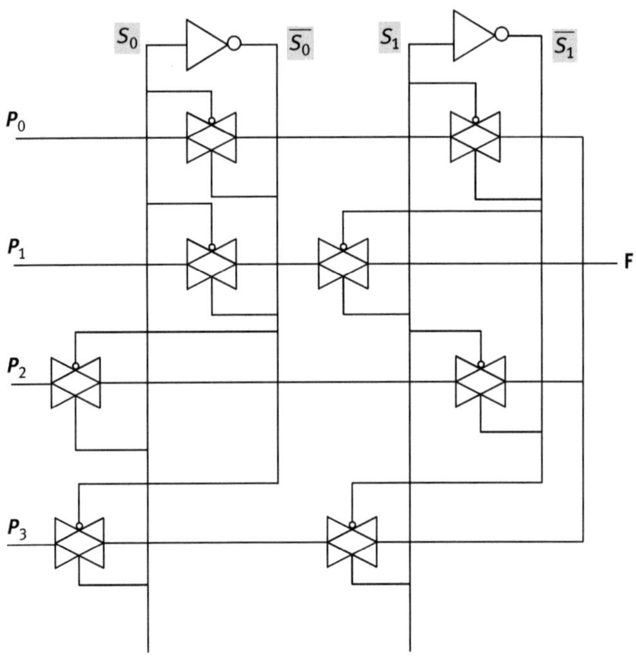

Figure 4.40: 4:1 MUX using TG.

discarded. Because a buffer output must always be strong. Actually the purpose of the buffer is to restore the logic.

4.12 Transmission Gates and Tristates

Transmission gate can be used to produce a buffer or tristate buffer. Figure 4.41 shows the symbols for a tristate buffer. When the enable input *EN* is "1", the output *Y* equals

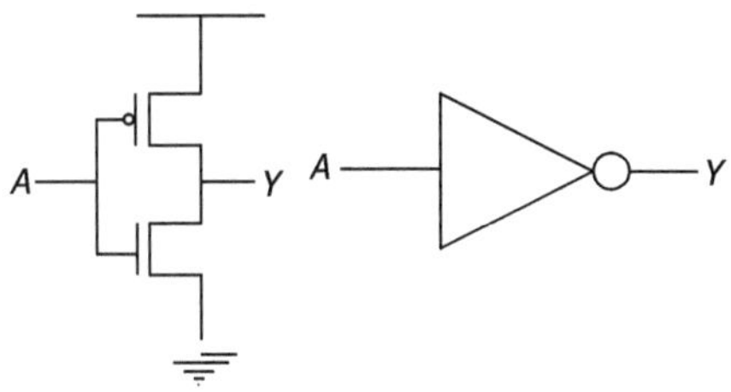

Figure 4.41: A bad signal degraded buffer with nMOS and pMOS interchanged in a standard CMOS static inverter.

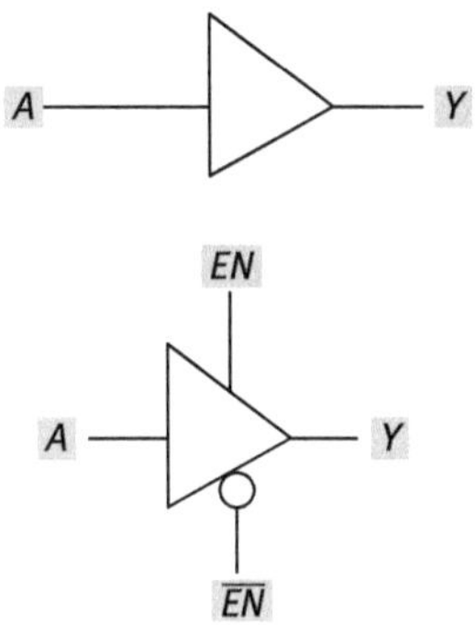

Figure 4.42: Tristate buffer both (a) with only enable shown (b) with enable and its complement shown.

Table 4.6: Truth table for tristate buffer.

$EN/\overline{EN}$	A	Y
0/1	0	Z
0/1	1	Z
1/0	0	0
1/0	1	1

the input A, just as in an ordinary buffer. When the enable input is "0", Y is left floating (a "Z" value) or high impedance state. The truth table and structure of tristate buffer is shown in Table 4.6 and Fig. 4.42 respectively.

Figure 4.43 shows a TG which has the same truth table as tristate buffer. It requires only two transistors but unfortunately it is a member of non-restoring logic family. If the input is a noisy or otherwise degraded signal, the output will receive the same noise. After several stages of non-restoring logic, a signal can become too degraded to recognize.

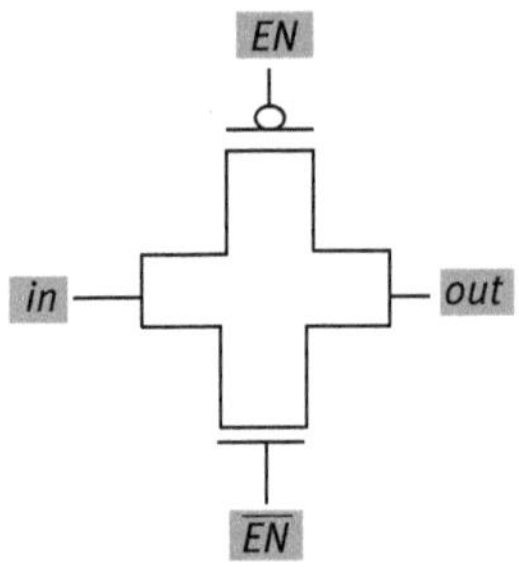

Figure 4.43: Transmission gate can be used as tristate buffer but will be a non-restoring logic.

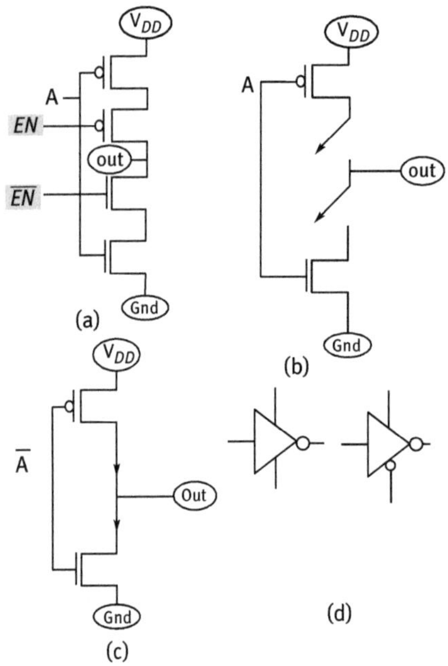

Figure 4.44: (a) Tristate inverter (b) when *EN* = "0", output is in high impedance state (c) when *EN* = "1", circuit effectively becomes an inverter (d) symbols of tristate inverter.

Figure 4.44 (a) shows a tristate inverter with logic restoring capability as the output is always connected to V_{DD} or ground. For some particular input combinations the output is floating, i.e. connected to neither V_{DD} nor ground, which is known as high impedance state. For example, when *EN* is "0" (Figure 4.44 (b)), both enable transistors are off, leaving the output floating. When *EN* is "1" (Figure 4.44 (c)), both enable transistors are on. They are conceptually removed from the circuit, leaving a simple inverter. Figure 4.44 (d) shows symbols for the tristate inverter. A tristate buffer can be built as a tristate inverter, following an ordinary inverter.

4.13 Implementation of Combinational Circuit Using DTMOS Logic for Ultralow Power Application

Subthreshold logic circuits are efficacious for ultralow power system design. In subthreshold logic circuits, as the supply voltage is scaled down below the threshold voltage of the MOSFET, gate overdrive voltage decreases, which also decreases the circuit speed. SO subthreshold logic circuits are advantageous where the low power has utmost importance instead of the speed of operation. Hence we cannot scale down the threshold voltage aggressively as the minimum value of the threshold voltage is fixed. Hence, device engineering becomes effective for the design of low power circuits. Dynamic threshold voltage MOSFET (DTMOS) is a class of subthreshold circuits which offer very less power dissipation [15]. Threshold voltage of nMOS can be written as $V_{Th} = 2\phi_B + V_{Fb} + y\left(\sqrt{|2\phi_B| - V_{BS}}\right)$. Hence φ_B is the inversion layer potential, V_{Fb} is the Fermi band voltage and y is the body effect parameter. Now, in DTMOS circuit gate and the body are tied up together. So in DTMOS (N-type) when the gate potential is low then the threshold voltage becomes $V_{Th} = 2\phi_B + V_{Fb} + y\left(\sqrt{|2\phi_B|}\right)$, i.e. maximum. Higher threshold voltage limits the leakage currents and also reduces the power dissipations due to the leakage currents at zero biasing. Also, when the gate potentials set to high, then the threshold voltage will be $V_{Th} = 2\phi_B + V_{Fb} + y\left(\sqrt{|2\phi_B| - V_{DD}}\right)$ i.e. minimum. Low threshold voltage results higher drive current than the normal condition for lower supply voltage. Threshold voltage variation can modify the inversion charge and also the carrier mobility. Reduction in threshold voltage increases the effective gate capacitance and the mobility of the carriers. So higher drive currents can be obtained.

From the designing point of view there is no basic differece with the conventional design, except that in DTMOS circuits gate and the body terminal will be tied up together. Hence the NAND and NOR structure is given for the analysis of the DTMOS logic. In DTMOS NAND gate, when $A = B = 0$, then the PUN will be on and the PDN will be off. So the output node potential will be raised up to the V_{DD}. Similarly, when $A = B = 1$, PUN will be turned off but the PDN will be turned on to pass the logic zero to the output node. Hence the operation is very much similar with the conventional operation. However, one should remember that due to the dynamic variation of the threshold voltage, leakage currrent and the leakage power dissipation will be very less compared to the conventional subthreshold logic operation.

The structure of DTMOS NAND and NOR structures is shown in Fig. 4.45. The output waveform of the NAND and the NOR gates using DTMOS structures is shown in Fig. 4.46 and 4.47, which is detailed in this section. It has been observed that completely full swings are obtained during the operations. Because of full swing, comparatively larger loads can also be driven by the DTMOS circuits. Hence pulse width of 100 μs is chosen during the operation. So the frequency is 10 KHz. If we increase the frequency, then the logic degradation will be observed, as in high frequency regime, and

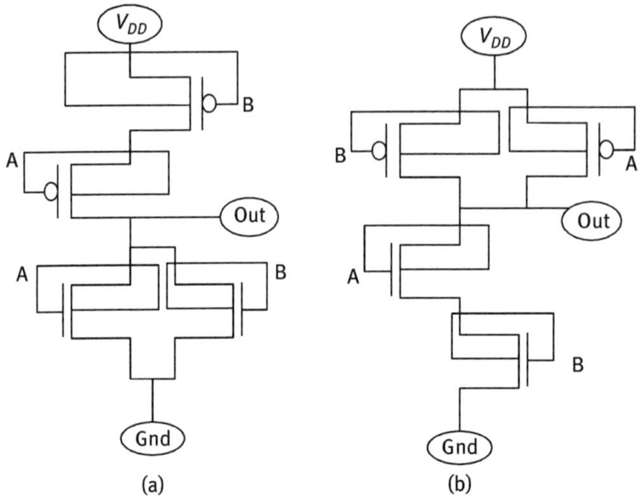

Figure 4.45: Subthreshold DTMOS logic circuit (a) NAND and (b) NOR.

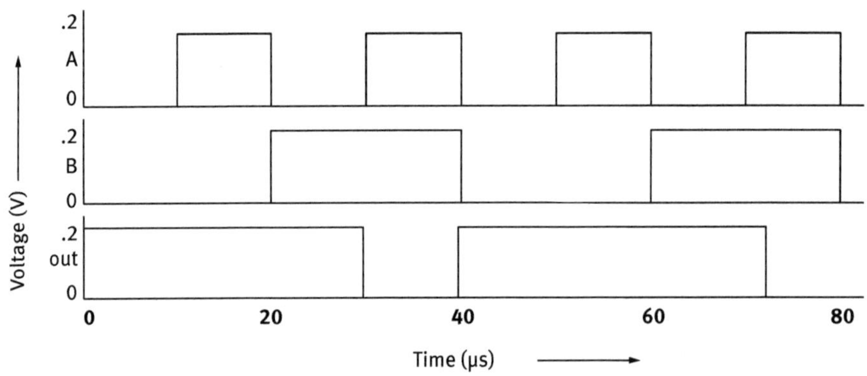

Figure 4.46: Output waveform of subthreshold DTMOS NAND.

the output load capacitor will not be able to charge or discharge completely. In other words, low operating currents will not be able to drive the capacitive load. DTMOS circuits are preferable in low frequency regime for better logic swing [16].

For the further design and the analysis of the DTMOS cicrcuits, two more examples are given here. In Figure 4.48, the implemented function is $Y = \overline{A + B(C + D)}$. Also it can be observed from the figure that same principles have been used during the implementation of the function. In PDN, the nMOS having gate inputs of C and D are connected in parallel. This is connected in series with the nMOS having gate input B. Whole combination will be in parallel with the nMOS having gate input A. Pull-up network will be complementary to the PDN. It must be noted that in case of every MOSFET, the gate and body terminals are tied up to make the DTMOS structure. In the

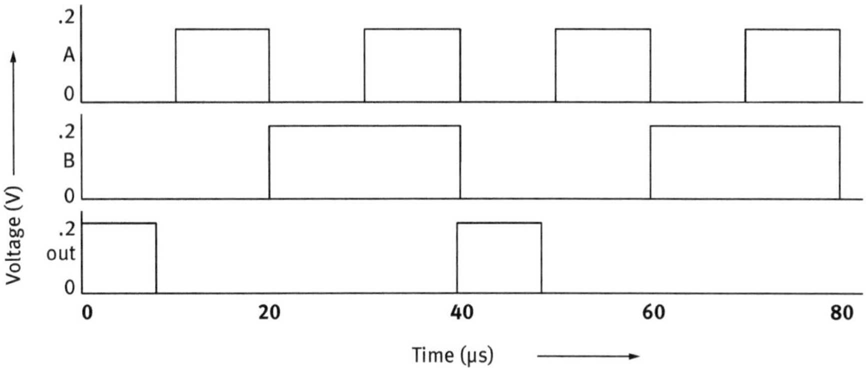

Figure 4.47: Output waveform of subthreshold DTMOS NOR.

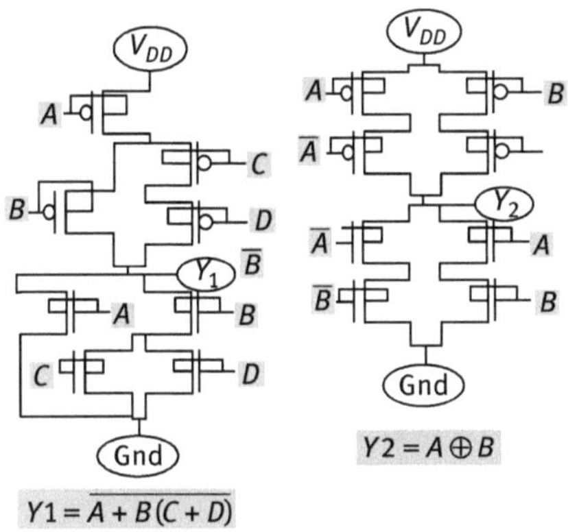

Figure 4.48: Schematic of the subthreshold DTMOS (a) $Y = \overline{A + B(C + D)}$ (b) XOR.

same fashion, XOR gates can be implemented. Using XOR and AND one can implement the full adder circuit, which is the basic building block of the multiplier. Hence Fig. 4.49 shows the output waveform of the $Y = \overline{A + B(C + D)}$ and the XOR gate is given for the ease of understanding the operations.

A full adder circuit in CADENCE SPICE Spectra and its output waveform in KHz regime of frequency are given for better understanding of the circuit. The transistor level schematic, test-bench setup and output waveform of the subthreshold DTMOS full adder circuit is shown in Fig. 4.50–4.52 respectively.

Parallel adders can add the binary strings of different length or of similar length in parallel. The schematic of a 4-bit parallel adder is given in Figure 3.8 for the ease of understanding. N bit parallel adder consists of N number of full adder block which has been discussed in the previous section. Hence we set $N = 4$. Similar length, i.e. 4 bits of augend and addend, has been chosen. Hence i-th position of augend bit will be added with the i-th position of the addend bit. Produced sum will be the output of

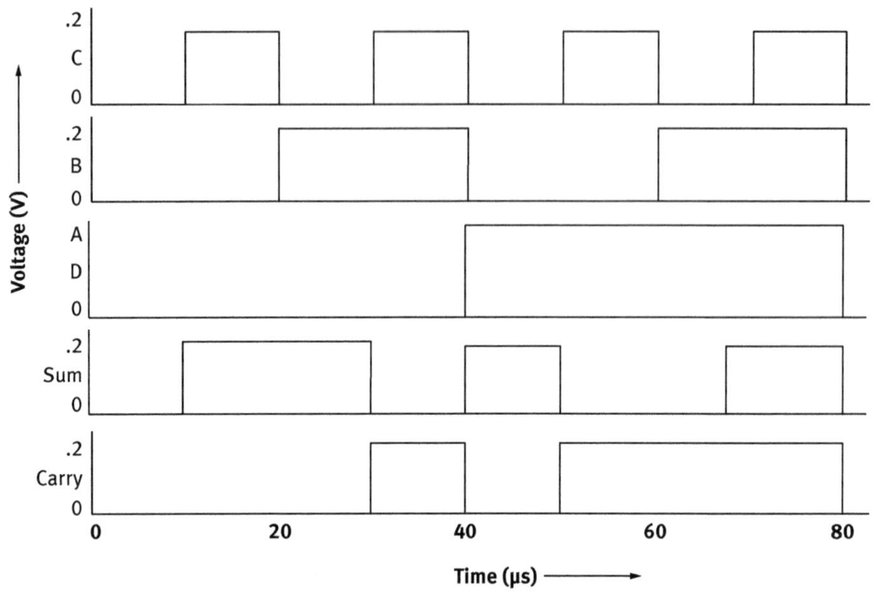

Figure 4.49: Output waveform of subthreshold DTMOS logic gate having output $Y = \overline{A + B(C + D)}$.

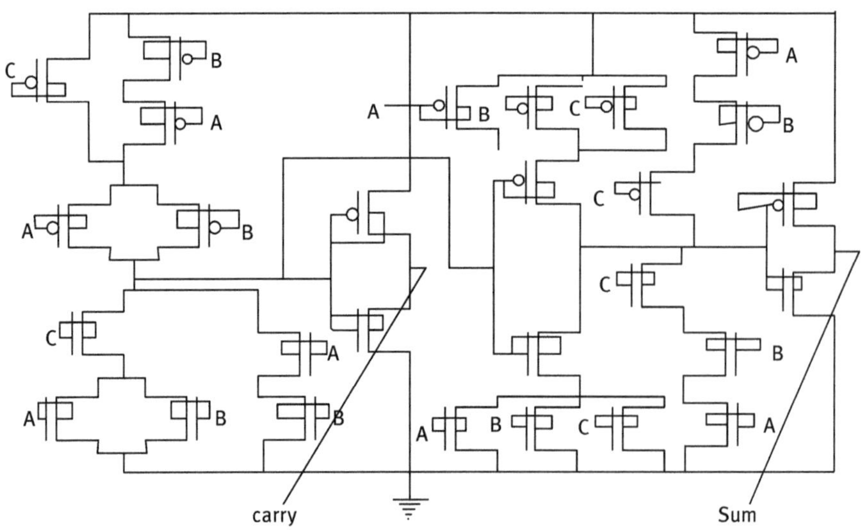

Figure 4.50: Transistor level schematic of Subthreshold DTMOS full adder in CADENCE.

i-th bit and the produced carry of the i-th stage will be propagated to the $(i+1)$th stage. Hence value of i will be from zero to $N-1$. Input carry will be fed to the 0th stage and the final carry will be produced at the $(N-1)$-th stage. The augend's bits of x are added to the addend bits of y, respectfully, of their binary position. Each bit 6 addition creates

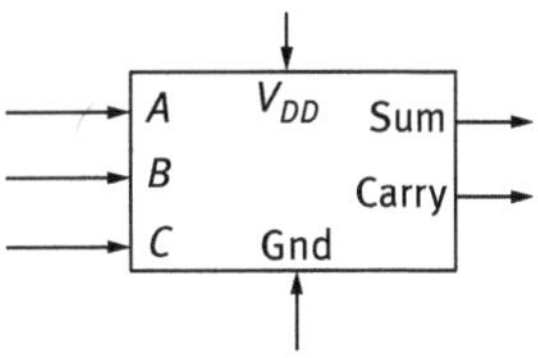

Figure 4.51: Test bench setup of subthreshold DTMOS full adder in CADENCE.

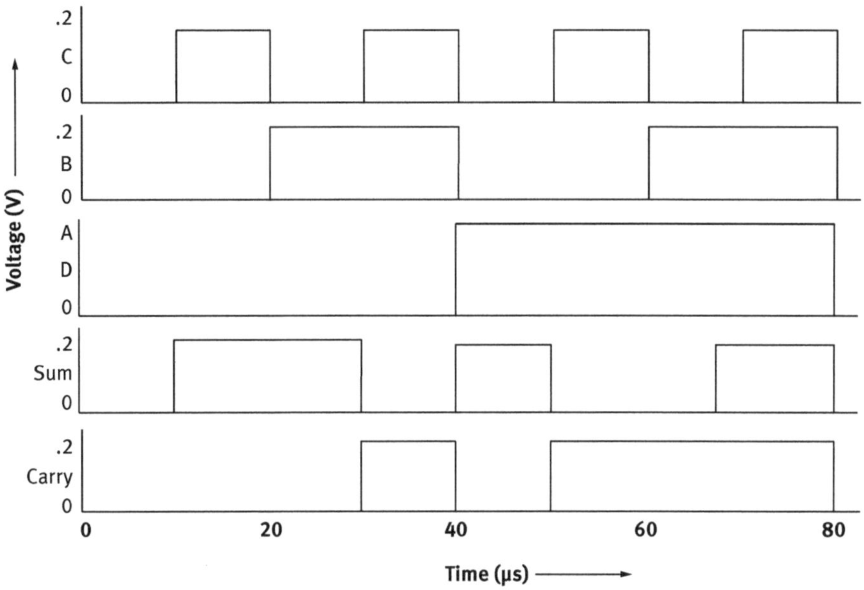

Figure 4.52: Output waveform of the subthreshold DTMOS full adder.

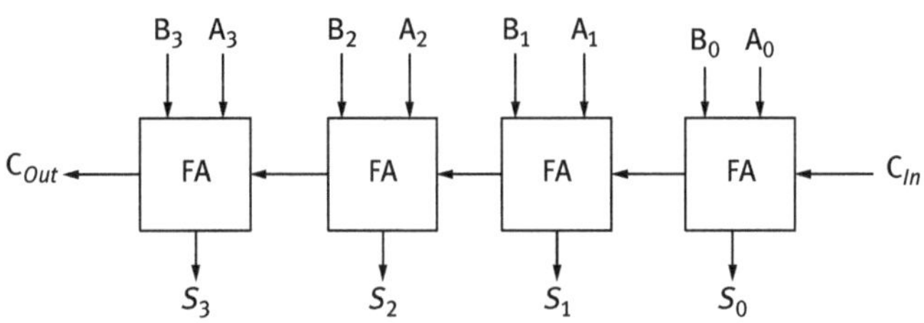

Figure 4.53: Block diagram of 4-bit parallel adder.

a sum and a carry out. The carry out is then transmitted to the carry in of the next higher-order bit. The final result creates a sum of 4 bits plus a carry out.

Using the full adder block given in Figure 4.53, one can easily implement an *n* bit parallel adder. A schematic of the 4-bit parallel adder is given in Figure 4.54 and its output waveform is shown in Fig. 4.55. The rectangular block is the presentation of the

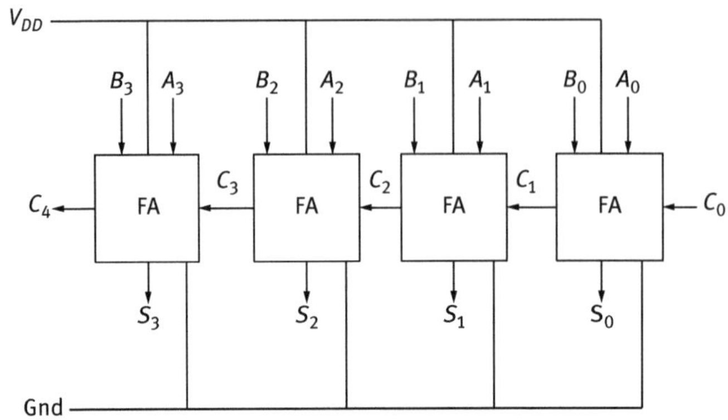

Figure 4.54: Schematic (block diagram) of 4-bit parallel adder.

DTMOS full adder. The input ports are given at the top left side. The output ports are given at the top from the right end, i.e. S_0, S_1, S_2 and S_3. Vertically rightmost four ports are A_0, A_1, A_2 and A_3; next set of four ports are B_0, B_1, B_2 and B_3; next ports are C_{In} and V_{DD} and the bottommost port is the ground node. Hence V_{DD} and ground node have been introduced separately for the ease of layout.

Hence the supply voltage is chosen as 400 mV in TSMC 180 nm technology node. The threshold voltage is almost 600 mV. So the circuit is operating in the subthreshold mode. The frequency is varied from 5 to 500 KHz. To check the workability of the 4-bit adder, high frequency regime has been selected. At low frequency the power dissipation of the circuit is less and the performance is also satisfactory. Hence a simple chart regarding the power dissipation of the 4-bit parallel adder is given to show the advantage of the DTMOS in terms of the ultralow power dissipations. The frequency and corresponding power dissipation are tabulated in Table 4.7 and plotted in Fig. 4.56.

If we would like to design more complex circuit then multiplier would be the best structure to check the workability of the DTMOS circuits in case of complex gate design. Hence the basic 2 × 2 multiplier circuit has been given as a basic building block of the higher order multiplier. The implementation of the 2 × 2 multiplier has been done by taking full adder block and AND gates. The AND gates and the full adders used in the 2 × 2 multiplier are also of DTMOS type, i.e. gate and body are tied up in case of nMOS and pMOS.

4.14 ECLR Structure

Efficient charge recovery adiabatic logic (ECRL) is one of the most simple approaches to implement low power VLSI circuits in superthreshold regime [17]. ECRL uses the cascode voltage swing logic (CVSL) network [18] having the differential inputs to produce complementary output. ECRL gates can be designed and implemented using single sinuosoidal, triangular or trapezoidal power clock. Inverter structure

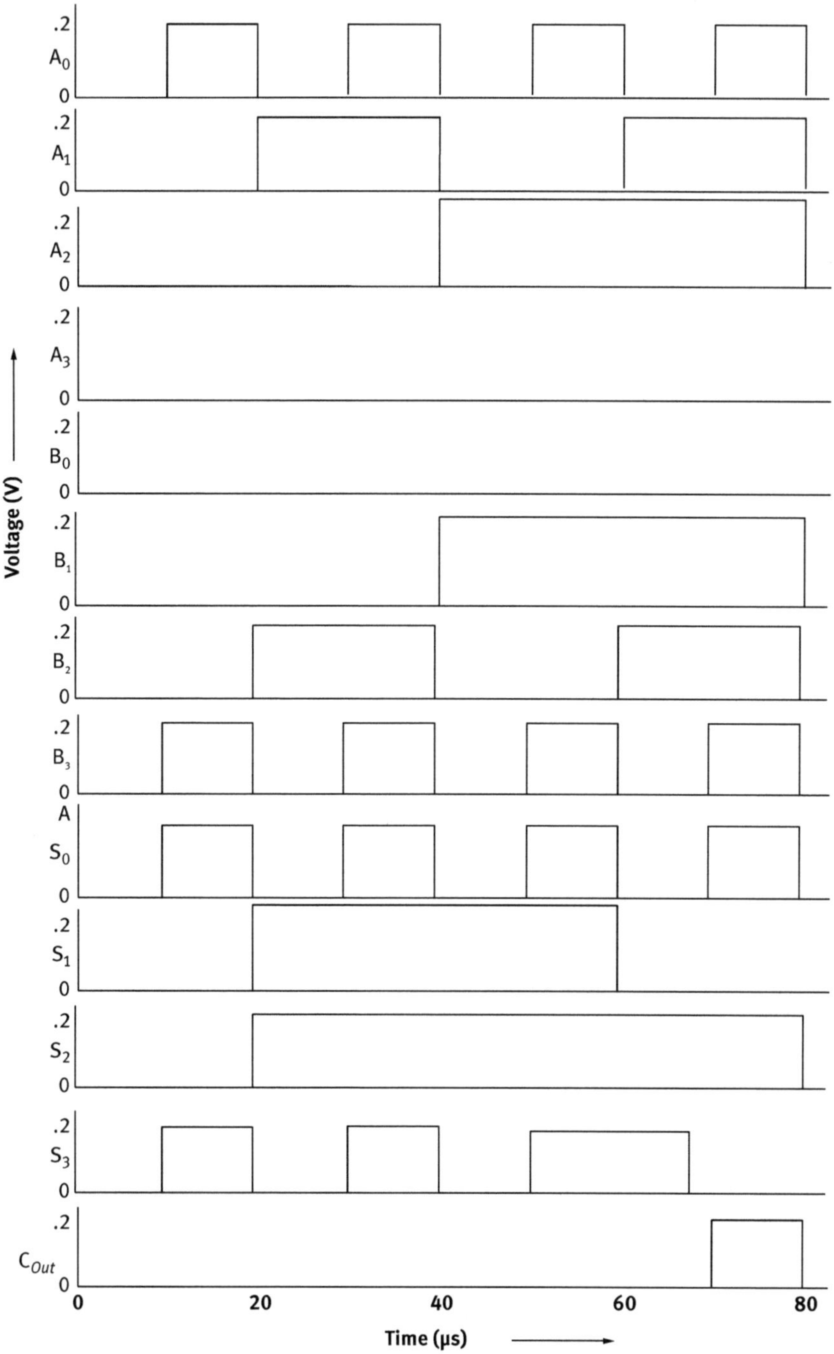

Figure 4.55: Output waveform of 4-bit parallel adder.

Table 4.7: Power dissipation of the 4-bit parallel adder in different frequency.

Operating frequency (KHz)	Power dissipation (nW)
5	0.33
10	0.42
25	0.66
50	1.19
125	2.2
250	4.29
500	8.37

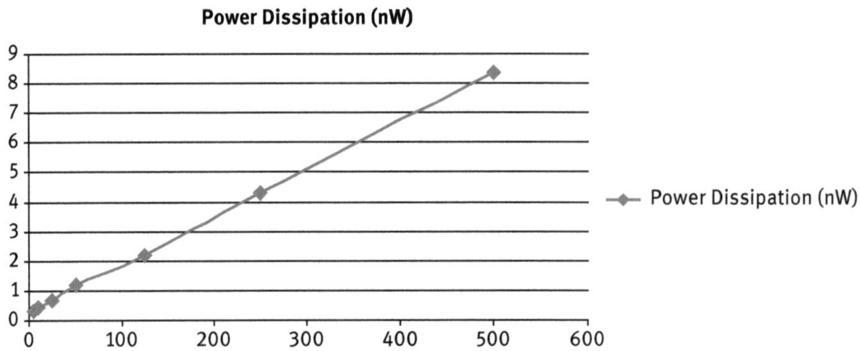

Figure 4.56: Power dissipation graph of the 4-bit parallel adder in different frequency.

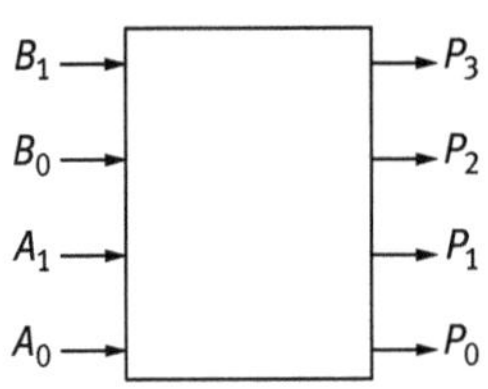

A1 A0 B1 B0	P0 P1 P2 P3
0000	0000
0001	0000
0010	0000
0011	0000
0100	0000
0101	0001
0110	0010
0111	0011
1000	0000
1001	0010
1010	0100
1011	0110
1100	0000
1101	0011
1110	0110
1111	1001

Figure 4.57: Block diagram of product generator and the truth table.

using the ECRL network is given in the previous chapter. Hence we will discuss the design and implementation of the combinational logic structures using the ECRL. Fig. 4.57 shows the truth table of the product generator. Fig. 4.58, 4.59 and 4.60 shows the gate-level schematics of 2×2 multiplier, its output waveform considering

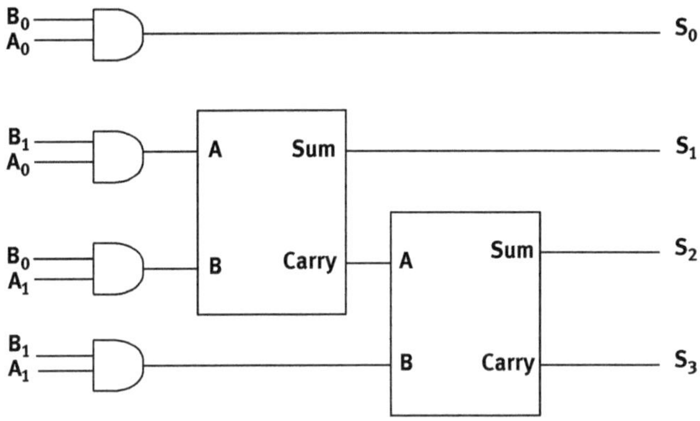

Figure 4.58: Block diagram of 2 × 2 multiplier.

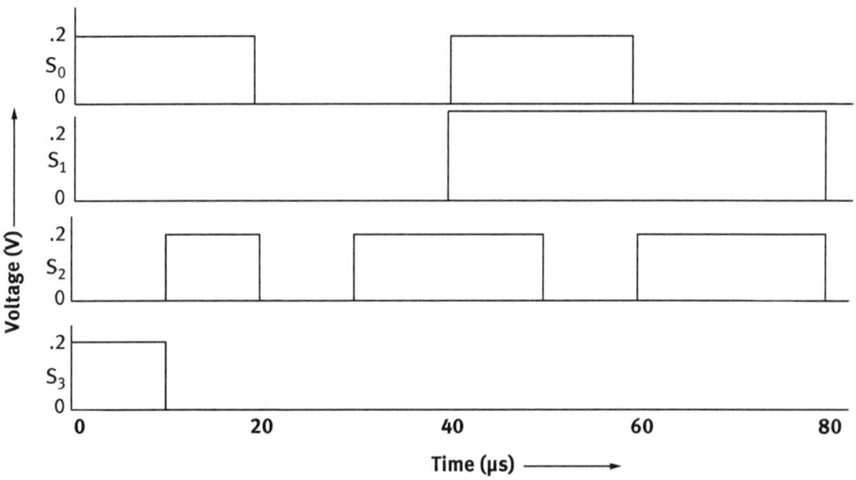

Figure 4.59: Output waveform of 2 × 2 multiplier considering subthreshold DTMOS logic.

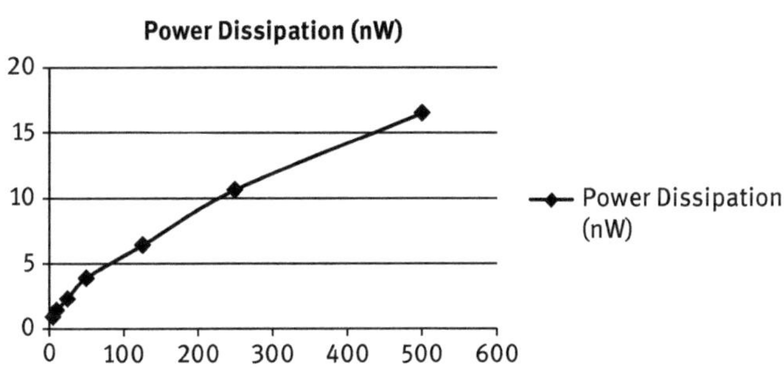

Figure 4.60: Power dissipation graph of the 2 × 2 multiplier in different frequency.

Table 4.8: Power dissipation of the 2 × 2 multiplier.

Operating frequency (KHz)	Power dissipation (nW)
5	0.9
10	1.4
25	2.3
50	3.9
125	6.4
250	10.6
500	16.5

subthreshold DTMOS logic implementation and power frequency curve. Table 4.8 shows the power-frequency curve of subthreshold DTMOS logic 2 × 2 multiplier.

ECRL NAND/AND and NOR/OR gates are given in Figure 4.61 and the output waveform of ECRL NAND/AND gate in Fig. 4.62 and ECRL NOR/OR gate in 4.63. The truth table of ECRL NAND/AND qate and NOR/OR gate is shown in Table 4.9 and 4.10 respectively. Hence in both gates, CVSL structure is used to design the pull-down part. Φ(t) is the supply clock which ramps up and down in between 0 and the V_{DD}. When the supply clock ramps up from 0 to V_{DD}, the phase can be turned as charging phase whereas when the supply clock ramps down from the V_{DD} to 0, the phase is termed as discharging phase. In ECRL gates, when the inputs are applied, any one of the two pull-down paths turns on or off. So any one of the two output nodes remains at ground potential and the other node follows the supply clock very closely to produce a supply clock like voltage waveform.

Hence in ECRL NAND left-sided PDN is basically the PDN of static CMOS NAND gate and the right-sided one is the complementary structure of the left one. The operation of the ECRL NAND/AND gate can be explained in terms of a specific input combination, i.e. $A = 1$ and $B = 0$, assuming the supply clock ramps up from 0 to V_{DD}. Under this input combination, M_3 and M_4 will be on and off. So NAND output node will be in floating condition for some time interval. Whereas M_6 will be on and will pull down the AND node to ground potential. Resultantly, the M_1 or pMOS will be turned of as the logic "0" will be applied to gate of M_1. So NAND node will be charged up by following the supply clock very closely during the charging phase. During discharging phase, AND node will remain at ground potential as there will be no changes in the input combinations. NAND node will also follow the supply clock closely just like the charging phase, and the stored charge in the NAND node will be sent back to the supply. Thus very less amount of energy will be dissipated during the total charging and discharging phase. Simplified truth table for the ECRL is given below for better understanding of the operations.

Similarly ECRL NOR/OR gate can be implemented by using the CVSL network. Hence pull-down and complementary PDN will be used to implement the NOR/OR structure. Operation of the NOR/OR gate can also be explained using some input

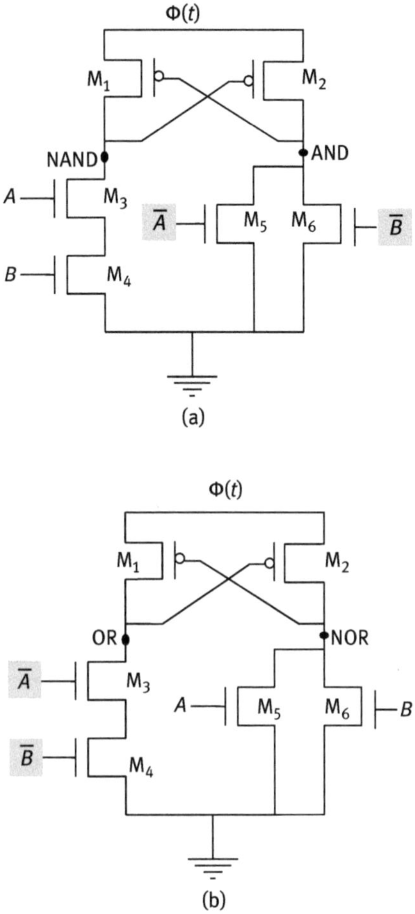

Figure 4.61: ECRL gate structure (a) NAND/AND (b) NOR/OR.

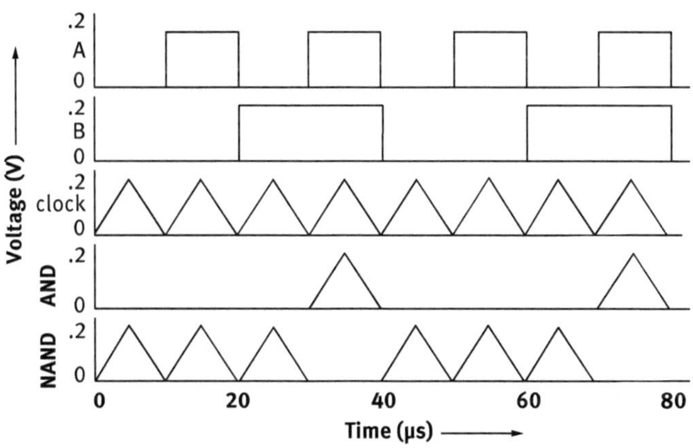

Figure 4.62: Output waveform of the ECRL NAND/AND gate.

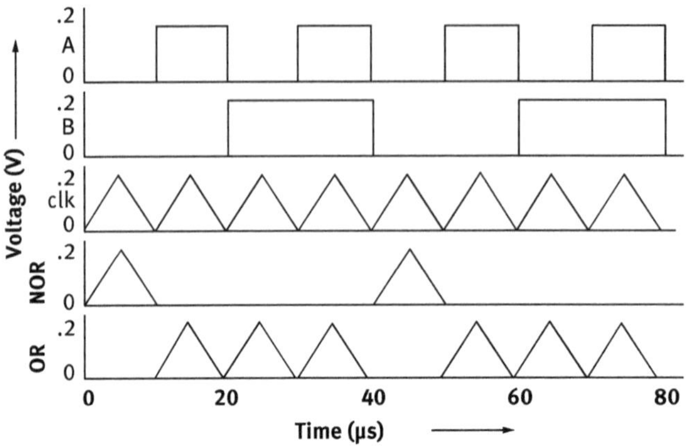

Figure 4.63: Output waveform of the ECRL NOR/OR gate.

Table 4.9: Truth table of ECRL NAND/AND gate.

Inputs									
k) A	***l) B***	M_1	M_2	M_3	M_4	M_5	M_6	**NAND**	**AND**
0	0	On	Off	Off	Off	On	On	1	0
0	1	On	Off	Off	On	On	Off	1	0
1	0	On	Off	On	Off	Off	On	1	0
1	1	Off	On	On	On	Off	Off	0	1

Table 4.10: Truth table of ECRL NOR/OR gate.

Inputs									
A	***B***	M_1	M_2	M_3	M_4	M_5	M_6	**OR**	**NOR**
0	0	Off	On	On	On	Off	Off	0	1
0	1	On	Off	On	Off	Off	On	1	0
1	0	On	Off	Off	On	On	Off	1	0
1	1	On	Off	Off	Off	Off	Off	1	0

combination, i.e. $A = 1$ and $B = 0$, assuming the supply clock ramps up from 0 to V_{DD}. Under this input combination, M_3 and M_4 will be off and on, repectively. So OR output node will be in floating condition for some time interval. Whereas, M_5 will be on because of higher logic input at the gate and will pull down the NOR output node to ground potential. Resultantly, the M_1 or pMOS will be turned of as the logic “0” will be applied to the gate of M_1. So OR output node will be charged up by following the supply clock very closely during the charging phase. During discharging phase, NOR node will remain at ground potential as there will be no changes in the input combinations. OR node will also follow the supply clock closely just like the charging phase, and the stored charge in the OR node will be sent back to the supply. Thus very less amount of energy will be dissipated during the total charging and discharging phases.

During the discharging phase, when the supply clock voltage approaches $|V_{Tp}|$ (threshold voltage of the pMOS), the pMOS transistor gets disconnected. So complete recovery of the charges is not possible. Hence instead of logic “0”, we will get few millivolts due to the stored charges. The associated energy loss due to the incomplete recovery of the charges is approximately $C_L|V_{Tp}|^2$. During the charging process, in the begining pMOS transistors will be in floating mode and charges leak away which cannot be retreived. This type of loss can be termed as *nonadiabatic* loss. Fig. 4.64 shows the output waveform of the ECRL XOR/XNOR gate considering sinusoidal source.

However, inspite of this nonadiabatic loss, ECRL is very much advantageous regarding the power dissipation. For example, it has been observed that the 16-bit ECRL carry look ahead adder consumes less than 50% of energy compared to the CMOS structure. Few combinational structures using ECRL have also been discussed here for the ease of understanding.

2 to 1 MUX and 4 to 1 multiplexers are given in Figures 4.65. and 4.66 and its truth table is shown in Table 4.11. Respectively. In 2 to 1 MUX, the output expression is

$$Y = A\overline{S} + BS.$$

When $S = 0$, A will be selected and when S = 1, B will be selected. Hence the pull-up structure consists of the cross-coupled pMOS transistors. In case of PDN, we are using complementary nMOS tree, which basically consists of CVSL network. In Figure 4.65, when $S = 0$ and $A = 1$, out node is pulled down at ground potential, which also turns on the right-sided pMOS to charge up the outb node in the charging phase. During

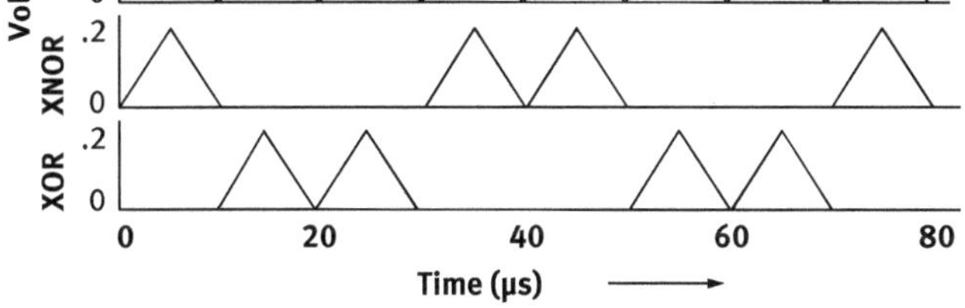

Figure 4.64: Output waveform of the ECRL XOR/XNOR gate considering sinusoidal source.

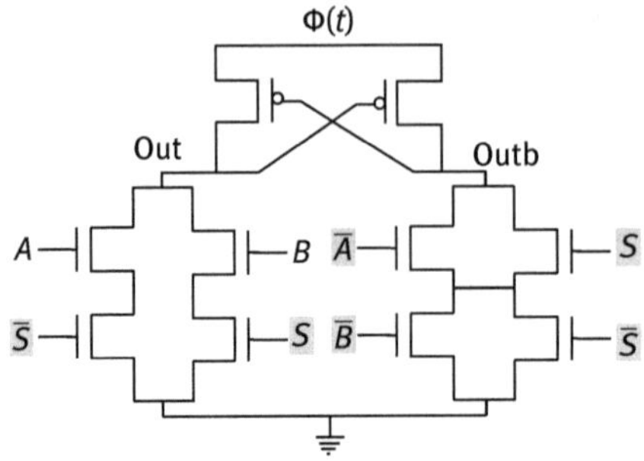

Figure 4.65: ECRL 2 to 1 multiplexer.

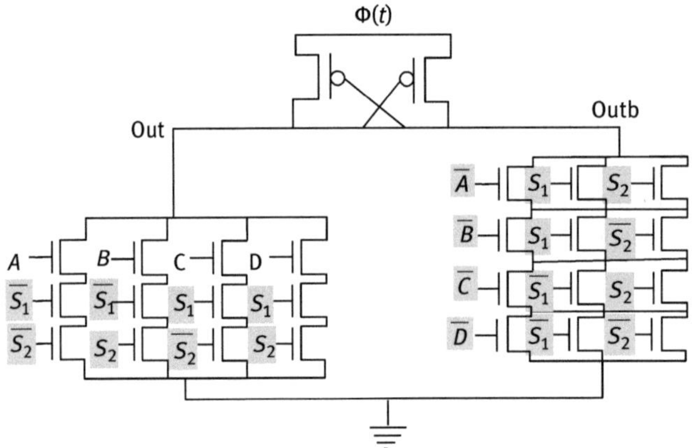

Figure 4.66: ECRL 4 to 1 multiplexer.

Table 4.11: Truth table of the ECRL multiplexer.

Select lines		Outputs	
S_1	S_2	**Out**	**Outb**
0	0	$\bar{A}$	A
0	1	$\bar{B}$	B
1	0	$\bar{C}$	C
1	1	$\bar{D}$	D

discharging phase, as the input condition remains same, stored charge in outb node will be sent back to the supply clock by the following the same path. So a full swing is observed at outb node, very similar to the supply clock. Similarly, when $S = 0$ and $A = 0$, outb node will be pulled down to the ground potential which also turns on the left pMOS transistor. Resultantly, out node will be charged up and later discharged by following the supply clock very closely. So it can be mentioned that the 2 to 1 MUX output will be obtained at outb node, and out node will produce the complementary logic.

Figure 4.66 shows the design of 4 to 1 MUX. The four inputs A, B, C and D will be selected by the select lines S_1 and S_2. The relationship is given below:

$$Y = A\overline{S_1}\overline{S_2} + B\overline{S_1}S_2 + CS_1\overline{S_2} + DS_1S_2$$

When S_1 and S_2 are set to logic 0, and $A = 1$, then the out node will be connected with ground node through the left-sided nMOS network (left most path out of the four paths). Ground potential of out node will turn on the pMOS transistor and the outb node will be charged up by following the supply clock through the turned-on pMOS. Also during the discharging phase, almost total stored charge of the outb node will be sent back to the supply clock through the same pMOS path and out node remains at the ground potential. Hence, in outb node we obtain the 4 to 1 MUX output, and out node produces the complementary logic output.

As we know that the adder is the basic building block of arithmetic blocks, a simple adder structure in ECRL logic is given in Figure 4.67. The adder consists of sum and carry structures. In both the structures, DCVS network is used to design the complementary logic paths. Hence the DCVS network for the sum/sumb implementation has been modified a bit to reduce the transistor count. This will also reduce the silicon area of the adder. This would be much more advantageous when we would design the multiplier circuits. Sum and sumb represent the complementary output nodes. To obtain a stable output waveform one can connect the load capacitors in the output nodes. Here one-stage full adder adiabatic logic system is implemented which will be used as a basic block in next designs.

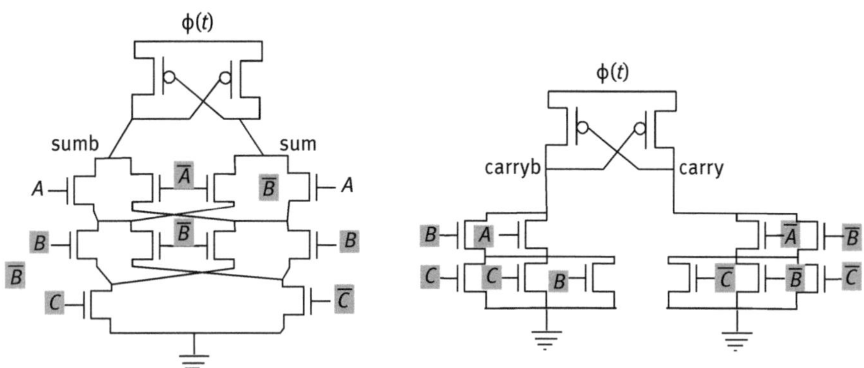

Figure 4.67: ECRL full adder schematic (a) Sum circuit (b) Carry circuit.

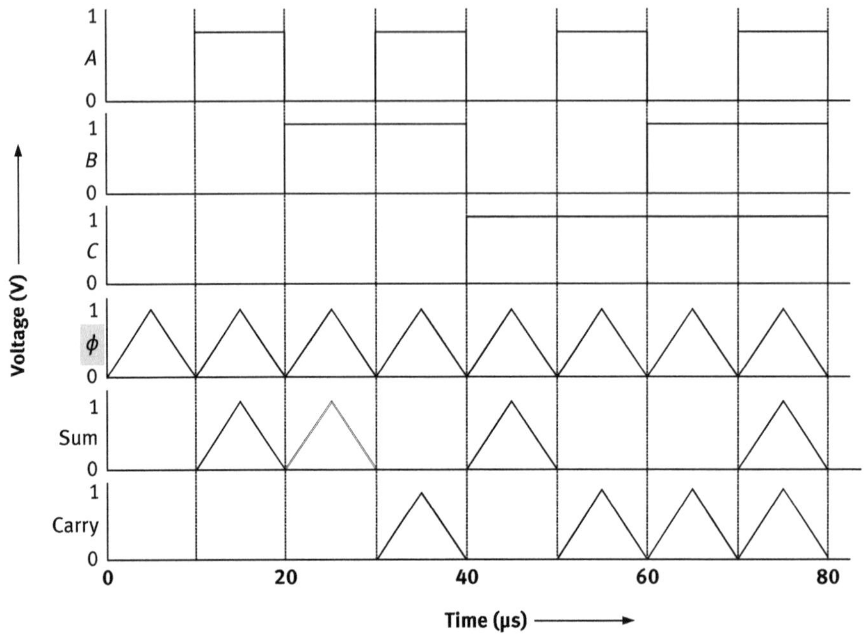

Figure 4.68: Output waveform of the ECRL full adder with a trapezoidal clock source.

Hence, $\varphi(t)$ is the clock supply source, here is a trapezoidal waveform. If we reduce the operating frequency of the ECRL, single sinusoidal clock source can be used as a supply clock. In that case buffer circuits are required after regular intervals to reduce the logic degradation in inner stage. 100 MHz clock frequency is used with a 2 V amplitude. 50 fF output loads have been set at the output nodes to obtain more stable output waveforms.

The output waveform is given in Figure 4.68. Output voltage will be measured at the $T/2$ time instant where T is the width of the clock pulse. So the distorted shape at the beginning or the end of the clock can be ignored. However, proper transistor sizing can minimize the distortion up to some extent. Here, as high frequency regime has chosen, the output nodes could not discharge properly. In case of low frequency operations, the output waveforms become better and also significant advantage can be observed in case of power savings.

However, the power dissipation of ECRL full adder for different frequencies is given in Table 4.12. It can be observed that with frequency, the power dissipation of the ECRL full adder increases significantly. Power dissipation for different load, keeping the frequency constant at 10 MHz, is given in Table 4.13. As the load increases, the energy dissipation will also increase as the energy dissipation is directly proportional to the square of the capacitive load. So the rate of change of energy or the power also increases.

Table 4.12: Power dissipation of ECRL full adder for different frequencies considering fixed load.

Frequency (MHz)	Average power consumption (µW)
10	0.53
25	2.06
50	5.73
100	16.76
200	46.90
250	64.75
500	135.82

Table 4.13: Power dissipation of ECRL full adder for different loads.

Capacitance (fF)	Average power consumption (µW)
10	1.491
25	3.016
50	7.212
75	13.10
100	20.42

4-2 compressor is also important logic structure for the implementation of the tree multiplier. A 4-2 compressor has four inputs and two outputs. The governing equation of the 4-2 compressor is given by

$$I_1 + I_2 + I_3 + I_4 + C_{In} = \mathrm{Sum} + 2(\mathrm{Carry} + C_{Out}).$$

The structure is given in Figure 4.69. Hence the inputs are I_1, I_2, I_3 and I_4 and two outputs are sum and carry. Also, carry-in (C_{In}) and a carry-out (C_{Out}) have been set. The input C_{In} is the output from the previous lower significant compressor. The C_{Out} is the output to the compressor in the next significant stage. Here, one point has to be remembered that C_{Out} should be independent of the input C_{In} as it can accelerate the carry, save summation of the partial products. Two different structures are given to implement the 4-2 compressor. As 4-2 compressor has higher capacitive load at the output, it consumes much more energy. Full adder-based implementation has simpler architecture compared to Figure 4.70(b). However, Figure 4.70(b) is an optimized one as the length of the critical path is smaller. So it provides much faster operation compared to the full adder-based implementation.

The output waveform of the 4-2 compressor is given in Figure 4.70. Hence sinusoidal clock source is used to show the workability of the ECRL logic block.

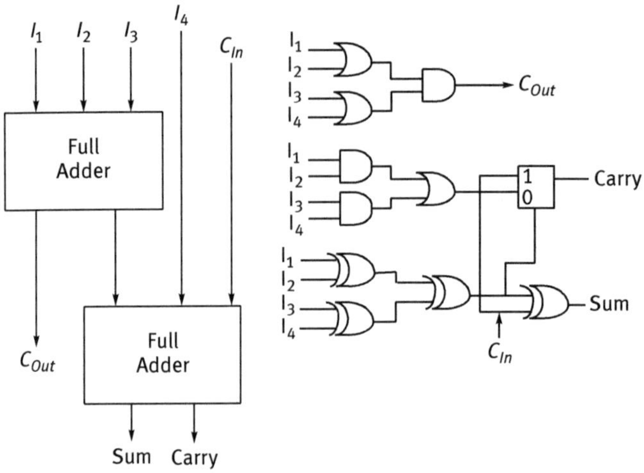

Figure 4.69: Schematic diagram of 4-2 compressor: (a) block diagram; (b) gate level presentation.

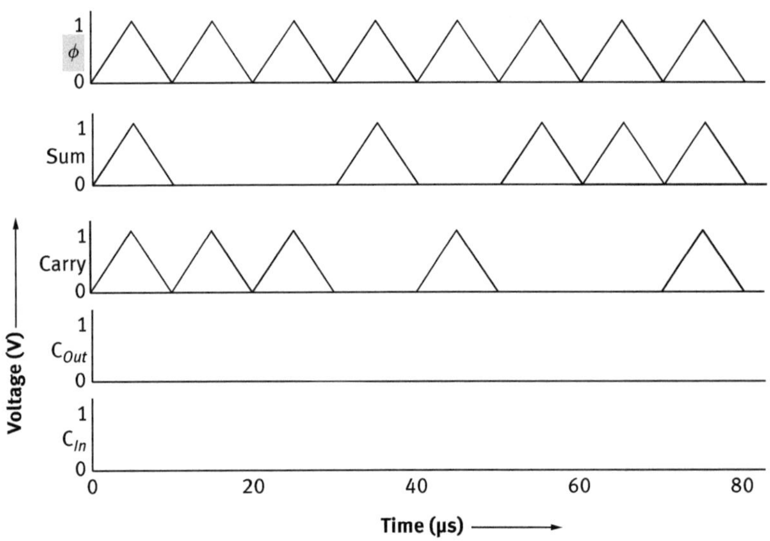

Figure 4.70: Output waveform of the ECRL 4-2 compressor.

1.8 Volt peak-to-peak voltage is chosen with 50 MHz frequency. Single sinusoidal clock can be generated using LC circuits and minimal power consumption can be enjoyed. Also single clock source can minimize the power dissipation and the silicon area, as it simplifies the distribution and the management of the clocks. Hence power dissipation is tabulated taking the load fixed at 50 fF.

For each bit in a binary sequence to be added, the carry look ahead logic will determine whether that bit pair will generate a carry or propagate a carry and is shown in Fig. 4.71. This allows the circuit to "pre-process" the two numbers being added to

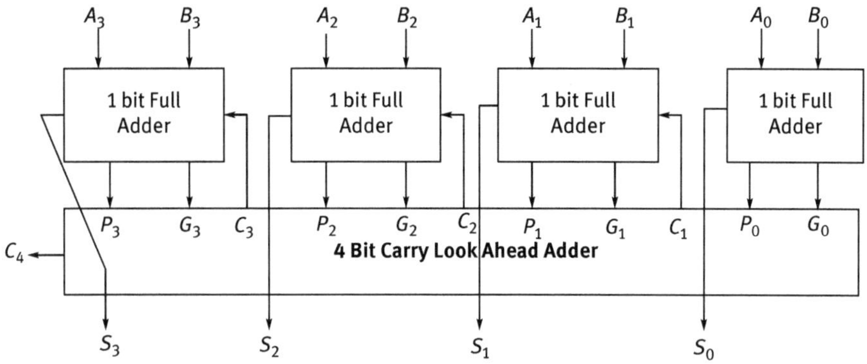

Figure 4.71: Block diagram of 4-bit carry look ahead adder.

determine the carry ahead of time. Then, when the actual addition is performed, there is no delay from waiting for the ripple carry effect (or time it takes for the carry from the first full adder to be passed down to the last full adder).

$$\begin{aligned} C_0 &= G_0 + P0C_{In} \\ C_1 &= G_1 + P_1G_0 + P_1P_0C_{In} \\ C_2 &= G_2 + P_2G_1 + P_2P_1G_0 + P_2P_1P_0C_{In} \\ C_3 &= G_3 + P_3G_2 + P_3P_1G_1 + P_3P_2P_1G_0 + P_3P_2P_1P_0C_{In} \end{aligned} \tag{4.3}$$

The carry look ahead 4-bit adder can also be used in a higher-level circuit by having each CLA logic circuit produce a propagate and generate signal to a higher level CLA logic circuit. The group propagate (PG) and group generate (GG) for a 4-bit CLA are

$$\begin{aligned} \mathrm{PG} &= P_0 \cdot P_1 \cdot P_2 \cdot P_3 \\ \mathrm{GG} &= G_3 + G_2 \cdot P_3 + G_1 \cdot P_3 \cdot P_2 + G_0 \cdot P_3 \cdot P_2 \cdot P_1 \end{aligned} \tag{4.4}$$

Using 4-bit CLAs together we can design 8-bit CLA. The schematic of adiabatic 8-bit carry look ahead adder is shown in Figure 4.72. We know the path from the input to the output signal that is likely to take the longest time, which is designated as "critical path". In Figure 4.72 circles and arrows highlight the fast carry computation tree. As opposed to ripple carry adder, the critical path in the CLA travels in vertical direction rather than a horizontal one as shown in Figure 4.72. Therefore the delay of CLA is not directly proportional to the size of the adder N, but to the number of levels used. This *log* dependency makes CLA one of the theoretically fastest structures for addition.

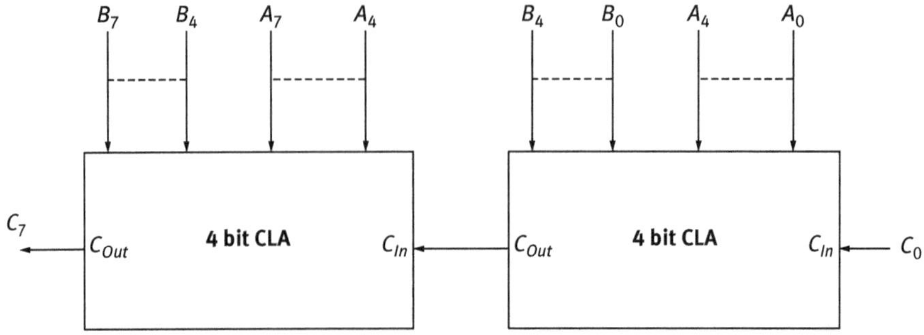

Figure 4.72: Schematics of 8-bit adiabatic CLA.

The schematic was simulated in cadence analog design environment [19]. All the inputs were buffered and the load capacitors were placed at the output nodes of each block. The simulation environment is given below.

Simulation environment
- Technology file: gpdk180
- V_{DD}: 1.8 V
- Load: 10 fF
- Clock: 10 MHz

The final simulation waveform of 16-bit CLA is given below. All the output nodes are loaded by 25 fF capacitors to get more stable waveform. A counting sequence $\{A\} = \{A_{32}, A_{31}, \ldots, A_2, A_1\} = \{000 \cdots 000, 000 \cdots 001, 000 \cdots 010, \ldots, 111 \cdots 111\}$ and $\{B\} = \{B_{32}, B_{31}, \ldots, B_2, B_1\} = \{000 \cdots 000, 000 \cdots 000, \ldots, 000 \cdots .000\}$ with $C_{in} = 0$ is assigned as the test patterns in case of CLA. The output results are $\{Z\} = \{S_{32}, S_{31}, \ldots, S_2, S_1\} = \{000 \cdots 001, 000 \cdots 010, 000 \cdots 011, \ldots, 100 \cdots 000\}$. Hence $Z_1 = 101010101\ldots$, $Z_2 = 01100110011\ldots$, $Z_3 = 0001111000011110000\ldots$, and so on. Waveforms of some output nodes in a 8-bit carry look ahead adder are shown in Figure 4.73.

The final layout was verified by running layout versus schematic (LVS) from Assura [20]. The log file is provided. However, the hardware-assisted emulation was not done due to technical limitations. The performance of the implementation was tested with the aim of determining the efficiency with which the system completes its tasks. Here we have provided the graphical output from the final simulation. The design methodology called design for manufacturability (DFM) includes a set of techniques to modify the design of integrated circuits (ICs) in order to make them more manufacturable, i.e. to improve their functional yield, parametric yield, or reliability. Detailed characterization of a statistically significant number of ICs is required before the project can be released to full production. This ensures that there are no inexplicable yield losses and the product will have a stable life once it is in full production. Preproduction characterization is generally conducted on automated test equipment

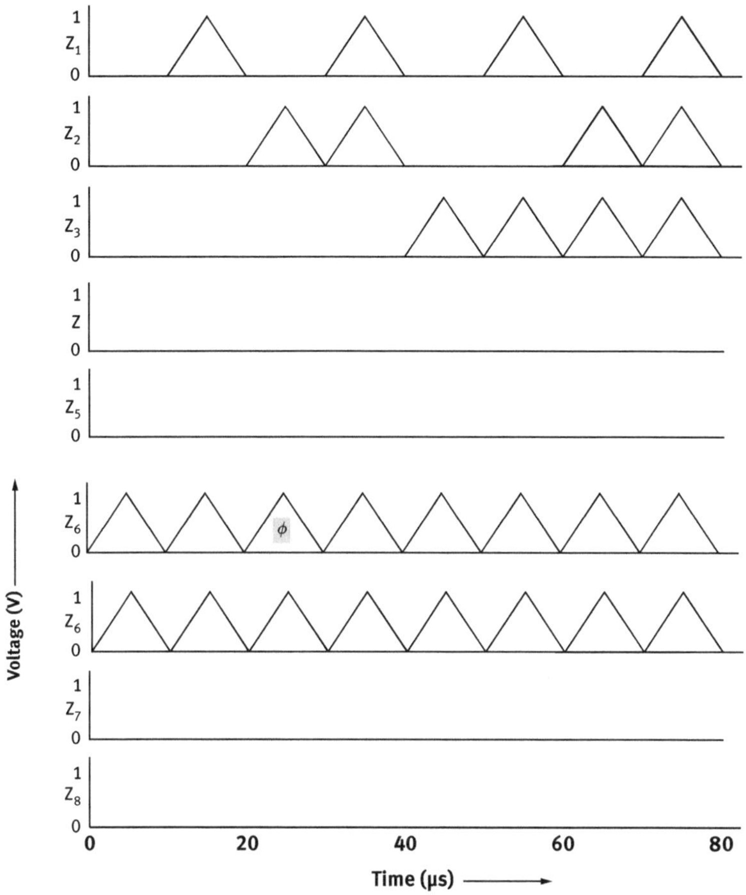

Figure 4.73: Output waveforms of first 8 nodes of 16-bit CLA based on Energy Efficient Adiabatic Logic (EEAL).

(ATE), usually concurrently with the development of the production test. The methods and the critical aspects of the preproduction characterization are documented by our engineers during the electrical design phase and we provide the level of support necessary to the test engineers for test development. For manufacturability, the design must be modified, where possible, to make it as easy and efficient as possible to produce. This is achieved by adding extra vias or adding dummy metal/diffusion/poly layers wherever possible while complying to the design rules set by the foundry. Since errors are expensive, time-consuming and hard to spot, extensive error checking is the rule, making sure the mapping to logic was done correctly and checking that the manufacturing rules were followed faithfully. The initial prototypes come from a matrix run to represent the expected variations of the IC manufacturing process. Then the design data is turned into photomasks in mask data preparation (MDP).

Mask data preparation is the step that translates an intended set of polygons on an IC layout into a form that can be physically written by the photomask writer. Usually this involves fracturing complex polygons into simpler shapes, often rectangles and trapezoids that can be written by the mask writing hardware. Typically a design is delivered to MDP in GDSII or OASIS format, and later fracturing is written out in a proprietary format specific to the mask writer. Next wafer fabrication, die test and packaging are done. Post-silicon validation and debug is the last step in the development of a semiconductor IC. During the pre-silicon process, engineers test devices in a virtual environment with sophisticated simulation, emulation and formal verification tools. In contrast, post-silicon validation tests occur on actual devices running at speed in commercial, real-world system boards using logic analyzer and assertion-based tools.

Table 4.14 shows the power dissipation for varying bit length for different types of multiplier circuits. The performance of multipliers is crucial for multimedia applications such as 3-D graphics and signal processing systems, which depend on the execution of large numbers of multiplications. Hence a Vedic Multiplier, based on UrdhvaTiryakbhyam Sutra (UTS) of Vedic Arithmetic, a promising approach due to less computational time, has also been given [21]. The multiplier is based on UTS of ancient Vedic Mathematics. The aim of this work is to demonstrate the advantages of concurrent generation and addition of partial products in Vedic Multiplier. It can be classified as a serial-parallel multiplier, which serves as a good trade-off between the time-consuming serial multipliers and area-consuming parallel multipliers.

UrdhvaTiryakbhyam Sutra: UrdhvaTiryakbhyam Sutra is a general multiplication formula applicable to all cases of multiplication. It literally means "*Vertically and Crosswise*". Let us explain UTS with an example. The numbers to be multiplied are written on two consecutive sides of the square as shown in Figure 4.74. The square is divided into rows and columns where each row/column corresponds to one of the digits of either a multiplier or a multiplicand. These small boxes are partitioned into two halves by the crosswise lines. Block diagram of the Vedic Multiplier is given in Figure 4.74. The schematics are given in Figures 4.75–4.77. The output waveform is shown in Fig. 4.78.

Table 4.14: Power dissipation of multiplier circuits with varying bit lengths at 100 MHz for 180 nm CMOS technology.

		Frequency (MHz)					
		10 M	25 M	50 M	100 M	200 M	500 M
16-bit CLA	CMOS	7.5	7.5	7.5	7.5	7.5	7.5
	ECRL	0.21	0.40	0.67	1.18	2.02	3.22
	Gain	35.7	18.7	12.0	6.3	3.7	2.32

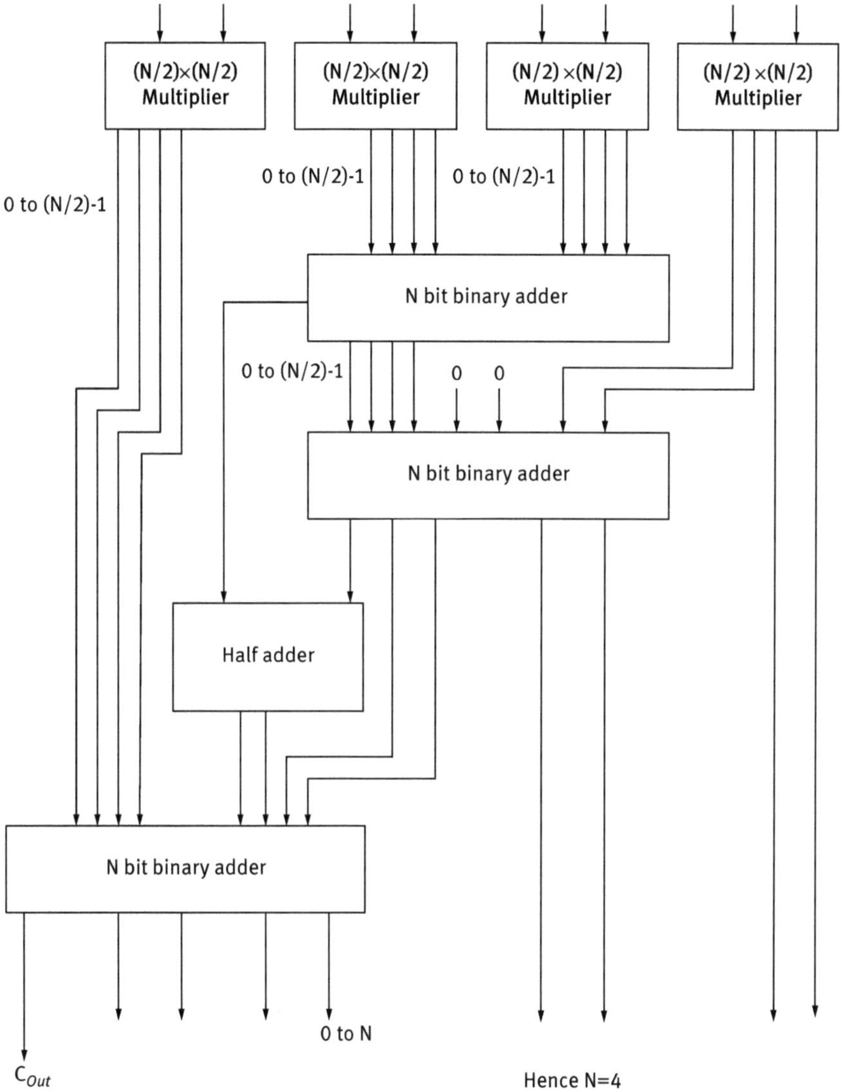

Figure 4.74: Diagram of $N \times N$ multiplier.

In this context, we would like to mention some of the innovative features in Cadence Schematic Editor that helped us complete this project smoothly.

- "Snap to Diamond" is an excellent feature that saves time and effort in large schematics.
- Multiple labels/pins addition by writing their names separated by space is a very handy feature. "Bus Expansion" also saves a lot of time when dealing with bus routing.
- User configurable rule is another great feature that helps finding bugs in early stages of the design.

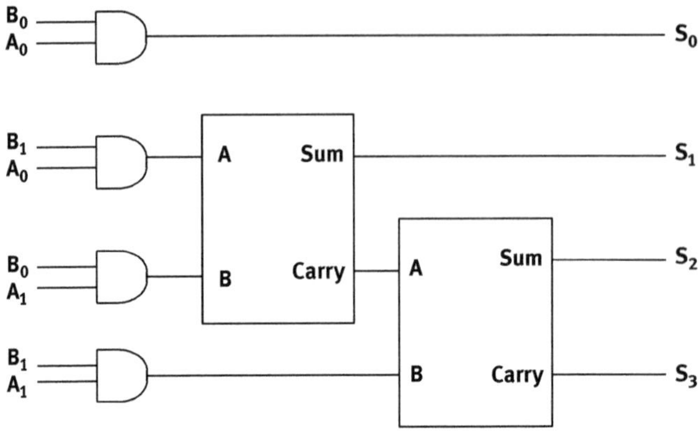

Figure 4.75: 2 × 2 Vedic Multiplier.

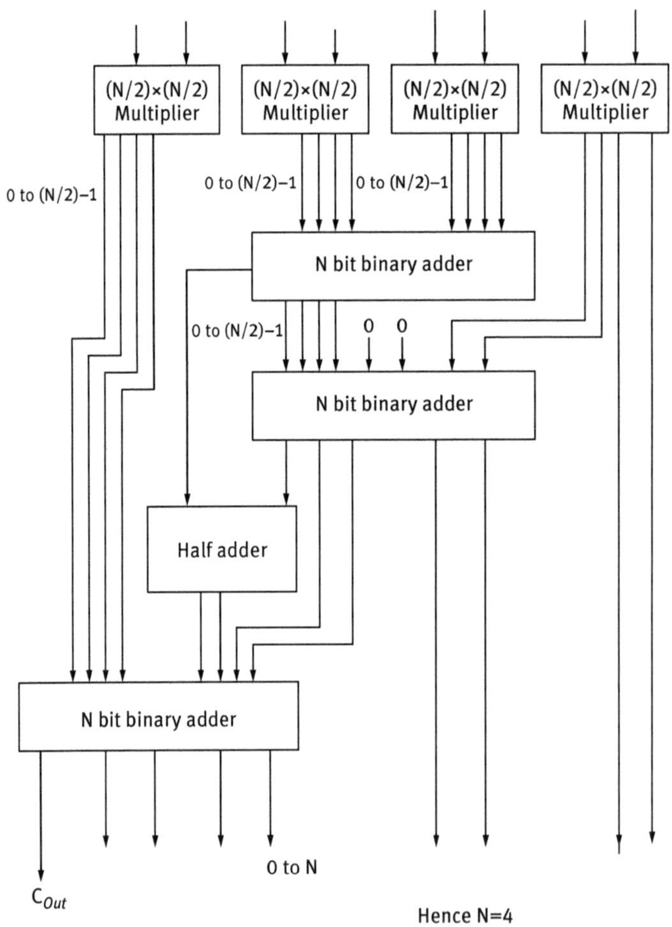

Figure 4.76: 4 × 4 Vedic Multiplier.

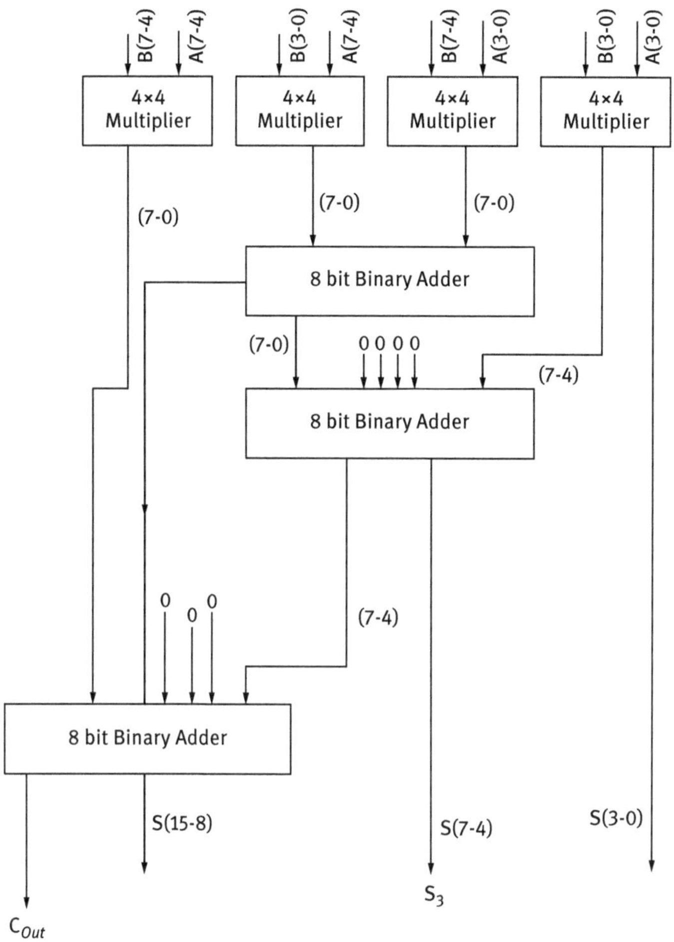

Figure 4.77: 8 × 8 Vedic Multiplier.

The creation of the mask layout is one of the most important steps in the full-custom design flow, because physical layout design is very tightly linked to overall circuit performance (area, speed and power dissipation). We have taken utmost care to design the layout in the smallest area possible while maintaining symmetry in the design. We have then eliminated all Design Rule Check (DRC) errors so that the design does not violate any of the layout design rules of the fabrication process, in order to ensure a high probability of defect-free fabrication. Then, using LVS comparison, we have verified that the layout actually implements the required functionality. After that, post-layout simulation can be performed taking into account the geometry of the circuit to include the parasitic effects from the layout view.

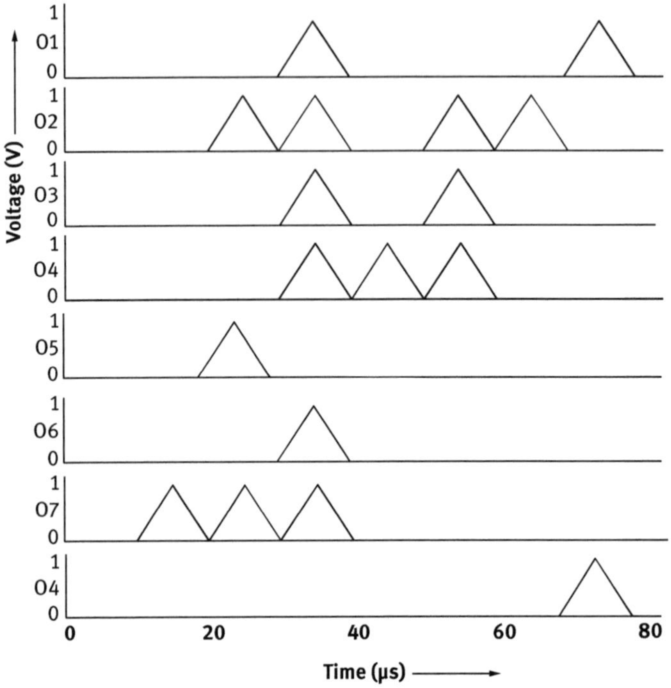

Figure 4.78: Output waveform of the Vedic Multiplier.

This report is meant to document the validation of the 8 × 8 Vedic Multiplier system. The validation of this system was subdivided into three distinct tasks:

- Functionality verification
- Layout verification
- Efficiency evaluation

The schematic was simulated in cadence analog design environment [19]. All the inputs were buffered and the load capacitor was placed at the output nodes of each block. The simulation environment is given below:

Simulation environment:

- Technology file: gpdk180
- V_{DD}: 1.8 V
- Load: 10 fF
- Clock: 10 MHz

The final simulation waveform of 8 × 8 multiplier for eight cycles is given below.

Table 4.15: Power dissipation of multiplier circuits with varying bit lengths at 100 MHz for 180 nm CMOS technology.

Bit length of multiplier	Vedic CMOS	Vedic adiabatic (ECRL)
2 × 2	29	11.64
4 × 4	107	48.76
8 × 8	899	36.28

Table 4.16: Delay comparison of multiplier circuits with varying bit lengths at 100 MHz for 180 nm CMOS technology.

Bit length of multiplier	Vedic CMOS	Vedic adiabatic
2 × 2	0.24	0.29
4 × 4	0.62	0.72
8 × 8	1.48	1.91

4.14.1 Power Consumption

Table 4.15 shows the comparison of delay as a function of varying bit lengths for different types of multiplier circuits.

4.14.2 Propagation Delay

Here the design and analysis of the ultralow power multiplier have been analyzed. Table 4.16 shows the comparison of delay as a function of varying bit lengths for different types of multiplier circuits. For the ease of understanding, starting from gate to the multiplier structure along with the waveform and the power dissipation, measurements are given to clear the true picture.

References

[1] J.M. Rabaey, *Digital Integrated Circuits*. Prentice Hall, Upper Saddle River, NJ, 1996.

[2] Smith Sedra, *Microelectronic Circuits*. Oxford University Press, New York, 1998.

[3] Forbes, Leonard, "Monotonic dynamic-static pseudo-nMOS logic circuit." US Patent No. 6,801,056. October 5, 2004.

[4] D.D. Gajski, *Principles of Digital Design*. Prentice Hall, Upper Saddle River, NJ, 1997.

[5] S. Brown and Z. Vranesic, *Fundamentals of Digital Logic with Verilog Design*. McGraw-Hill, New York, 2002.

[6] Reto Zimmermann and Wolfgang Fichtner, "Low-power logic styles: CMOS versus pass-transistor logic." IEEE Journal of Solid-State Circuits, 32.7 (1997): 1079–1090.

[7] M. Michael Vai, *VLSI Design*. CRC Press, FL, 2001.

[8] John. P. Uyemura, *CMOS Logic Circuit Design*. Kluwer Academic Press, MA, 1999.

[9] Ken Martin, *Digital Integrated Circuits*. Prentice Hall, New Jersey, 1996.

[10] Robert F. Sproull and Ivan E. Sutherland., "Logical effort: Designing for speed on the back of an envelope." IEEE Advanced Research in VLSI (1991): 1–16.

[11] R. C. Jaeger, and T. N. Blalock, (1997). *Microelectronic circuit design* (Vol. 97). New York: McGraw-Hill.

[12] Neil Weste and Kamran Eshraghian, *Principles of CMOS VLSI Design*, 2nd edn. Addison-Wesley, MA, 1993.

[13] R. Jacob Baker W. Li Harry and David E. Boyce, *CMOS Circuit design, layout and simulation.* IEEE Press, Piscataway, NJ, 1988.

[14] Hendrawan Soeleman, Kaushik Roy and Bipul C. Paul, "Robust subthreshold logic for ultra-low power operation." IEEE Transactions on Very Large Scale Integration (VLSI) Systems, 9.1 (2001): 90–99.

[15] F. Assaderaghi, D. Sinitsky, S. A. Parke, J. Bokor, P. K. Ko, and C. Hu, (1997). Dynamic threshold-voltage MOSFET (DTMOS) for ultra-low voltage VLSI. Electron Devices, IEEE Transactions on, 44(3), 414–422.

[16] Hu, Chenming, Ping K. Ko, Fariborz Assaderaghi, and Stephen Parke. "Dynamic threshold voltage mosfet having gate to body connection for ultra-low voltage operation." U.S. Patent 5,559,368, issued September 24, 1996.

[17] Yong Moon and Deog-Kyoon Jeong, "An efficient charge recovery logic circuit." IEEE Journal of Solid-State Circuits, 31.4 (1996): 514–522.

[18] William R. Griffin and Lawrence G. Heller, "Clocked differential cascade voltage switch logic systems." US Patent No. 4,570,084. February 11, 1986.

[19] A. Vachoux, J. M. Bergé, O. Levia, and J. Rouillard, (Eds.). (2012). *Analog and Mixed-Signal Hardware Description Language* (Vol. 10). Springer Science & Business Media, Berlin, Germany.

[20] Ul, New LVS Debugging. "LVS: Required." Assura™ 2.0 to 3.0 Migration Guide, 2003.

[21] Honey Durga Tiwari, et al. "Multiplier design based on ancient Indian Vedic Mathematics." IEEE International SoC Design Conference, 2008. ISOCC '08. Vol. 2, 2008.

[22] Harpreet Singh Dhillon and Abhijit Mitra, "A reduced-bit multiplication algorithm for digital arithmetic." International Journal of Computational and Mathematical Sciences, 2.2 (2008). Vol. 2, No. 2, pp. 719–724.

5 Advanced Energy-reduced Sequential Circuit Design

5.1 Introduction to Sequential Circuit

Combinational circuits, as we have seen in the last chapter, cannot store any state information. They have the property that the output of a logic block is only a function of the current input values [1–3]. Thus, the output levels at any given time point are directly determined as Boolean functions of the input variables applied at that time. They are also known as non-regenerative circuits, since there is no feedback relationship between the output and the input. But virtually all useful systems require storage of state information, leading to another class of circuits called sequential logic circuits. In these circuits, the output depends not only upon the current values of the inputs but also upon preceding input values [4–6]. In other words, a sequential circuit remembers some of the past history of the system – it has memory. A sequential circuit consisting of a combinational circuit and a memory block in the feedback loop is shown in Figure 5.1. The regenerative behavior resulting in memory function of sequential circuits is due to a feedback connection between the output and the input.

Figure 5.2 shows the block diagram of a finite state machine (FSM), which is a sequential circuit having the capability of holding the system state. The outputs of the FSM are a function of the current inputs and the current state. The next state is determined based on the current state and the current inputs and is fed to the inputs of registers [7, 8]. On the rising edge of the clock, the next state bits are copied to the outputs of the registers (after some propagation delay), and a new cycle begins. The register then ignores changes in the input signals until the next rising edge. The register can be either positive or negative edge triggered.

5.2 Basics of Regenerative Circuits

The logic circuits in general can be classified into two types: (a) non-regenerative (combinational) and (b) regenerative (sequential) circuits as shown in Figure 5.3. Regenerative operation or memory function based on positive feedback falls under the class of elements called multivibrator circuits.

The basic regenerative circuits can be classified into three groups: (a) bistable; (b) monostable; and (c) astable.

Bistable circuits have two stable states or operational modes as the name suggests. These stable states can be attained under certain input and output combinations. In contrast, the monostable circuits have only one stable operating point (state). Even if the circuit experiences an external perturbation, the output eventually returns to the single stable state after a certain time period. Finally, in astable circuits (ring

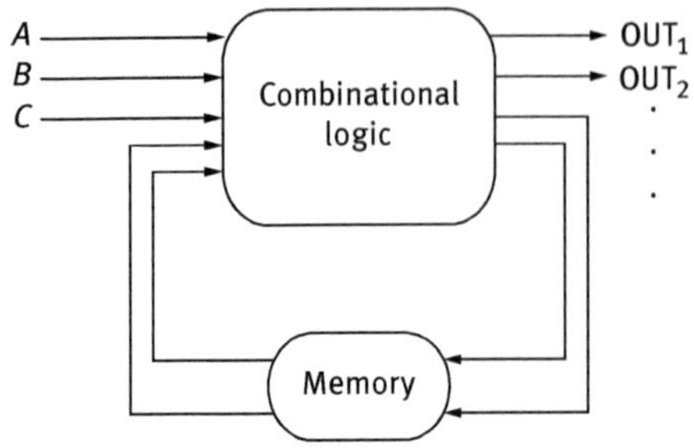

Figure 5.1: Block diagram of a sequential circuit consists of a combination logic with memory connected in feedback loop.

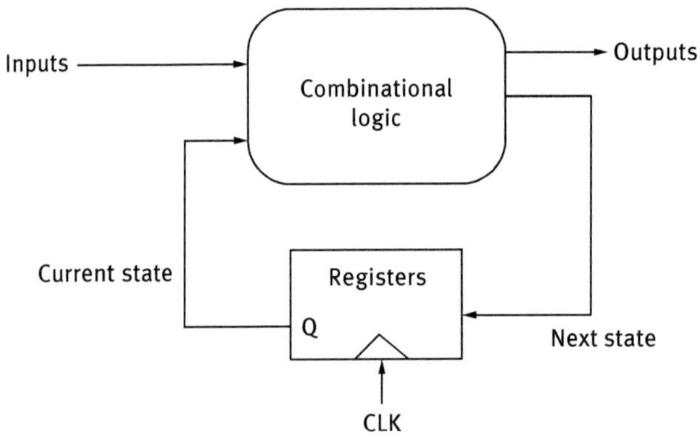

Figure 5.2: A finite state machine (FSM) with positive edge-triggered register.

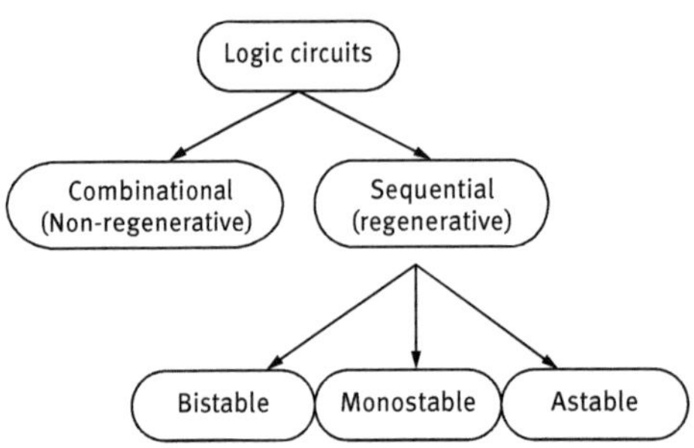

Figure 5.3: Types of logic circuits.

oscillator), there are no stable states; as a result, the circuit oscillates without settling to a stable state. The bistable element is its most popular and important representative circuit. All basic latch and flip-flop circuits, registers and memory elements used in digital systems fall into this category.

In this section, we will discuss about the basic behavior and application of bistable circuits. Static memories use positive feedback to create a bistable circuit– a circuit having two stable states that represent 0 and 1.

The basic bistable element to be examined in this section consists of two identical cross-coupled inverter circuits, as shown in Figure 5.4(a). Here, the output voltage of inverter no. 1 is equal to the input voltage of inverter no. 2, i.e. $V_{01} = V_{i2}$ and $V_{02} = V_{i1}$.

In order to investigate the static input–output behavior of both inverters, we must plot the voltage transfer characteristic (VTC) of both the inverters as shown in Figure 5.4(b) separately. Consequently, we can also plot the VTC of inverter no. 2 using the same axis pair of inverter no. 1, as shown in Figure 5.4(c), resulting in butterfly plots [9–11].

It can be seen that the two VTCs intersect at three points. The resulting circuit has only three possible operation points (*A*, *B* and *C*), as demonstrated on the combined VTC.

Simple reasoning can help us to make the conjecture that two of these operating points are stable. Under the condition that the gain of the inverter in the transient region is larger than 1, only *A* and *B* are stable operation points, and *C* is a metastable

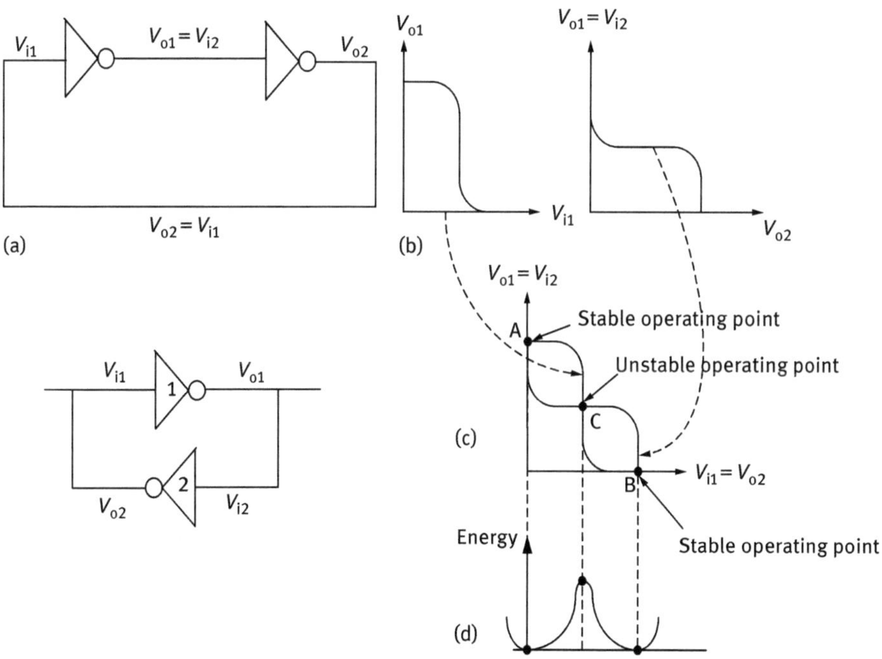

Figure 5.4: (a) Back-to-back connected inverters and (b) their VTC separately. (c) Butterfly plots: VTC superimposed on same axis pair shows intersecting voltage transfer curves of the two inverters, demonstrating the three possible operating points. (d) Qualitative view of the potential energy levels corresponding to the three operating points.

operation point. If the circuit is initially operating at one of these two stable points, it will preserve this state unless it is forced externally to change its operating point. Note that the gain of each inverter circuit, i.e. the slope of the respective voltage transfer curves, is smaller than unity at the two stable operating points. Thus, in order to change the state by moving the operating point from one stable point to the other, a sufficiently large external voltage perturbation must be applied so that the voltage gain of the inverter loop becomes larger than unity.

The circuit has two stable operating points; hence, the name is bistable. Let us assume that the cross-coupled inverter 1 is biased at point C. Noise may cause deviation from this bias point and is amplified and regenerated around the circuit loop [12–14]. As a result, the gain around the loop is larger than 1. A small deviation is applied to V_{i1} (biased in C). This deviation is amplified by the gain of the inverter. The enlarged divergence is applied to the second inverter and amplified once more. The bias point moves away from C until one of the operation points A or B is reached. In conclusion, C is an unstable operation point. Every deviation (even the smallest one) causes the operation point to run away from its original bias. The chance is indeed very small that the cross-coupled inverter pair is biased at C and stays there. Operation points with this property are termed metastable.

Figure 5.4(d) demonstrates qualitatively that the potential energy is minimum at two stable operating points, where the voltage gains are less than "1". In contrast, the energy is maximum at the metastable point, where the voltage gains of the inverters are maximum.

Thus, the circuit has two stable operating points corresponding to the two energy minima, and one unstable operating point corresponding to the potential energy maximum.

In summary, a bistable circuit has two stable states. In the absence of any triggering, the circuit remains in a single state (assuming that the power supply remains applied to the circuit), and hence remembers a value. A trigger pulse must be applied to change the state of the circuit. Another common name for a bistable circuit is flip-flop (sometimes an edge-triggered register is also referred to as a flip-flop).

Figure 5.4 below shows two back-to-back connected inverters. Assuming initially the inverters are in unstable operating point. Therefore, all four transistors are working in saturation region resulting in a maximum loop gain. A small voltage perturbation will cause significant changes in the operating modes of the transistors. Thus, the output voltage of the two inverters will settle in either V_{Oh} or in V_{Ol}, respectively, as shown in 5.5. The direction in which each output voltage diverges is determined by the initial perturbation polarity.

In brief, every sequential logic circuit has two stable stages that are complementary to each other. In the absence of triggering, circuit will remain in the stable stage. In the presence of triggering, circuit may switch or flip into another stable stage. These bistable circuits can also be termed as flip-flop. Flip-flop is the main block of sequential architectures. Here, design and analysis of low power flip-flop

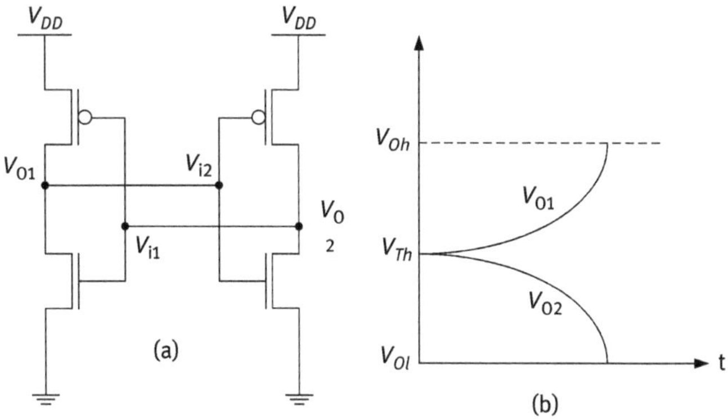

Figure 5.5: (a) Schematic of CMOS bistable circuit and (b) expected behavior of output voltage after the application of a small input voltage perturbation if the circuit is initially biased in a metastable point.

will be discussed in depth for the implementation of ultra-low power sequential logic architecture.

5.3 Basic SR Flip-flop/Latch

As we have learnt from the previous section that a bistable circuit consists of two cross-coupled inverters, the circuit can preserve its state (acting as a memory) as long as power supply is present. However, it is not possible in a two-inverter circuit to change the state. To do so, we must add simple switches to the bistable element, i.e. extra circuitry must be added to enable control of the memory states. Figure 5.6(a) shows a simple *SR* latch with two triggering inputs *S* (set) and *R* (reset). *SR* latch is also called an *SR* flip-flop, since two stable states can be switched back and forth. This circuit is similar to the cross-coupled inverter pair with NOR gates replacing the inverters. The second input of the NOR gates is connected to the trigger inputs (*S* and *R*), which make it possible to force the outputs Q and $\overline{Q}$ to a given state. If a positive (or 1) pulse is applied to the *S* input, the Q output is forced into the 1 state (with $\overline{Q}$ going to 0). Vice versa, a 1 pulse on *R* resets the flip-flop and the Q output goes to 0. It can easily be seen that when both input signals are equal to logic “0”, the *SR* latch will operate exactly like the simple cross-coupled bistable element examined earlier, i.e. it will preserve (hold) either one of its two stable operating points (states) as determined by the previous inputs. Finally, consider the case in which both of the inputs *S* and *R* are equal to logic “1”. In this case, both output nodes will be forced to logic “0”, which conflict with the complementarity's of Q and $\overline{Q}$, i.e. this input combination does not correspond to the definition of flip-flop that Q and $\overline{Q}$ must be complementary.

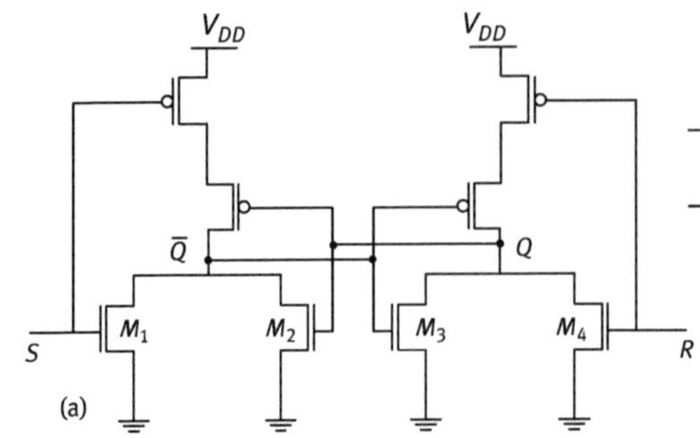

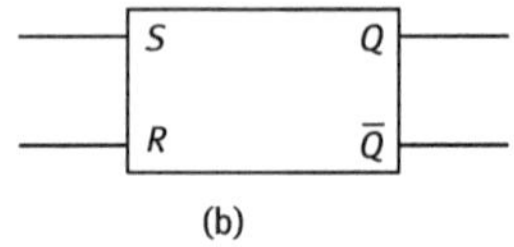

(c)

S	R	Q_{n+1}	$\overline{Q_{n+1}}$	Operation
0	0	Q_n	$\overline{Q_n}$	Hold
1	0	1	0	Set
0	1	0	1	Reset
1	1	0	0	Not allowed

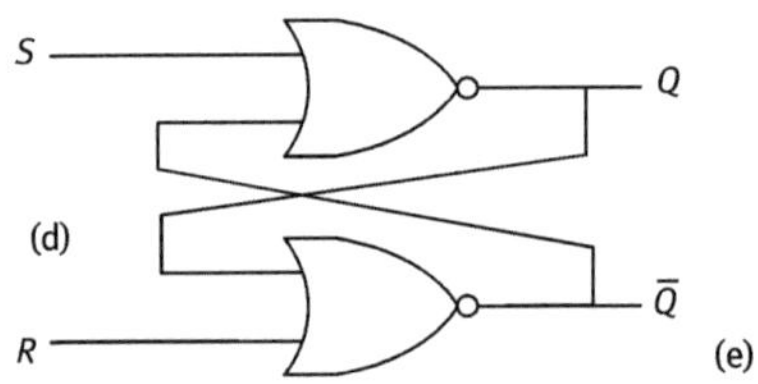

(e)

S	R	Q_{n+1}	$\overline{Q_{n+1}}$	Operation
V_{Oh}	V_{Ol}	V_{Oh}	V_{Ol}	M_1 and M_2 ON, M_3 and M_4 OFF
V_{Ol}	V_{Oh}	V_{Ol}	V_{Oh}	M_1 and M_2 OFF, M_3 and M_4 ON
V_{Ol}	V_{Ol}	V_{Oh}	V_{Ol}	M_1 and M_4 OFF, M_2 ON or,
V_{Ol}	V_{Ol}	V_{Ol}	V_{Oh}	M_1 and M_4 OFF, M_3 ON

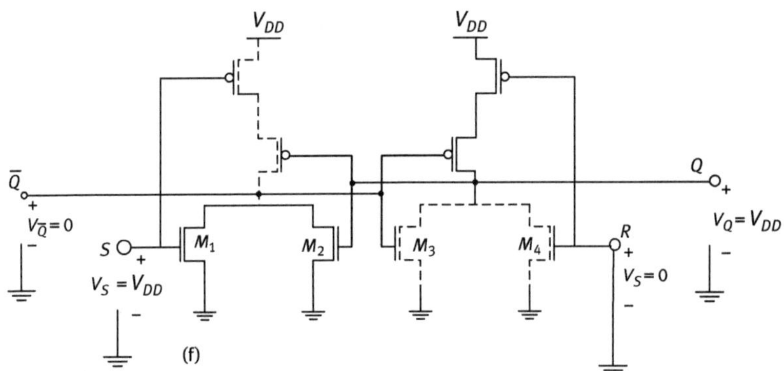

Figure 5.6: (a) CMOS *SR* latch realized with two input NOR gate; (b) symbol of *SR* latch; (c) truth table of *SR* latch; (d) gate diagram of *SR* latch; (e) operational modes of the nMOS transistor in two input NOR-based *SR* latch; and (f) voltages for set operation.

Therefore, this input combination is not permitted during normal operation and is considered to be a not-allowed or forbidden condition. The symbol of *SR* latch is shown in Figure 5.6(b). The truth table of *SR* latch is shown in Figure 5.6(c).

The operational modes of the transistor in NOR-based *SR* latch can be found. When S = "1" = V_{Oh} and R = "0" = V_{Ol}, both M_1 and M_2 are ON; at the same time, M_3 and M_4 become OFF. As a result, the voltage on Q is equal to V_{Oh} = "1" and output voltage on $\overline{Q}$ equals to "0" = V_{Ol}. In contrast when R = "1" = V_{Oh} and S = "0" = V_{Ol}, the situation will be reversed (M_1 and M_2 turned OFF and M_3 and M_4 turned ON).

When both S and R = "0" = V_{Ol}, there are two possibilities. Depending on the previous state of the *SR* latch, either M_2 or M_3 will be ON, while both of the trigger transistors M_1 and M_4 are OFF. This will generate a logic low level of $V_{Ol} = 0$ at one of the output nodes, while the complementary output node is at V_{Oh}.

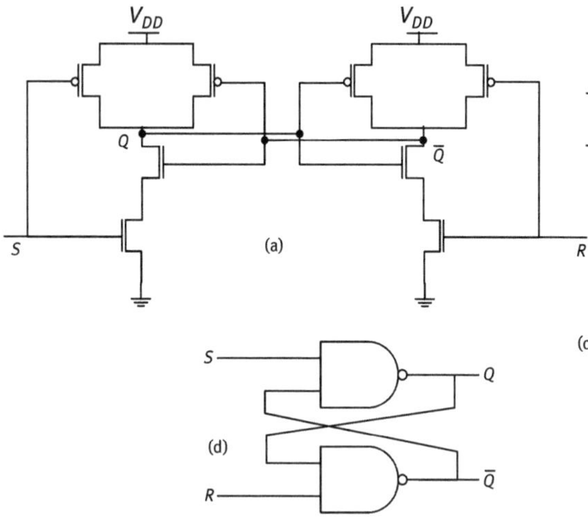

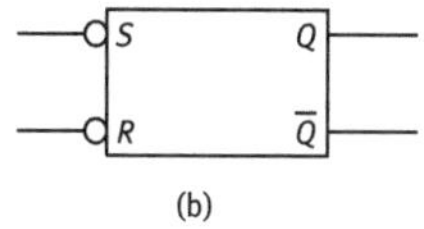

(c)

S	R	Q_{n+1}	$\overline{Q_{n+1}}$	Operation
0	0	1	1	Not allowed
0	1	1	0	Set
1	0	0	1	Reset
1	1	Q_n	$\overline{Q_n}$	Hold

Figure 5.7: (a) Circuit; (b) truth table; (c) gate-level diagram; (d) circuit symbol of two input NAND gate-based *SR* latch.

5.3.1 NAND Gate-based Negative Logic *SR* Latch

Figure 5.7(a) shows a NAND gate-based *SR* latch. Here two input NAND gates are cross-coupled with one input with the other input terminal connected to external trigger inputs *S* and *R*. To preserve a state one must apply an input combination of "1" to both *S* and *R*. If *S* = "0" and *R* = "1" the output *Q* attains a voltage equals V_{OL}. Thus, we can observe that to set the flip-flop we must have *S* = "0" and *R* = "1". Similarly, to reset the flip-flop *S* = "1" and *R* = "0" must be applied [15, 16]. The conclusion is that the NAND-based *SR* latch responds to active low input signals, as opposed to the NOR-based *SR* latch, which responds to active high inputs. For *S* = *R* = "0", both the output wants to attain a logic high, which is a forbidden condition. The truth table, gate-level schematic and symbol are shown in Figure 5.7(b) and 5.7(c) and 5.7(d), respectively. The small circles at the *S* and *R* input terminals indicate that the circuit responds to active low input signals.

5.3.2 Clocked *SR* Latch

The *SR* flip-flops discussed so far are asynchronous and do not require a clock signal. Most systems operate in a synchronous fashion with transition events referenced to a clock. In a clocked latch, the outputs will respond to the input levels only during the active period of a clock pulse. For simple reference, the clock pulse will be assumed to be a periodic square waveform, which is applied simultaneously to all clocked logic gates in the system.

The gate-level diagram of a clocked *SR* latch is shown in Figure 5.8(a). One can observe that when the clock signal is equal to "0", the output of the AND gate is equal

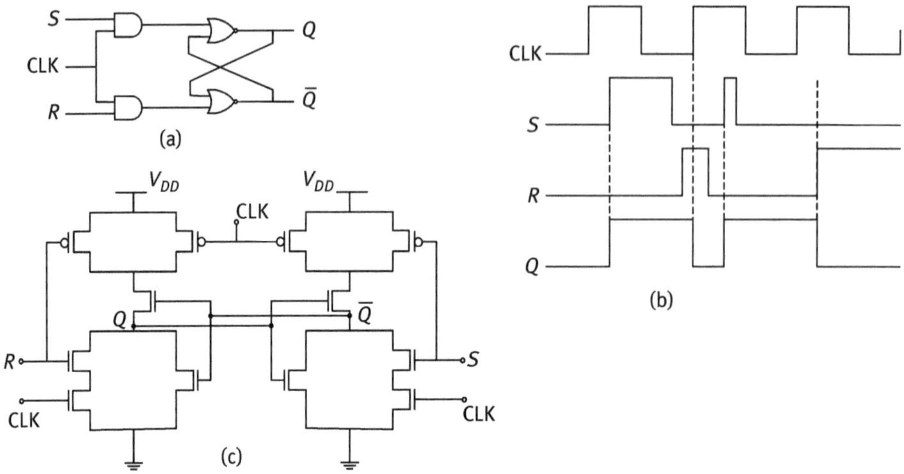

Figure 5.8: (a) Gate-level diagram of clocked *SR* latch; (b) sample waveform demonstrating the level sensitivity of the clocked *SR* latch; and (c) AOI-based implementation of clocked *SR* latch.

to “0”; thus, the input signals have no influence to the circuit output. As long as the output of the AND gates is “0”, the *SR* latch holds its current state without getting influenced by the current *S* and *R* inputs. When the clock becomes high equal to “1”, then the *S* and *R* values are permitted to reach the *SR* latch and may change the state of the latch. Notice that *S* = *R* = “1” remains still a not-allowed input combination in clocked *SR* latch. When *S* = *R* = 1 and clock is high, then both the outputs try to go to level “0”. When the clock pulse becomes “0”, the state of the latch cannot be predicted. The output can be in either state caused by slight difference in delay between the output signals.

Figure 5.8(b) shows a sample waveform of clocked *SR* latch. It is evident from the waveform that the latch is strictly level sensitive during active edge of the clock pulse. Any change in the input when the clock is high is reflected in the output. Even a narrow spike or glitch in the input during active period of click causes changes in the output.

Figure 5.8(c) shows a neither clocked NOR-based *SR* latch using two simple AOI gates, resulting in a smaller transistor count.

The gate-level diagram of a NAND gate-based clocked *SR* latch is shown in Figure 5.9(a). Notice that the inputs *S*, *R* and clock are all active low signals, i.e. when the clock is high (“1”) the input signals cannot change the state of the latch. The inputs will influence the outputs only when the clock is active, i.e. clock = “0”.

A different realization of clocked NAND-based *SR* latch is shown in Figure 5.9(b). Notice here that the clock, *S* and *R* are all active high signals, i.e. latch is set when *S* = “1”, *R* = “0” and clock = “1” and latch is reset when *S* = “0”, *R* = “1” and clock = “1”. The latch preserves its state as long as the clock signal is inactive, i.e. clock = “0”.

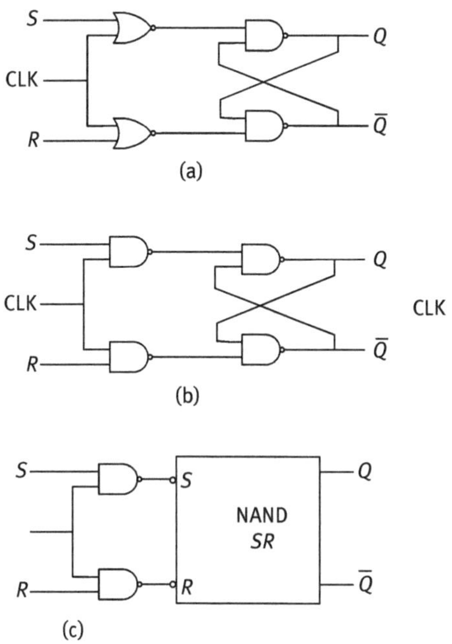

Figure 5.9: (a) Gate-level diagram of a NAND-based *SR* latch with active low inputs; (b) alternative gate-level diagram of a NAND-based *SR* latch with active high inputs; and (c) block diagram representation of the circuit shown in (b).

5.4 Clocked *JK* Latch

The *SR* latch and gated *SR* latch both are having problems with not-allowed/forbidden input combinations. The state of the latch is indeterminate when both are active at the same time. This problem can be solved by drawing a feedback line from the output to the input, resulting in a circuit called *JK* latch. It was named after Jack Kilby, who first produced a digital IC (*JK* flip-flop).

Figure 5.10(a) shows the clocked NAND gate-based *JK* latch gate-level schematic diagram. Figure 5.10(b) shows the symbol of *JK* latch. Figure 5.10(c) shows all NAND implementation of *JK* latch circuit. Figure 5.10(d) shows the truth table of *JK* latch. A *JK* latch is also sometimes known as *JK* flip-flop.

When the clock is active, to set the latch we must apply J = “1” and K = “0”. To reset the latch we must apply J = “0” and K = “1”. When J = “0” and K = “0”, the latch holds its current state. When the clock is inactive the latch preserves its state. Therefore, J and K resemble the set and reset inputs of *SR* latch. Now if both J = K = “1” during active clock period, the latch switches its state due to feedback. *JK* latch does not have a not-allowed input combination like in *SR* latch circuits.

Still there is problem in a *JK* latch. When J = K = “1” in active phase of clock, the output oscillates (toggles) due to feedback until the clock becomes inactive or any

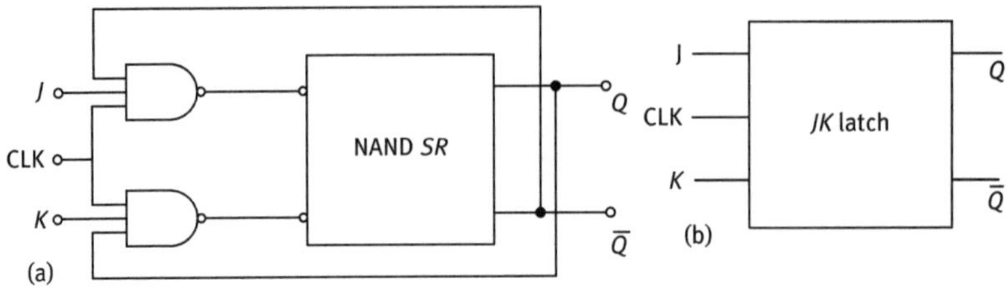

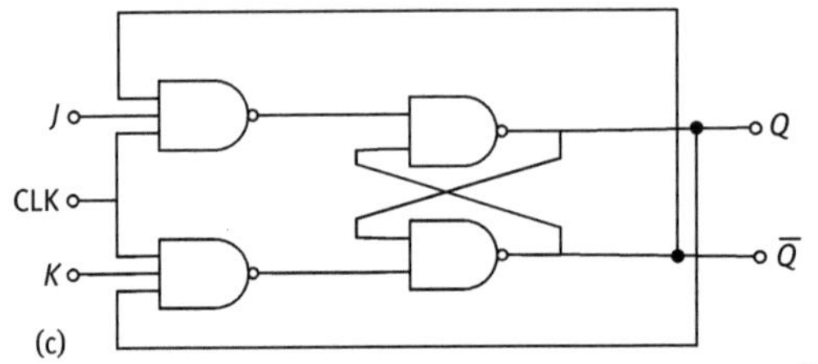

J	K	Q_n	$\overline{Q_n}$	S	R	Q_{n+1}	$\overline{Q_{n+1}}$	Operation
0	0	0	1	1	1	0	1	Hold
		1	0	1	1	1	0	
0	1	0	1	1	1	0	1	Reset
		1	0	1	0	0	1	
1	0	0	1	0	1	1	0	Set
		1	0	1	1	1	0	
1	1	0	1	0	1	1	0	Toggle
		1	0	1	0	0	1	

(d)

Figure 5.10: (a) Gate schematic of NAND-based *JK* latch showing the feedback connection from the output to the input. (b) Symbol of *JK* latch. (c) All NAND implementation of a *JK* latch. (d) Truth table of *JK* latch.

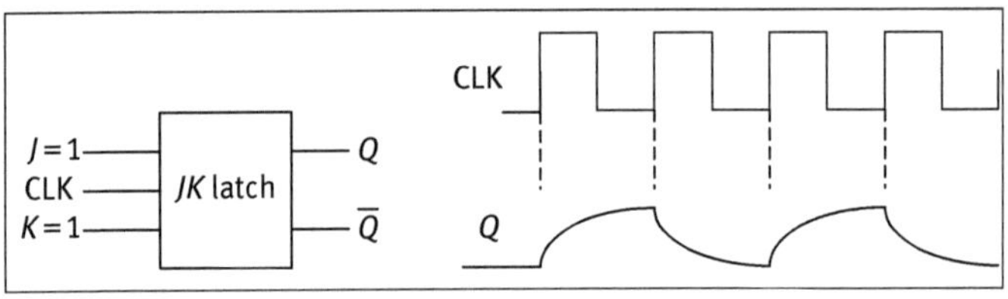

Figure 5.11: *JK* latch working as a toggle switch.

of the input signal becomes “0”. To get rid of this problem, the clock width must be smaller than input-to-output propagation delay of the latch. In that case, the clock will switch its state before the output level has an opportunity to switch again. However, it is difficult to impose this restriction in a real situation.

5.4.1 Toggle Switch

If the time period of the clock signal is less than the input-to-output propagation delay, and if $J = K =$ “1”, then for every clock ticking, the output will change (toggle) its state. The resulting switch is known as toggle switch and is shown in Figure 5.11.

5.5 Master–slave Flip-flop

The master–slave *JK* Flip-Flop removes the timing problem associated with ordinary clocked flip-flops. There are two cascaded stages of latches working with opposite phase of clock. The first stage is called master, which is activated with the high time

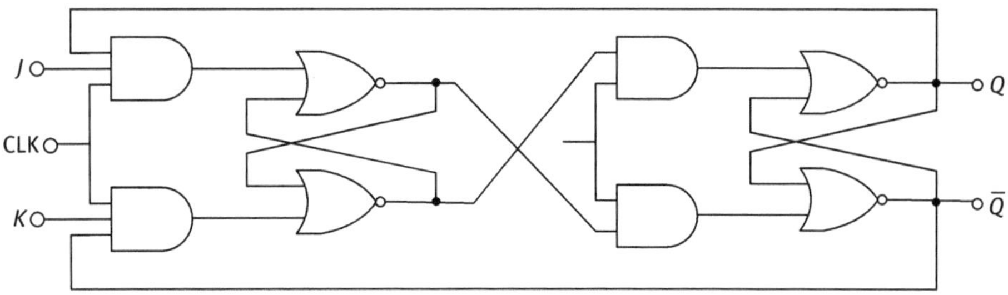

Figure 5.12: NOR gate-based master–slave *JK* Flip-Flop.

period of the clock. The second stage is called the slave, which is activated with the low time phase of the clock. When clock is high, the master is active, allowing the inputs *J* and *K* to pass through the latch. The output of the first stage is set according to the inputs given in high period of the clock. At the same time, the slave stage is deactivated, holding the previous value. When the clock pulse goes to zero, the master latch becomes inactive and the second-stage latch called the "slave" becomes active. The output levels of the flip-flop circuit are determined during this second phase, based on the master-stage outputs set in the previous phase.

The whole latch is never transparent. During the first phase of the clock, the first phase is transparent, while during second phase of the clock, the second stage is transparent. Therefore, change in input never reflected directly to output in the same phase of clock. The two stages are effectively decoupled from each other by the opposite phase of clock. When $J = K =$ "1", the circuits allow toggling, but uncontrolled oscillations are eliminated as two stages are not coupled and they are not transparent at any given time. Figure 5.12 shows a NOR gate-based master–slave *JK* flip-flop.

Still master–slave *JK* latch has one problem. When the clock pulse is high, any narrow spike or glitch in *J* or *K* may set or reset the master latch, which may cause unwanted state transitions, which again may get propagated to the slave stage in the next phase of the clock. This problem is known as ones catching. This problem can be eliminated by edge-triggered flip-flop discussed in the next section.

5.6 D Latch

The flip-flops designed using CMOS logic are straightforward, but they consume a large number of transistors. In this section, we will investigate some structures that require less number of transistors than conventional sequential circuit structures. Let us first consider the simple *D* latch circuit shown in Figure 5.13. This circuit is obtained by modifying the NOR gate-based *SR* latch. The circuit shows a single input *D* is connected to *S* input and its inverted version is connected to *R* input. When the clock is active, the output follows the input with a propagation delay. When the clock signal goes to zero, the output will simply preserve its state. Thus, the CLK input acts as an enable signal, which allows data to be accepted into the *D* latch. *D* latch finds its use in digital application for temporary storage and delay element.

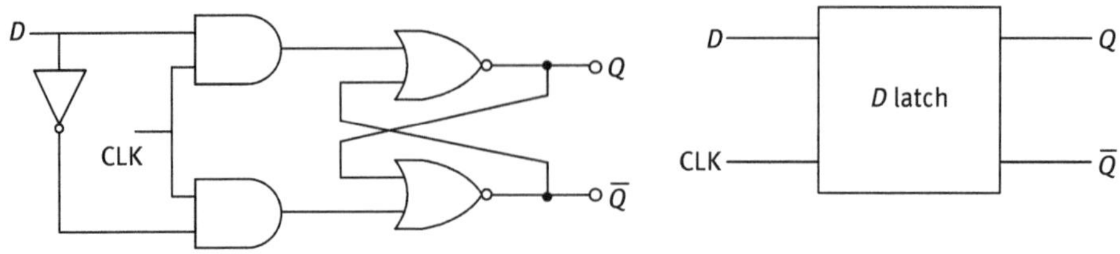

Figure 5.13: *D* latch obtained from NOR-based *SR* latch.

5.6.1 Positive and Negative Latch

A latch is a level-sensitive circuit. A *D* latch passes the input *D* to the *Q* output when the clock is high. This latch is said to be in transparent mode. When the clock is low, the input data sampled on the falling edge of the clock is held stable at the output for the entire phase, and the latch is in hold mode. The inputs must be stable for a short period around the falling edge of the clock to meet setup and hold requirements. A latch operating under the above conditions is a positive latch. Similarly, a negative latch passes the *D* input to the *Q* output when the clock signal is low.

The signal waveforms for a positive latch and negative latch are shown in Figure 5.14.

5.6.2 Multiplexer-based Latch

There are many ways to build latches, out of which one common technique uses transmission gate multiplexers. Multiplexer-based latches can provide similar functionality to the static *SR* latch, but has the important added advantage that the sizing of devices affects only performance and does not affect the functionality. Figure 5.15 shows the

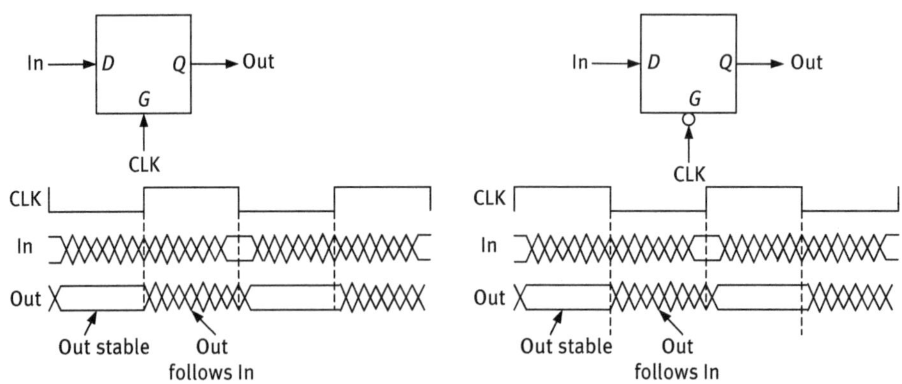

Figure 5.14: Timing of positive and negative latches.

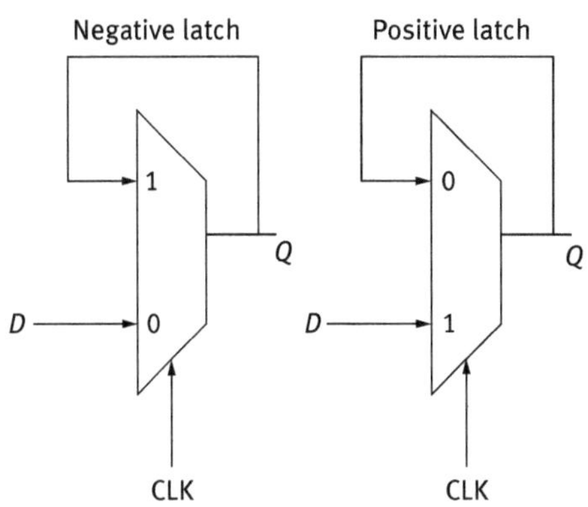

Figure 5.15: Positive and negative latches using multiplexer.

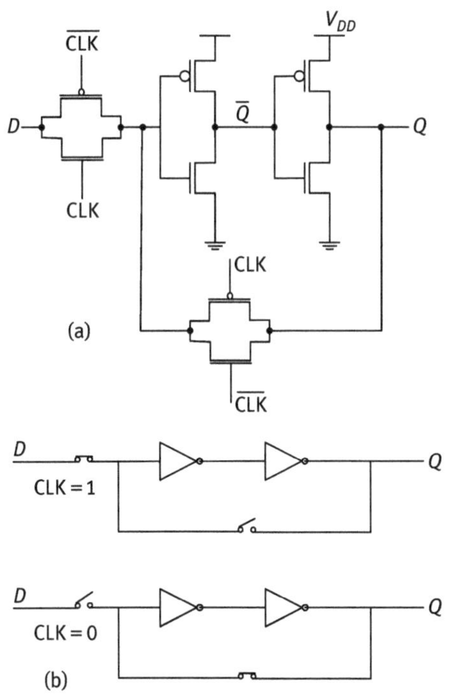

Figure 5.16: (a) Positive latch built using transmission gate and (b) operation of the *D* latch circuit during two phases of the clock.

negative and positive latches based on multiplexers. For a negative latch, when clock is low, it selects the input 0, and input *D* is connected to output *Q*. When the clock is high, it selects the input 1, and output *Q* is connected by feedback. Thus for low clock, the latch is transparent and for high clock the latch is in hold state, providing a stable

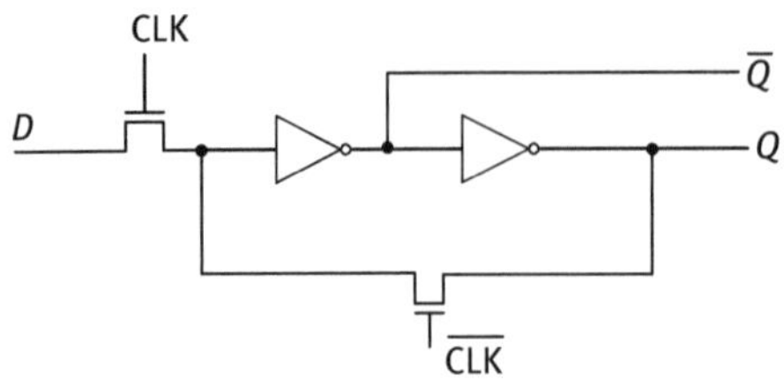

Figure 5.17: Latch implemented by nMOS pass transistors.

output. Similarly in the positive latch, the *D* input is selected when clock is high, and the output is held (using feedback) when clock is low.

The transmission gate implementation of a positive latch using multiplexer is shown in Figure 5.16. When CLK is high, the input transmission gate is ON and the latch is transparent – that is, the *D* input is copied to the *Q* output. When the clock is low, the state of *D* latch in current state is preserved as input switch (TG) is OFF and loop switch (TG) is ON. Unlike the *SR* FF, the feedback does not have to be overridden to write the memory and hence sizing of transistors is not critical for realizing correct functionality. The number of transistors that the clock touches is important. This particular latch implementation is not particularly efficient as it presents a load of four transistors to the CLK signal.

It is possible to implement the same latch with clock load of two transistors using pass transistors as shown in Figure 5.17.

The advantage of this type of latch is reduced clock load and simplicity. But nMOS pass transistors result in a degraded high voltage of $V_{DD}-V_{Thn}$ to the input of the first inverter. This impacts noise margins and switching performance especially for a low V_{DD} and high V_{Thn}. It also causes static power dissipation in first inverter, since the maximum input voltage to the inverter equals $V_{DD}-V_{Thn}$, and the pMOS device of the inverter is never turned off, resulting in a static current flow.

5.7 Master–slave Edge-triggered Flip-flops

The *D* latch shown in Figure 5.17 is not an edge-triggered storage element because the output changes according to the input, i.e. the latch is transparent while the clock is high. The transparency property makes the applications of this *D* latch unsuitable for counters and some data storage implementations.

Edge-triggered flip-flops are built using primitive latches. They sample the input only in the time of clock transition: 0→1 for positive and 1→0 for negative edge-triggered flip-flops. Positive and negative latches are connected in cascade in master–slave configuration to form edge-triggered flip-flops. Positive latch followed by a negative latch forms negative edge-triggered flip-flop, whereas negative latch followed by a positive latch forms positive edge-triggered flip-flop.

Figure 5.18 shows a positive edge-triggered flip-flop implemented by cascading first negative and then positive latches in master–slave configuration. A multiplexer-based latch is used in this particular implementation, although any latch could be used. The master stage is transparent, and the D input passes to master-stage output Q_M when the clock is low. During this period, the slave stage is in the hold mode, keeping its previous value using feedback. On the rising edge of the clock, the master–slave stops sampling the input, and the slave stage starts sampling. During the high phase of the clock, the slave stage samples the output of the master stage (Q_M), while the master stage remains in a hold mode. Since Q_M is constant during the high phase of the clock, the output Q makes only one transition per cycle (Figure 5.19). The value of Q is the value of D right before the rising edge of the clock, achieving the positive edge-triggered effect. A negative edge-triggered register can be easily obtained by cascading positive latch first and then negative latch.

The transmission gate realization of the positive edge-triggered flip-flop is shown in Figure 5.20. When the clock is low, T_1 is ON and T_2 is OFF in the master stage, so that input D is sampled into Q_M. Also during this phase, T_3 is OFF and T_4 is ON in the slave stage to complete the feedback loop of the cross-coupled inverters (I_5 and I_6) to hold the state of the slave latch. When the clock goes high, the master stage stops sampling the input and goes into a hold mode. T_1 is OFF and T_2 is ON, and the cross-coupled

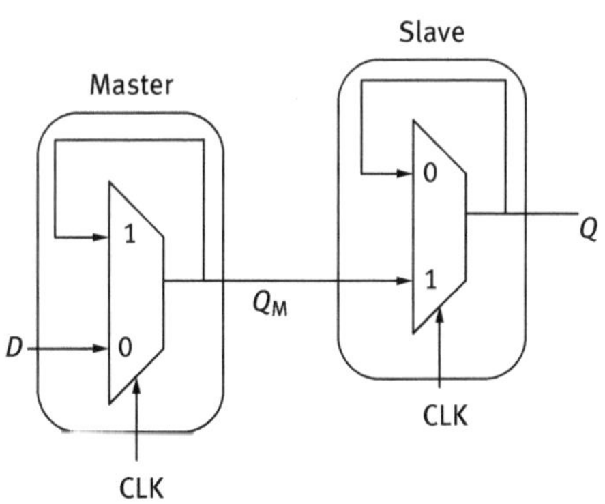

Figure 5.18: Positive edge-triggered flip-flop using master–slave configuration of latches using multiplexers.

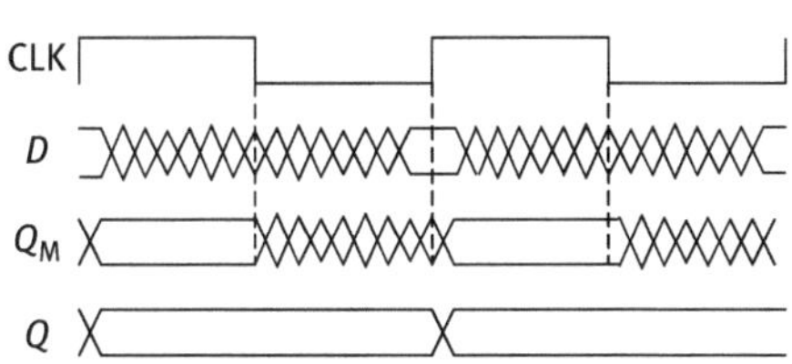

Figure 5.19: Timing of a positive edge-triggered flip-flop.

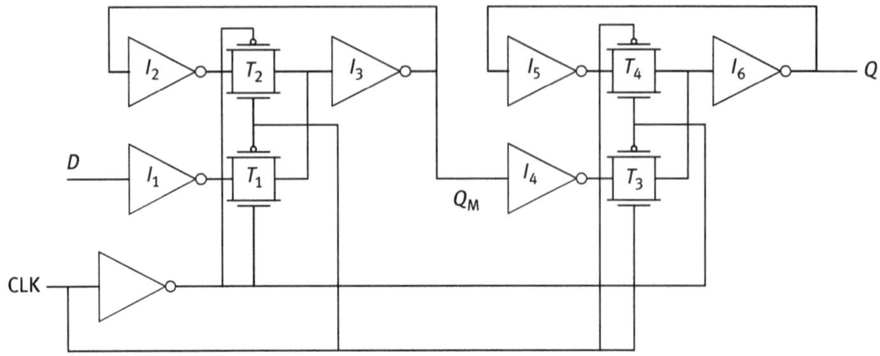

Figure 5.20: Implementation of master–slave positive edge-triggered flip-flop using multiplexers.

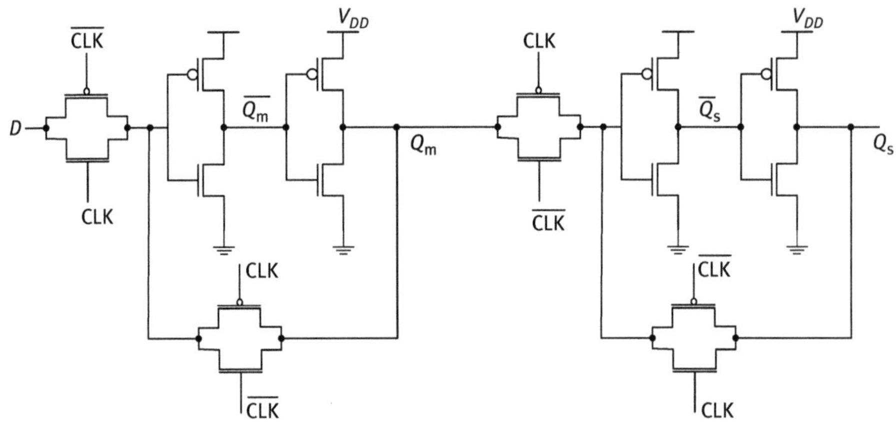

Figure 5.21: Negative edge-trigged master–slave D flip-flop.

inverters I_3 and I_4 hold the state of Q_M. Also, T_3 is ON and T_4 is OFF, and Q_M is copied to the output Q.

Finally, by cascading a positive level-sensitive master stage D latch with a negative level-sensitive slave stage D latch, a negative edge-triggered master–slave D FF is constructed and is shown in Figure 5.21. This circuit is a negative edge-triggered D flip-flop by virtue of the fact that it samples the input at the falling edge of the clock pulse.

5.8 Timing Parameters for Sequential Circuits

There are three important timing parameters for a sequential circuit like flip-flop or register.

The setup time (t_{Su}) is the time that the data inputs (D input) must be valid before the clock transition. (For a positive edge-triggered flip-flop that is "0" to "1", transition 0→.) The hold time (t_{Hold}) is the time the data input must remain valid after the clock edge. Assuming that the setup and hold times are met, the data at the D input is copied to the Q output after a worst-case propagation delay (with reference to the clock edge) denoted by t_{c-q}. In other words, D input must be stable before (setup time, t_{Setup}) and after the negative clock transition to allow time for the input and loop switch (TG) to OFF and ON. Once the loop switch is closed and input switch is open, the output is preserved. The constraint of setup and hold time must be fulfilled; otherwise, metastability problem can cause chaotic behavior leading to an indeterminate (unpredictable) state after transitional period. This situation is illustrated in Figure 5.22, where the input D switches from "0" to "1" immediately before the clock transition occurs (setup time violation). As a result, the master stage fails to latch the correct value, and the slave stage produces an erroneous output. The timing should be properly synchronized to avoid this kind of problem.

Assume that worst-case propagation delay of the logic circuit is denoted as t_{Plogic}, and minimum delay or contamination delay is denoted as t_{Cd}.

The minimum clock period T, required for proper operation of the sequential circuit, is given by

$$T \geq t_{c-q} + t_{Plogic} + t_{Su}$$

The hold time of the register imposes an extra constraint for proper operation:

$$t_{Cdregister} + t_{Cdlogic} \geq t_{Hold}$$

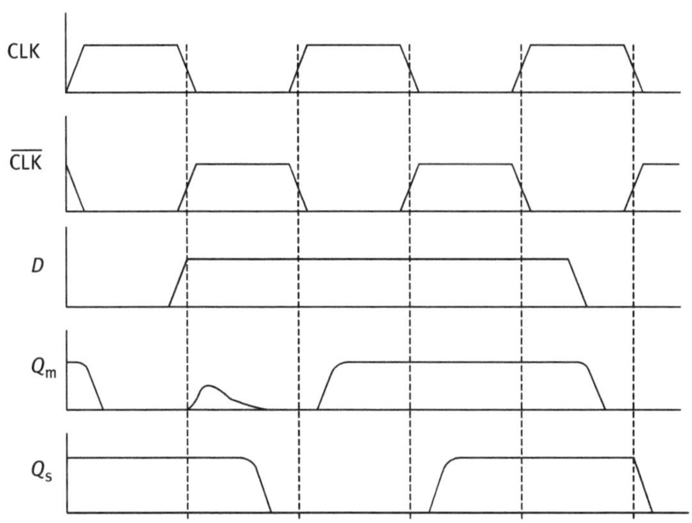

Figure 5.22: Waveform of D FF showing that setup time violation can cause erroneous output.

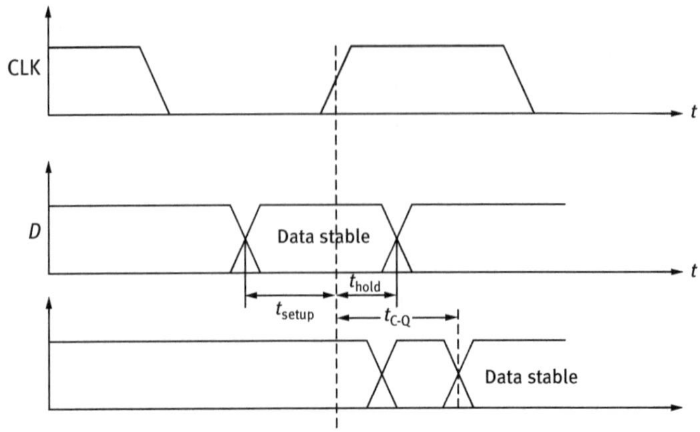

Figure 5.23: Definition of setup time, hold time and propagation delay of a flip-flop.

where $t_{Cdregister}$ is the minimum propagation delay (or contamination delay) of the register.

Figure 5.23 shows the various timing constraints of D-type flip-flop.

5.8.1 Timing of Multiplexer-based Master–slave Flip-flop

Assume that the propagation delay of each inverter is t_{Pd_inv}, and the propagation delay of the transmission gate is t_{Pg_tx}. Also assume that the contamination delay is 0 and the inverter delay to derive $\overline{CLK}$ from CLK has a delay equal to 0. The setup time is the time before the rising edge of the clock that the input data D must become valid. Another way to ask the question is how long before the rising edge does the D input have to be stable such that Q_M samples the value reliably. For the transmission gate multiplexer-based register, the input D has to propagate through I_1, T_1, I_3 and I_2 before the rising edge of the clock. This is to ensure that the node voltages on both terminals of the transmission gate T_2 are at the same value. Otherwise, it is possible for the cross-coupled pair I_2 and I_3 to settle to an incorrect value. The setup time is therefore equal to $3 \times t_{Pd\_inv} + t_{Pd\_tx}$.

The propagation delay is the time for the value of Q_M to propagate to the output Q. Note that since we included the delay of I_2 in the setup time, the output of I_4 is valid before the rising edge of clock. Therefore, the delay $t_{c\text{-}q}$ is simply the delay through T_3 and I_6 ($t_{c\text{-}q} = t_{Pd\_tx} + t_{Pd\_inv}$).

The hold time represents the time that the input must be held stable after the rising edge of the clock. In this case, the transmission gate T_1 turns off when clock goes high and therefore any changes in the D input after clock going high are not seen by the output. Therefore, the hold time is 0.

To obtain the setup time of the register using SPICE, we progressively skew the input with respect to the clock edge until the circuit fails. In a similar fashion, the

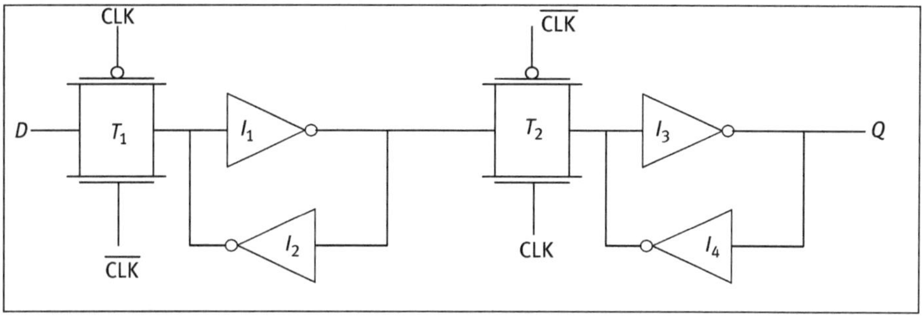

Figure 5.24: Static master–slave flip-flop.

hold time can be simulated. The D input edge is once again skewed relative to the clock signal till the circuit stops functioning.

The drawback of the transmission gate register is the high capacitive load presented to the clock signal. Each register has a clock load of eight transistors. Figure 5.24 shows the approach to reduce the clock load by eliminating the feedback transmission gate by directly cross-coupling the inverters.

The penalty for the reduced clock load is increased design complexity. The transmission gate (T_1) and its source driver must overpower the feedback inverter (I_2) to switch the state of the cross-coupled inverter.

5.8.2 The Sizing Requirements for the Transmission Gates

The input to inverter I_1 must be brought below its switching threshold in order to make a transition. Another problem with this scheme is the reverse conduction– that is, the second stage can affect the state of the first latch. When the slave stage is on (Figure 5.25), it is possible for the combination of T_2 and I_4 to influence the data stored in I_1–I_2 latch. As long as I_4 is a weak device, this is fortunately not a major problem.

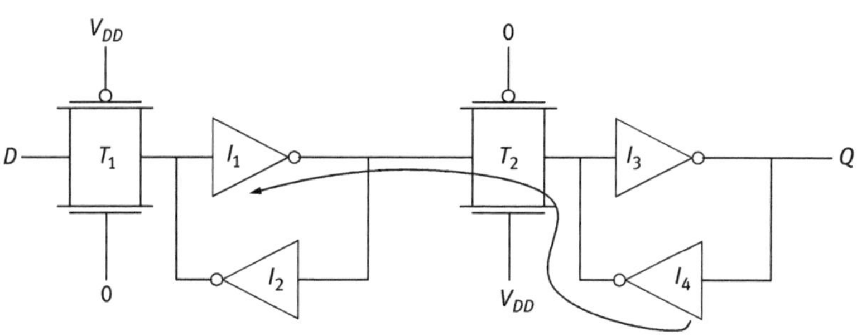

Figure 5.25: Reverse conduction in a static master–slave flip-flop.

5.9 Clock Skews due to Nonideal Clock Signal

For a master–slave flip-flop we need $\overline{CLK}$ and CLK signal. So far we have assumed that $\overline{CLK}$ is a perfect inversion of CLK. The delay generated by the inverter is zero. But this is not a good assumption. Also the variations can exist in the wires used to route the two clock signals, or the load capacitances can vary based on data stored in the connecting latches. This effect is known as clock skew, where two opposite phase clock signals overlap as shown in the figure below. Clock skew/overlap causes failures in master–slave flip-flop as illustrated below.

If both $\overline{CLK}$ and CLK become high for a short period of time due to clock skew, both the master stage and slave stage sampling pass transistors become ON. Thus, there is a direct path between input D and output Q. As a result, data at the output can change with positive edge of the clock, which is undesirable for a negative edge-triggered flip-flop. This is known as a race condition in which the value of the output Q is a function of whether the input D arrives at node X before or after the falling edge of $\overline{CLK}$. If node X is sampled in the metastable state, the output will switch to a value determined by noise in the system. Due to the clock overlap between $\overline{CLK}$ and CLK, node A can be driven by both D and B, resulting in an undefined state in a master–slave flip-flop shown in Figure 5.26.

This problem of clock skew can be solved by using nonoverlapping clocks as shown in Figure 5.27 and by keeping the nonoverlap time $t_{Nonoverlap}$ between the clocks large enough such that no overlap occurs even in the presence of clock-routing delays. During the nonoverlap time, the flip-flop is in the high-impedance state – the feedback loop is open, the loop gain is zero and the input is disconnected. Leakage will destroy the state if this condition holds for too long a time. Hence, the name pseudo-static: the register employs a combination of static and dynamic storage approaches depending upon the state of the clock.

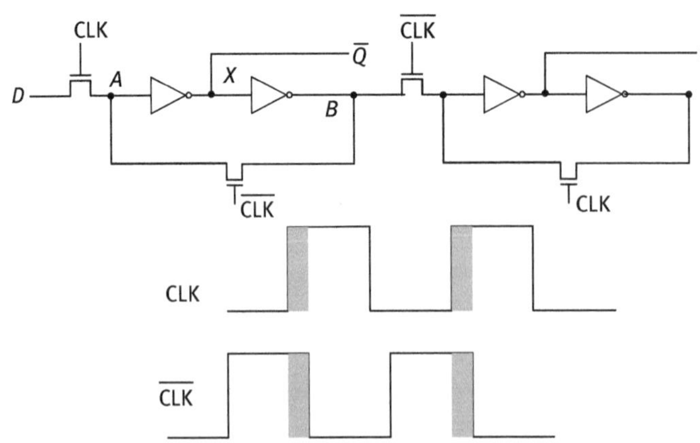

Figure 5.26: (a) Schematic diagram of master–slave registers based on nMOS pass transistors and (b) overlapping clock pairs responsible for clock skew.

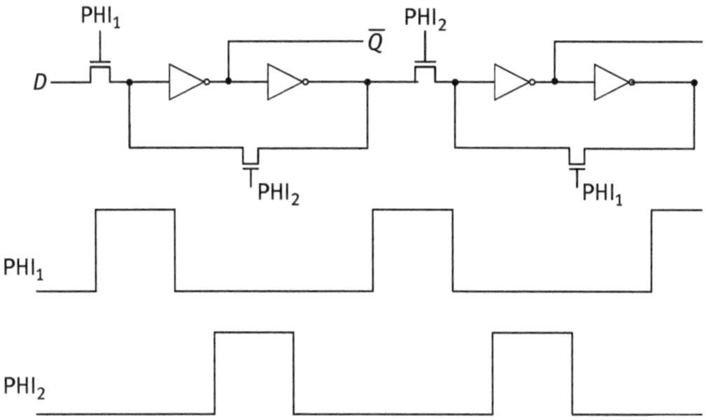

Figure 5.27: (a) Pseudo-static two-phase *D* register and (b) two-phase nonoverlapping clock signal.

5.10 Design and Analysis of the Flip-flops Using DTMOS Style

5.10.1 *SR* Latch and Flip-flop

SR latch is one of the simplest latch circuits having two triggering inputs "*S*" and "*R*" for set or reset operation of the outputs. These inputs can set or reset the output nodes externally. Hence, the *SR* latch circuit has two complementary stable outputs, *Q* and Qb. When $Q = 250$ mV, the latch is "set" and when the $Q = 0$, the latch is reset. *SR* latch can be formed by cross-coupling two NAND or NOR gates. Hence, first NOR-based *SR* latch shown in Figure 5.28 is discussed.

In case of NOR-based latch, when both the inputs *S* and *R* are set to 0, the output nodes hold the previous state. Simply it behaves like a bistable Boolean logics as both the elements. It is known as hold operation. When $S = 250$ mV and $R = 0$, the output *Q* is forced to 250 mV, and the latch is set. When $S = 0$ and $R = 250$ mV, the output *Q* is forced to 0, and the latch is reset. If we apply $R = S = 250$ mV, both the

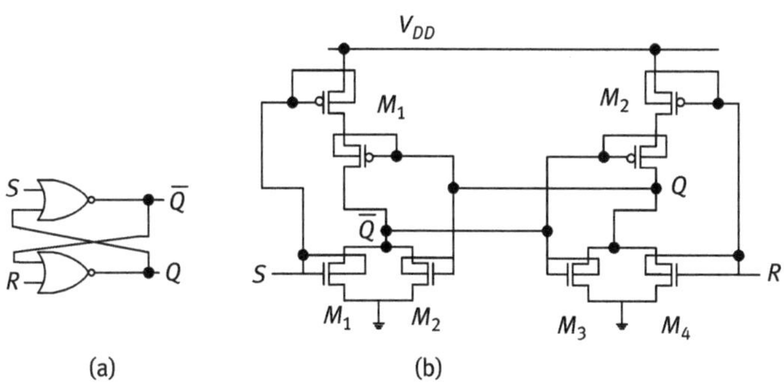

Figure 5.28: NOR-based *SR* flip-flop using DTMOS.

outputs are forced to 0. This operation simply violates the Boolean logic as both the complementary outputs cannot remain in one stable state. So this input combination is not allowed for NOR-based *SR* latch circuit, also termed as forbidden state. The truth table of the *SR* latch is given as follows:

Inputs		Outputs		
S	***R***	**$Q(n)$**	**Qb(n)**	**Operation**
0	0	$Q(n-1)$	Qb($n-1$)	Hold
0	250 mV	250 mV	0	Set
250 mV	0	0	250 mV	Reset
250 mV	250 mV	250 mV	250 mV	Forbidden

When $S = V_{DD}$ and $R = 0$, M_1 and M_4 will be ON and OFF, respectively, Qb node will be set because of M_1 and Q node will be charged up through the series pMOS paths. As a result, M_2 turns ON and M_3 remains OFF. So "250 mV" and "0" are obtained at Qb and Q nodes, respectively. The latch is said to be in reset mode. Reverse operations will occur if we reverse the input combinations. When we set both the inputs to "250 mV", both the M_1 mV and M_4 will be turned ON and both Q and Qb will be forced to be at zero. This operation is not allowed as during the operation, Boolean principles will be violated. Assuming both the inputs are set to "0", and if Q and Qb are initially set to "0" and "250 mV", because of inputs M_1 and M_4 will remain OFF. However, because of the Q and Qb nodes, M_2 and M_3 will be turned OFF and ON, respectively. So Q and Qb remain in the previous state.

In case of NAND-based latch, when both the inputs S and R are set to 0, both the outputs are forced to 250 mV. This operation simply violates the Boolean logic as both the complementary outputs cannot remain in one stable state. So this input combination is not allowed for NAND-based *SR* latch circuit, also termed as forbidden state. If we apply $R = S = 250$ mV, then the output nodes hold the previous state. Simply it behaves like a bistable element. It is known as hold operation. When $S = 250$ mV and $R = 0$, the output Q is forced to 0, the latch is reset. When $S = 0$ and $R = 250$ mV, the

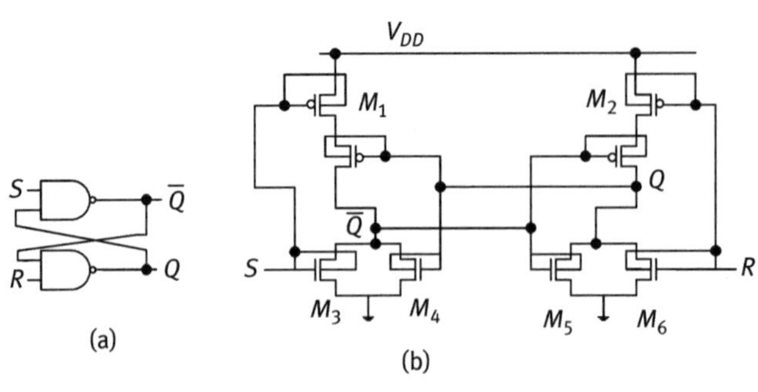

Figure 5.29: NAND-based *SR* flip-flop using DTMOS.

output Q is forced to 250 mV, the latch is set. The truth table of the *SR* latch is given as follows:

Inputs		Outputs		
S	***R***	**Q(*n*)**	**Qb(*n*)**	**Operation**
0	0	250 mV	250 mV	Forbidden
0	250 mV	250 mV	0	Set
250 mV	0	0	250 mV	Reset
250 mV	250 mV	$Q(n-1)$	$Qb(n-1)$	Hold

Implementation of NAND-based *SR* latch circuit using the transistor is also given in Figure 5.29. The operation will be very much similar to the NOR-based circuit. The basic difference between the NAND- and NOR-based circuits lies in response to the inputs. NAND-based latch responds to the active low input signal, whereas the NOR-based latch circuit responds to the active high input signals.

Until now we have discussed the latch circuit, which is asynchronous in nature. In case of latch circuit, clock is not required as it is asynchronous in nature, whereas flip-flop circuits are driven by the clock input for the synchronization purpose. So flip-flop circuits are not transparent like the latch circuits. Flip-flop circuits can be made idle by setting clock to logic "0" or "250 mV". When the flip-flop becomes idle, then unwanted transitions in the inputs do not affect the circuit as the disturbance cannot switch the output swing. So, in noisy environment, flip-flop circuits are much more efficacious compared to the latch circuits. However, introduction to clocking scheme, distribution and management of the clock tree add more power in case of larger blocks. However, it is acceptable as the flip-flops behave more accurately in noisy environment. Hence, clocked *SR* latch circuit or *SR* flip-flop is discussed in detail. *SR* flip-flop is driven by a clock and synchronous in nature. Hence, if CLK = 0, M_1 and M_6 both will be turned OFF. Triggering inputs, like S and R, can turn ON or OFF the M_2 and M_5 but it does not affect the circuit operations. Depending on the output nodes M_3 and M_4 will be turned ON or OFF, i.e. previous states will hold. When CLK = 0, input transitions would not flip the output nodes and simply previous stages will hold. When CLK = 250 mV, M_1 and M_6 will turn ON. The triggering inputs can set or reset the *SR* flip-flops. When CLK = 250 mV and S = 250 mV, $R = 0$, then Qb node will be set to zero, which forces the Q node to be at logic "1", i.e. the flip-flop will be set. When both the inputs are zero and CLK = 250 mV, the flip-flop holds the previous states. The problem arises when both the inputs are set to 250 mV and CLK = 250 mV, Q and Qb nodes will be forced to zero. These input combinations cannot be allowed and can be considered as the forbidden stage in the operation of flip-flop. AOI representation of clocked *SR* flip-flop is given in Figure 5.30 and its timing diagram is shown in Figure 5.31. In an alternative approach, a NOR-based SR flip-flop is shown in Figure 5.32.

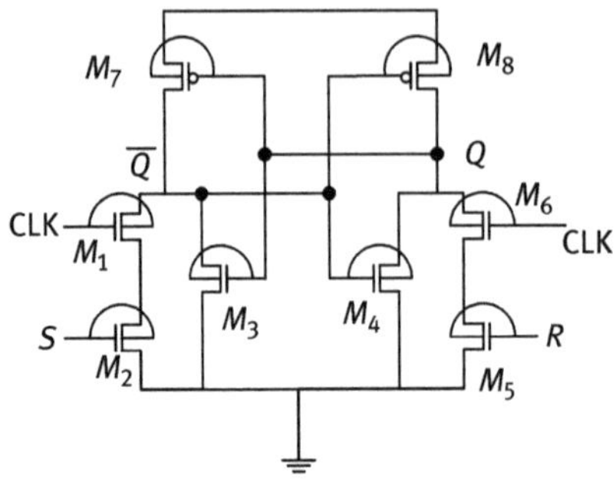

Figure 5.30: AOI-based *SR* flip-flop using DTMOS.

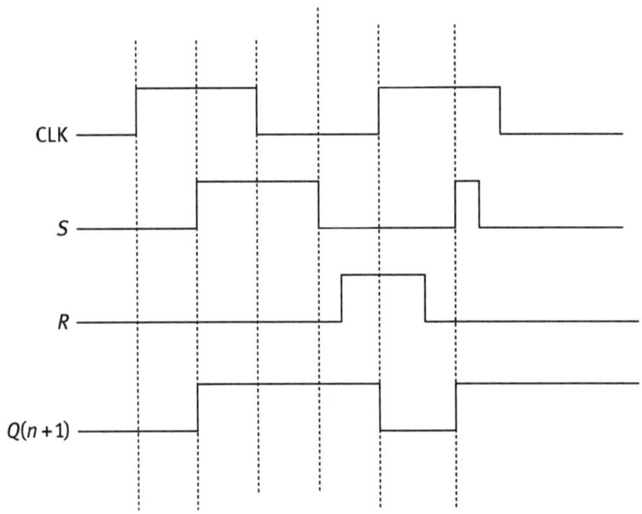

Figure 5.31: Timing diagram of AOI-based *SR* flip-flop.

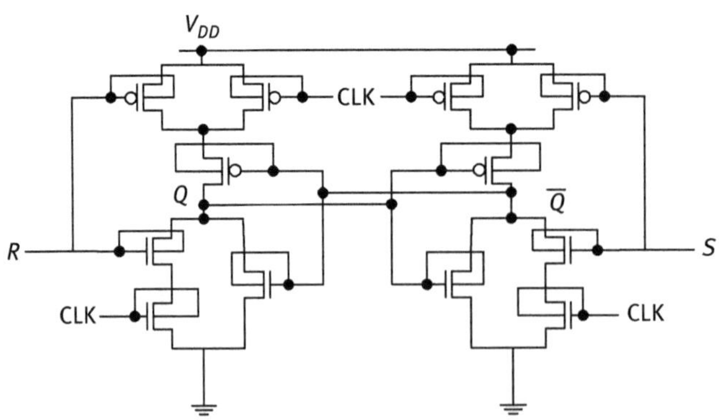

Figure 5.32: NOR-based *SR* flip-flop using DTMOS.

5.10.2 *JK* Latch and *JK* Flip-flop

The main problem associated with the *SR* flip-flop is the forbidden state. In case of NOR-based *SR* latch, if we set $S = R = 250$ mV or in case of NAND-based *SR* latch if $S = R = 0$, then the outputs will be forced to either 0 or 1 state, which are basically not allowed during the operations of the *SR* latch or in any bistable circuit element. This problem can be solved using a feedback path from the outputs to the inputs. Conversion of *SR* latch into *JK* latch or *JK* flip-flop is given in Figure 5.33. Hence, also implementation of *JK* flip-flop using the NAND gates that respond to the active high inputs is shown in Figure 5.34. The forbidden stage of the *SR* flip-flop becomes the toggling mode in the *JK* flip-flop.

In all NAND-based *JK* flip-flops when we set CLK = 0, the outputs will preserve the previous state. When the clock is set to 250 mV, the *JK* flip-flop can be set, reset or can perform in toggle mode of operation depending on the triggering inputs. Hence, the operation of *JK* flip-flop when the clock is set to 250 mV is discussed in brief. When $J = 250$ mV and $K = 0$, the *JK* flip-flop is set, i.e. Q is set to 250 mV. When $J = 0$ and

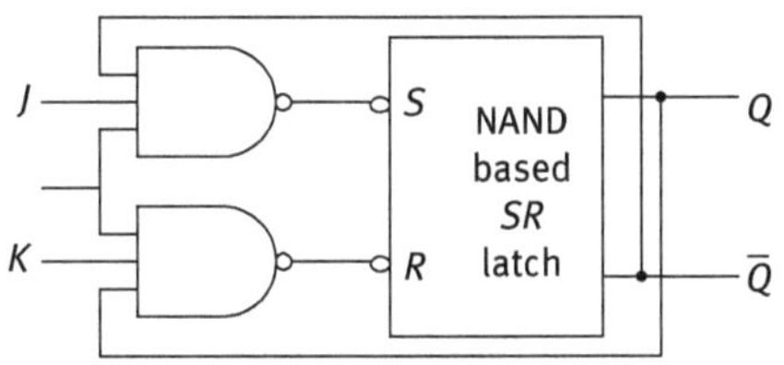

Figure 5.33: *SR* latch to *JK* latch in subthreshold regime.

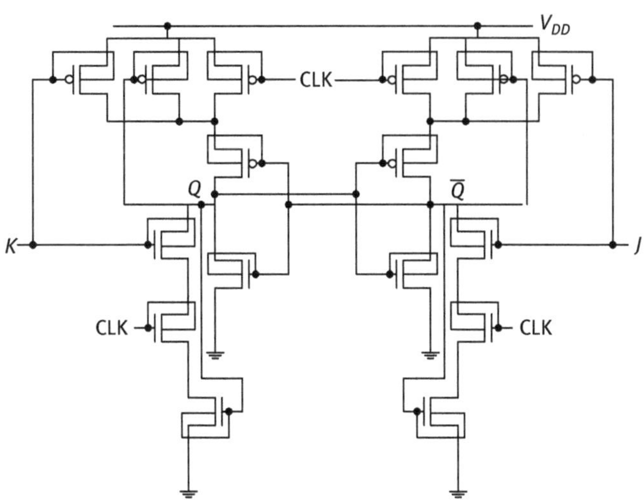

Figure 5.34: *JK* flip-flop using DTMOS.

K = 250 mV, the *JK* flip-flop is reset, i.e. Q is set to 0. When both the inputs are set to 0, the outputs hold the previous state, i.e. hold operations will be performed by the flip-flop. If both the inputs are set to logic "1", then the outputs switch in between "0" and "250 mV" or oscillate due to the feedback path. This toggling operation will be continued until the clock phase becomes inactive or one input goes to "0" to set or reset the flip-flop. This operation is termed as toggling. So, in *JK* flip-flop there is not any forbidden state like the *SR* flip-flop. Truth table of the *JK* flip-flop is given below to analyze the operation of the *JK* flip-flop. In the truth table, the relationship between J, K and S, R is also given for the ease of understanding.

J	*K*	*Q*(*n*)	Qb(*n*)	*S*	*R*	*Q*(*n* + 1)	Qb(*n* + 1)	Operation
		0	250 mV	250 mV	250 mV	0	250 mV	
0	0	250 mV	0	250 mV	250 mV	250 mV	0	Hold
		0	250 mV	250 mV	250 mV	0	250 mV	
0	250 mV	250 mV	0	250 mV	0	0	250 mV	Reset
		0	250 mV	0	250 mV	250 mV	0	
250 mV	0	250 mV	0	250 mV	250 mV	250 mV	0	Set
		0	250 mV	0	250 mV	250 mV	0	
250 mV	250 mV	250 mV	0	250 mV	0	0	250 mV	Toggle

Hence, also the AOI-based implementation of the *JK* flip-flop is given for in-depth analysis of the *JK* flip-flop. AOI implementation of the *JK* flip-flop needs comparatively less number of transistors, so the silicon area is reduced in this case. In addition, less transistor count consumes less power without sacrificing the noise immunity and the voltage swing. The basic operations of the *JK* flip-flop will be same, which we have mentioned earlier. Hence, when CLK = 250 mV and $J = K =$ 250 mV, the circuit will toggle between "0" and "250 mV". Therefore, we can use a *JK* flip-flop as a toggle switch. The problem is the toggling operation will be continued until one input goes low or the clock becomes inactive. This undesirable timing problem in the *JK* flip-flop can be eliminated by setting comparatively smaller width of the clock pulse compared with the input-to-output propagation delay of the circuit. Also to avoid the timing problem, clock signal will be low before the output level switches. Then the toggling will be done once in one clock pulse for $J = K =$ 250 mV as shown in Figure 5.35.

5.10.3 *D* Flip-flop

In case of static circuit, the output nodes produce the true logic as long as the supply voltage is ON. The main disadvantages of the static latches and the flip-flops are their

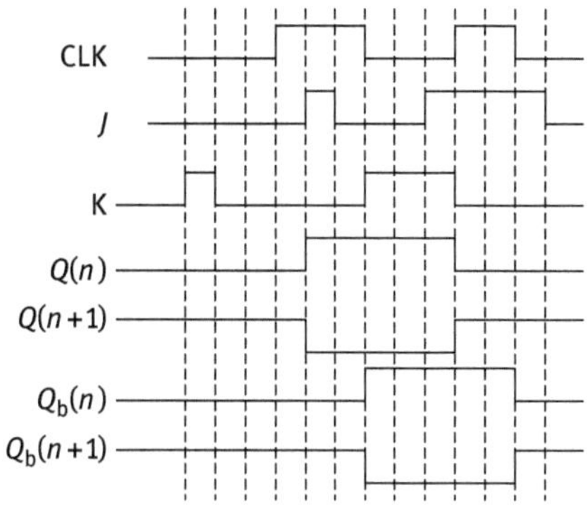

Figure 5.35: Output waveform of *JK* flip-flop using DTMOS.

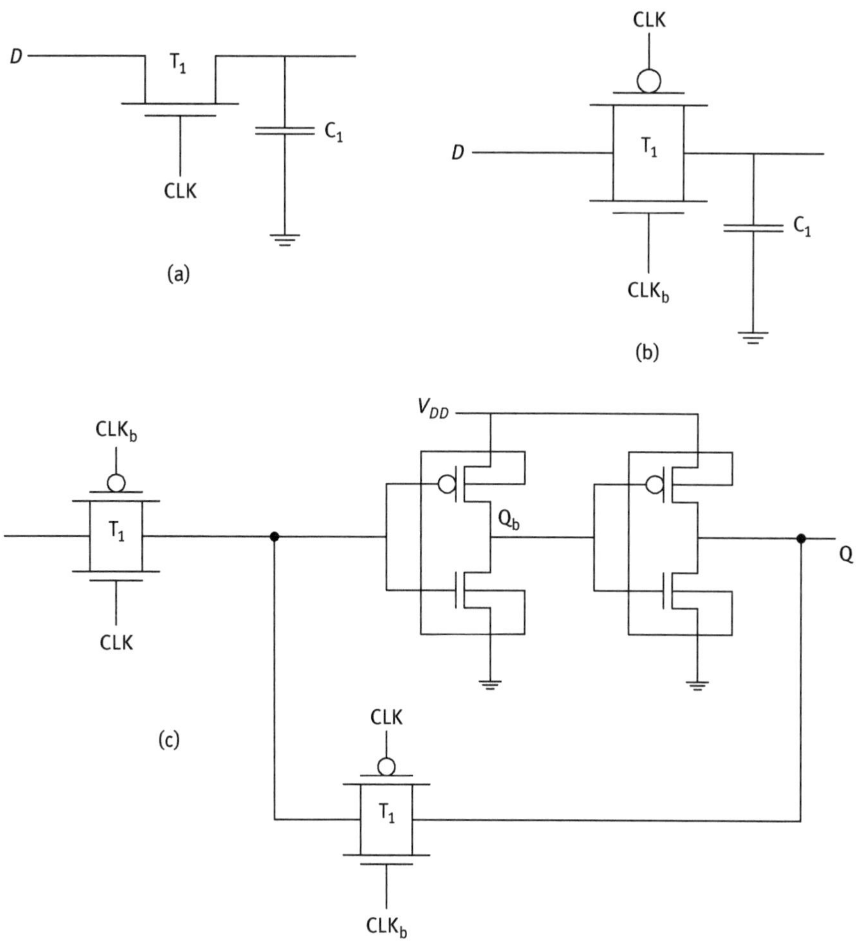

Figure 5.36: *D* flip-flop using DTMOS.

higher complexity and power dissipation, and less performance. Dynamic latches are much more efficient in terms of performance and power dissipation. Moreover, dynamic latches require less silicon area compared to the static counterpart. However, in larger sequential architectures, many times we need to store the data on a temporary basis. Hence, dynamic logic circuits are also advantageous over the static logic counterparts. In dynamic circuits, the stored charge across the output capacitors represents the logic swing. Presence of significant amount of charge can be treated as logic "1", whereas the absence of charges can be treated as logic "0". As there is always some provision of the charge leakage through the capacitors, the stored charge can be kept significantly for a certain amount of time, in the order of milliseconds. Moreover, in dynamic circuits, periodic refreshing (charging or discharging) of the output node capacitance is required to obtain the true logic for long time.

Figures 5.36 and 5.37 show two alternative implementation of D flip-flop, whereas Figure 5.38 shows four types of implementation of D flip-flop for low power applications. Hence, a positive edge-triggered dynamic *D* flip-flop architecture is also detailed for the ease of understanding. Master consists of gate T_1 and the next inverter, whereas

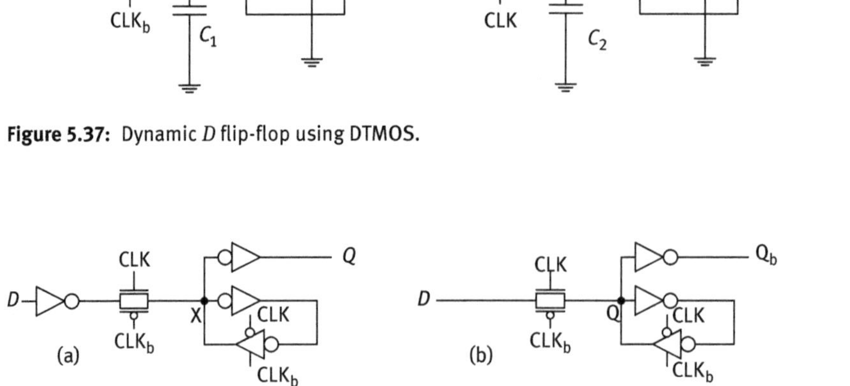

Figure 5.37: Dynamic *D* flip-flop using DTMOS.

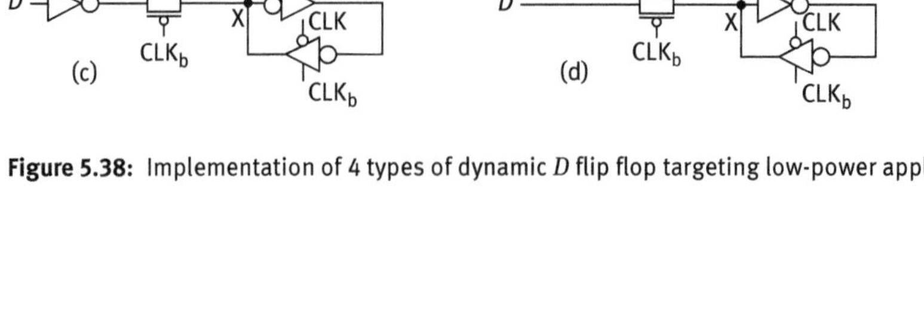

Figure 5.38: Implementation of 4 types of dynamic *D* flip flop targeting low-power applications.

the slave consists of gate T_2 and the next inverter. When master is ON, the slave will be OFF and vice versa. The operation of the master–slave *D* flip-flop can be explained as follows: When CLK = 0, T_1 will be ON and the C_1 node capacitance will be charged or discharged by following the voltage swing of input signal *D*. So the output can be obtained at node 1 and the complemented *Q* can be obtained from the output of the next inverter stage. As CLK = 0, so master will be ON and the slave will be OFF, and the output will not be able to propagate through the slave stage. Now when CLK = 250 mV, master becomes inactive and the slave turns ON. Hence T_1 turns OFF and T_2 turns ON to pass the output of node 1 to node 2. Inverted version of node 1 will be stored in node 2. So the final output will be obtained from the output of the next inverter stage. This architecture needs only eight transistors that significantly reduce the power dissipations and increase the performances (Figure 5.38).

5.11 Adiabatic Flip-flop

The structures of the adiabatic flip-flops are given in Figure 5.39. *D*, *T* and *JK* flip-flops are used to show the workability of the adiabatic logic styles for ultra-low power application. Also the pre-settable flip-flops are shown in Figure 5.40, which are very much efficacious for ultra-low power application.

Figure 5.41 depicts the output waveforms of adiabatic *D* flip-flop. Transistor-level representations of the above-mentioned ECRL gates have been discussed in earlier sections. These flip-flops can be used further for the implementation of the larger sequential blocks for low power applications.

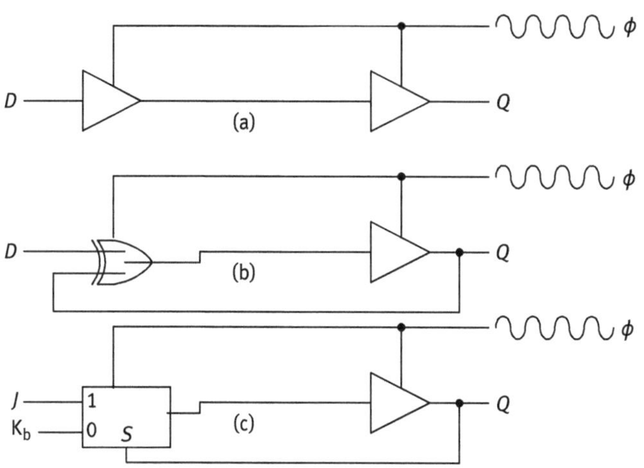

Figure 5.39: Adiabatic (a) *D* flip-flop; (b) *T* flip-flop; and (c) *JK* flip-flop.

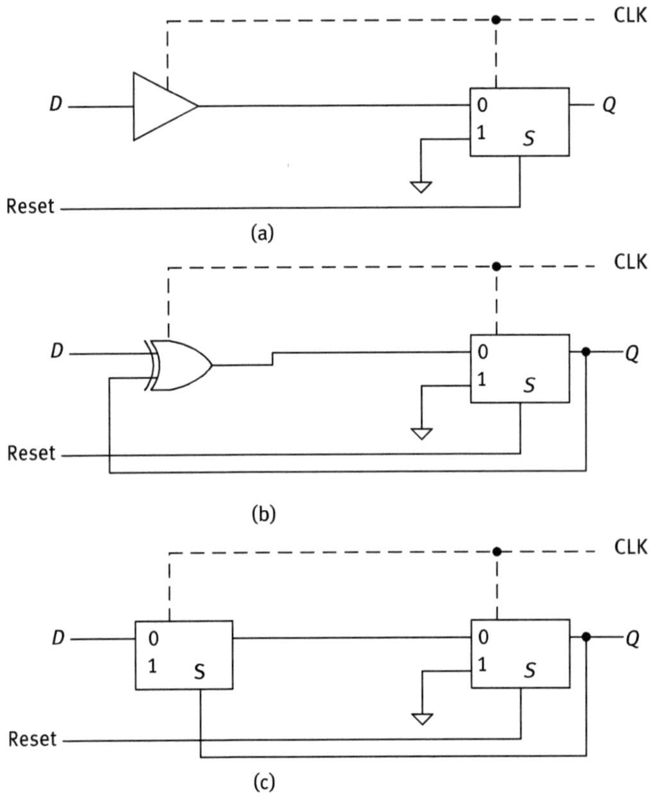

Figure 5.40: Pre-settable adiabatic flip-flops: (a) D flip-flop; (b) T flip-flop; and (c) JK flip-flop.

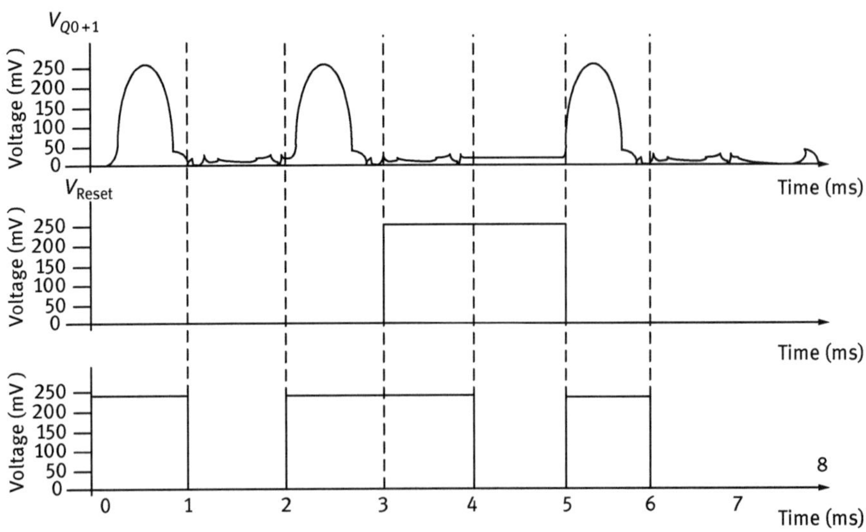

Figure 5.41: Output waveform of the D flip-flop.

References

[1] J.M. Rabaey, *Digital Integrated Circuits*. Prentice-Hall, Upper Saddle River, NJ, 1996.
[2] N. Weste and K. Eshraghian, *CMOS VLSI Design*, 2nd edn., Addison-Wesley, Reading, MA, 1994.
[3] M. Mano, *Computer System Architecture*, Prentice Hall, Upper Saddle River, New Jersey, USA, 1982.
[4] N. Goncalves and H. De Man, "NORA: A racefree dynamic CMOS technique for pipelined logic structures," IEEE JSSC, SC-18.3 (1983): pp. 261–266.
[5] John P. Hayes, *Computer Architecture and Organization*. McGraw-Hill, Inc., New York, USA, 1988.
[6] Randall L. Geiger, Phillip E. Allen and Noel R. Strader. *VLSI Design Techniques for Analog and Digital Circuits*. McGraw-Hill, Inc., New York, USA, 1990.
[7] C. Mead and L. Conway, *Introduction to VLSI Systems*. Addison Wesley, Boston, USA, 1980.
[8] A Mukherjee, *Introduction to nMOS and CMOS VLSI Design*. Prentice Hall, NJ, 1996.
[9] John P. Uyemura. *Circuit Design for CMOS VLSI*. Kluwer Academic Publishers, Berlin, Germany, 1992.
[10] S. Muroga, *VLSI System Design*. Wiley, New York, 1983.
[11] Sung-Mo Kang and Yusuf Leblebici, *CMOS Digital Integrated circuits*. McGraw Hill, New York, 1996.
[12] L. J. Herbst, (1996). Integrated circuit engineering: establishing a foundation (No. 4). Oxford University Press, USA.
[13] K. Martin, *Digital Integrated Circuit Design*. Oxford University Press, New York, 2000.
[14] M. Elmasry (ed.), *Digital MOS Integrated Circuits*. IEEE Press, New York, USA, 1991.
[15] Kiat-seng Yeo, Samir S. Rofail and Wang-Ling Goh, *CMOS/BiCMOS ULSI*, Pearson Education, Upper Saddle River, New Jersey, USA, 2002.
[16] Dougals A. Pucknell and Kamran Eshraghian, *Basic VLSI Design*. Prentice Hall, Upper Saddle River, New Jersey, USA, 1984.

6 Introduction to Memory Design

Introduction

In the last few decades, miniaturization and their corresponding reliability issues have been emerged as a major research topic. In this frame, low cost memory chips with low power consumption are of paramount interest and are indispensible components of modern life. The memory remains the driving force in the thrust for higher transistor counts per chip and faster operation. It spearheads development and sets the pace for VLSI in general. The aim of this chapter is to provide a broad overview of the principles and technology of memory systems.

6.1 Types of Semiconductor Memory

Based on the type of data storage and data access, semiconductor memories can broadly be classified into two categories, RAMs and ROMs, as shown in Figure 6.1.

As the name implies, the RAM allows entry and extraction at random. In a RAM, the accession of any cell is random in nature with almost equal accession time, in contrast to sequential access memories (magnetic tapes, cassettes, etc.) where memory accession is purely on sequential basis; i.e. data is randomly accessible, which mean that if any address is given, read and write operation can be performed using the following memory functions:

```
ReadData = Memory[ Address ];
Memory[ Address ] = WriteData;
```

This implies, data is written into a desired location, overwriting the information it contains. In a RAM, data extraction (readout) procedure needs to be nondestructive; i.e. the data at the location being read is retained. A RAM is volatile; it means that all its information is lost when the power is removed or fails, unless battery backup is used to maintain power. In contrast, the ROM is a fixed content store and information is extracted nondestructively. Unlike RAM it is nonvolatile.

In general, RAMs can be clustered broadly into two groups such as dynamic RAMs (DRAM) and static RAMs (SRAM). Until the evolution and adaptation of dynamic circuit techniques the RAMs were static.

Storage capacity increased dramatically with the adaptation of dynamic logic design techniques in the late 1960s. All dynamic logic relies on temporary charge storage across a nodal capacitor. Static devices need more power but can hold the information for an indefinite period of time (as long as power is supplied). The dynamic devices use low power, but need periodic refreshment of the stored charge, to compensate the charges lost due to leakage. The charge tends to leak away with time

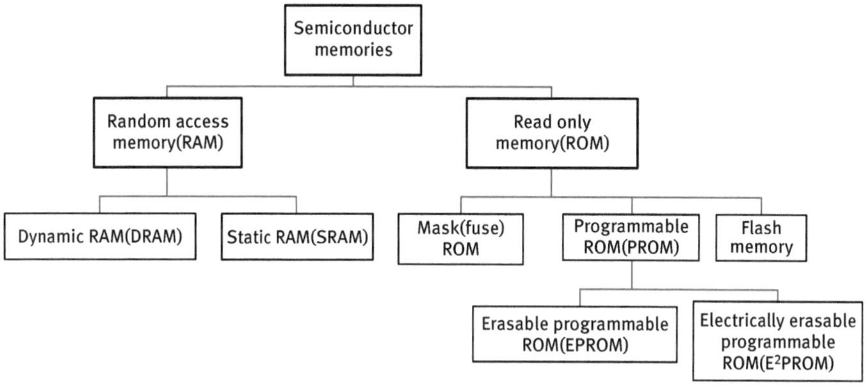

Figure 6.1: Types of semiconductor memory.

even when the transistor driving the node is off (subthreshold conduction). In DRAM technology, the leakage demands a refresh operation at fixed intervals of time (few microseconds with current technology). Because of the one-transistor cell, DRAMs exceed the storage capacity of SRAMs for identical die size; their cost per bit (of storage) is consequently much less. DRAMs maintain the highest RAM storage, but the SRAM is unquestionably faster. As a result, SRAMs must not however be discounted. Their (SRAMs) superior speed combined with the absence of overhead circuits and refresh management ensures their continued existence [1].

The term *ROM* on its own stands for a mask (mask-programmed) ROM. Mask ROMs are suited to equipment manufactured on a large scale and being operated with fixed programs already burnt (programmed) in computers. Equipment with smaller volumes of production and operated with software undergoing development and changes are satisfied far more economically with user-programmable (field-adjustable) PROMs (programmable ROMs). Photolithographic techniques with photo masks are used to write the data at some stage of the fabrication process in a masked ROM and data is written electrically after the chip is fabricated in PROM. The various categories of such ROMs are shown in Table 6.1. User-programmable PROMs are more popular than mask-programmable ROMs, because of their high consumption and requirement by large number of customers. The various categories of PROMs are also shown in Table 6.1. It is not possible to modify or rewrite the data once that is programmed in a ROM having fuse technology. In contrast, it is possible to rewrite the data electrically in an EPROM or E^2PROM. However, it is worth mentioning that the number of the rewrite to be performed is constrained to 10^4–10^5. In EPROMs, ultraviolet rays that can penetrate through the crystal glass on the package are used to erase whole data in the chip simultaneously, while high electrical voltage is used to erase data in 8-bit units in E^2PROMs. A flash memory uses a very high electrical voltage to wipe out the data present inside a block.

Table 6.1: Categories of ROM.

Type	Technology	Operation
EPROM	CMOS	Electrically programmable. UV light erasable.
OTP EPROM (one-time programmable ROM)	CMOS	EPROM without erase facility
PROM	Bipolar	Electrically programmable. Cannot be erased.
E^2PROM (E^2PROM)	CMOS	Electrically programmable and erasable. Usually erasable byte by byte
Flash EPROM	CMOS	E^2PROM that cannot be erased byte by byte, but only by erasing the entire chip or large sections thereof.

6.2 Memory Organization

Figure 6.2 depicts a generalized block diagram of a typical memory chip organization. Semiconductor memories are universally organized as matrix or 2-D arrays of 1-bit storage cells. These arrays are delimited by address decoding logic and interface circuitry connected to external signals. Each memory cell is connected with the other cells present in the same row using a common connection (WordLine). Similarly they are connected with the other cells in the same column using another common connection (BitLine). In this structure, there are 2^N rows, same as the number of word lines, and 2^M columns, same as the number of bit lines. Thus, the array contains $2^N \times 2^M$ cells of memory.

The purpose of a row decoder is to select a particular row by enabling the row address provided to all cells along the selected row. As a result, the contents of these cells become available along the column lines. The column address is provided to select the particular column containing the desired data bit. Therefore, in order to access (read/write) a specific individual cell for gaining access to the data bit stored in the memory array, the word line and the bit line correspond to the memory array needs to make active as per the addresses coming from the processor. There are some memory organizations where n bits can be accessed simultaneously. For these memories, the data from n columns are selected and routed to n data output pins simultaneously. Normally, signal levels are different inside the memory chip (CMOS) and outside the memory chip in board Transistor-Transistor Logic (TTL). The purpose of the input and output address buffer is to make the compatible [2].

The row and column decoders complete the tasks of row and column address selection [3]. Once a particular row is selected by the row decoder, a data read or write can be performed by selecting a single or multiple bits on particular row [4]. The charge on one of the bit lines from each pair of bit lines on each column will discharge through the enabled memory cell, representing the state of the active cells on that column. The column decoder will enable columns and will connect the bit lines

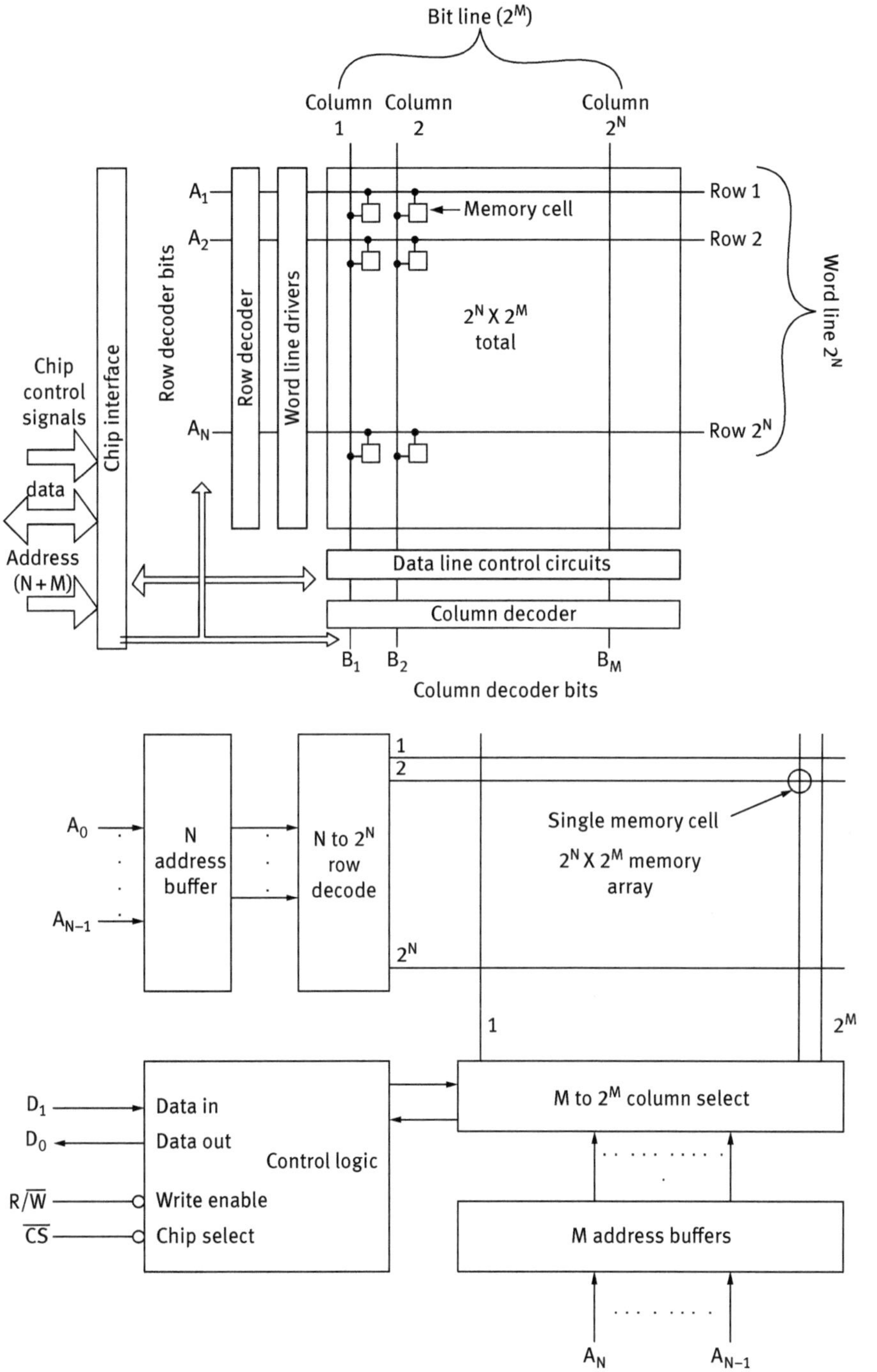

Figure 6.2: Generalized memory organization.

to the I/O lines, which are passed to the output buffer. Note that the memory cells are small and can only sink current, not source it. The sense amplifier has driving capability and is now enabled. As we explained earlier, the unbalanced bit lines will affect the balanced sense amplifier to make a transition just before the state of the bit lines when it is enabled. Output buffers are required to force relatively large current load on the board [5].

A write operation is performed by setting $\overline{CE} = 0$ and $R/\overline{W} = 0$. The input data will be amplified by the superbuffers and applied to the I/O bit lines which will cause one-bit line high and one-bit line low. The column decoder will select a particular column connecting the bit lines to the I/O lines. The row decoder will select a particular row followed by the information present in the hit lines to be written to the cell at the intersection of the row and column.

The internal timing starts with the chip-enabled ($\overline{CE}$) signal going low. The precharge (in case of DRAM) is turned off immediately after this and is not turned on again until the entire operation is done. The column and row decoders are then activated, followed by the activation of the sense amplifier. The sense amplifier is deactivated along with $\overline{CE}$ becoming low after the completion of the read/write process. This is followed by disabling the decoder and then the precharge line comes on again [6].

6.3 Introduction to DRAM

In a computer system, the main memory typically consists of a high quantity of DRAM. In VLSI circuit design, the quest for low power device design and implementation are always in demand. The VLSI circuit efficiency utilizing either memory or processors can achieve a significant improvement using a combination of device scaling, along with novel device structures and the usage of new material properties. For the last few decades, memory chips have been demonstrated more than 1 million hikes in its storage capacity [7]. One-transistor (1-T) dynamic RAM (DRAM) has initiated this tremendous increase of capacity [8] to find its use in almost every computing machines and other electronic gadgets and equipment [9]. One single capacitor and a single transistor are required to construct A 1-T DRAM for storing one bit of information, as compared to six transistors (6-T) to store one bit using SRAM technology. Thus, DRAM technology provides high integration density and very high storage capacity [10, 11]. The term *dynamic* is used to denote the leakage phenomenon inherent to DRAM that degrades the charge stored in the memory cell with time and requires periodic refreshment necessary in order to retain the data stored in it, which makes DRAM less power efficient [12]. However, such power consumption can be reduced and more speed can be achieved by reducing the leakage through the MOSFET [13, 14]. As scaling of conventional MOSFET beyond 100 nm is facing severe roadblocks due to the various short channel effects (SCEs), various novel MOSFET device structures have been considered experimentally and theoretically in recent times. Among the several alternative technologies attempted for mitigating SCEs, silicon-on-insulator (SOI) technology presented itself as a major competitor for the next-generation emerging CMOS technology due to its several inherent advantages [15] like higher speed, large acceptable tolerance from radiation effects, reduced parasitic capacitance, reduced SCEs, better current deliverability, less leakage current and manufacturing compatibility with the existing technology [16].

6.4 One-transistor DRAM Cell

The memory cell uses only one transistor, and a storage capacitor to store one bit is shown in Figure 6.3. The basic requirement to construct a single unit of dynamic random-access memory (DRAM) cell is an access transistor M_A and a storage capacitor C_S. The word line signal and the bit line present in the input/output route are required to govern the access to the capacitor via the access transistor M_A. The simplicity of the circuit makes it very attractive for high density storage. For proper operation of a DRAM, the first step is to validate the row address and the second step is to validate the column address.

Step 1: To validate the row address: $\overline{RAS}$ (row address access clock) is responsible for the internal validation of the row addresses arriving from the external address pads. The row access requires a large time as rows exhibit a large time constant (due to the connection of the several gates of memory cells).

Step 2: To validate the column address: In a similar manner, addresses present in address pads are internally validated in a sense amplifier by the $\overline{CAS}$ (column address access clock). In contrast to row access, column accession is faster to move the data sensed by the sense amplifier to the peripheral output pin through the column decoder and output buffer. The typical DRAM access timing (Step 1 and Step 2) is shown in Figure 6.4.

As with any RAM, there are three feasible functions:

- Write – In order to store a data bit;
- Hold – In order to preserve the value of the data undamaged in the cell;
- Read – In order to pass on the value of the data to a peripheral circuit.

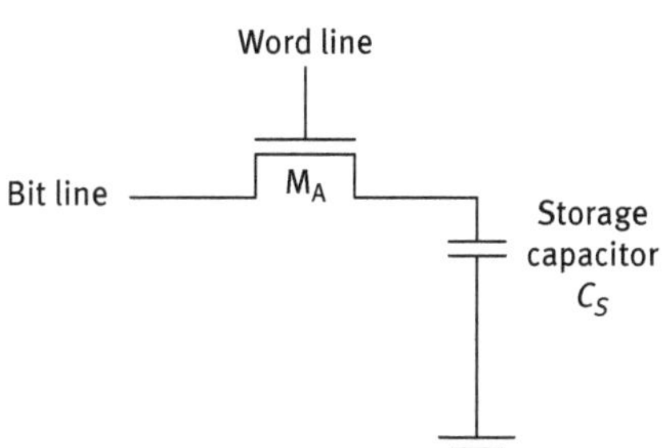

Figure 6.3: Basic DRAM cell.

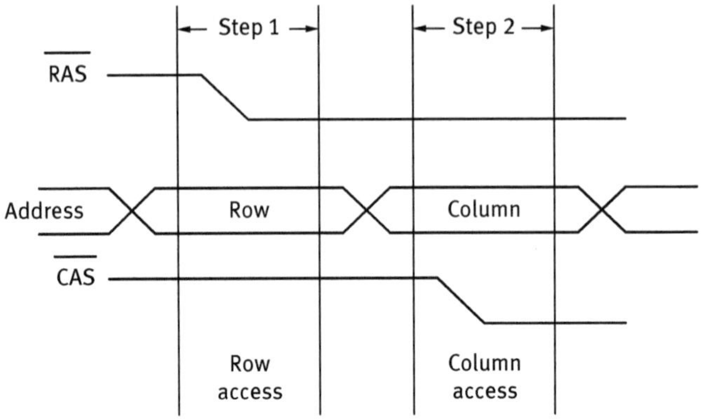

Figure 6.4: DRAM access timings.

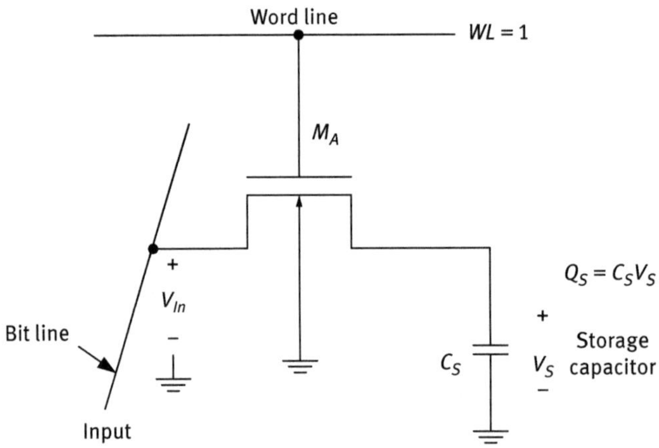

Figure 6.5: Write operation in a DRAM cell.

6.4.1 Write

Figure 6.5 exemplifies the write process. To turn on access transistor M_A the word line is activated high so that WL = "1". Now to write a logic "1" also the bit line is charged to the desired level (logic "1") so that capacitor C_S charges via M_A. An input voltage V_{In} = "0" causes the discharge of the capacitor, thus ensuring the storage of logic "0".

6.4.2 Hold

Figure 6.6 exemplifies the hold operation to protect the charges stored in a DRAM capacitor. In order to achieve this, the access transistor is turned off by providing WL = 0. As a result, the capacitor storage node becomes ideally isolated from other part of the

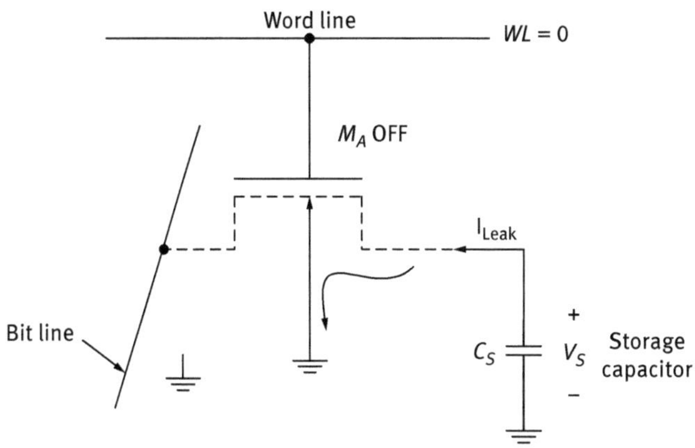

Figure 6.6: Hold operation in a DRAM cell.

DRAM circuit. However, it is important to notice that, in practice, charges tend to leak away from the capacitor storing logic "1". Many experimental and theoretical works have been devoted to increase the charge retention time. In order to store the data for a longer duration, refreshing circuitry responsible for reading the data, then amplifying it and finally rewriting the data is provided [17].

6.4.3 Read

To read the stored data in a DRAM cell, the first step is to activate the WL. Then the charge in C_S is put in the bit line and is detected by a specially designed sense amplifier. The reading is destructive. During a read operation, the bit lines containing the data are coupled with the input of a high-gain sense amplifier that is designed to provide amplification of the voltage level. In a typical DRAM architecture shown in Figure 6.7, the data bit line is coupled with the gates of MOSFET; as a result the bit lines exhibit capacitive characteristics. Figure 6.8 exhibits this by the inclusion of a line capacitance C_{Line}. In the course of a read operation, a small voltage change in the output line appears due to the sharing of charges among the DRAM storage cell and the output line capacitance. A comparator present within the sense amplifier is responsible to identify the change in voltage. It is worth mentioning that the same bit line is exploited for reading and writing steps together, with the reverse current direction. To read, the bit line is first precharged to $V_{DD}/2$, when the word line rises and the capacitor shares its charges with bit line, causing a voltage change ΔV that can be sensed by a sense amplifier. It indicates that for a faithful operation, the DRAM capacitor must have a large capacitance value. However, at the same time they must be physically very small to achieve large integration density. Typically, the bit line is connected to a large number of DRAM cells and thus got a relatively large capacitance (C_{Line}) value. Therefore

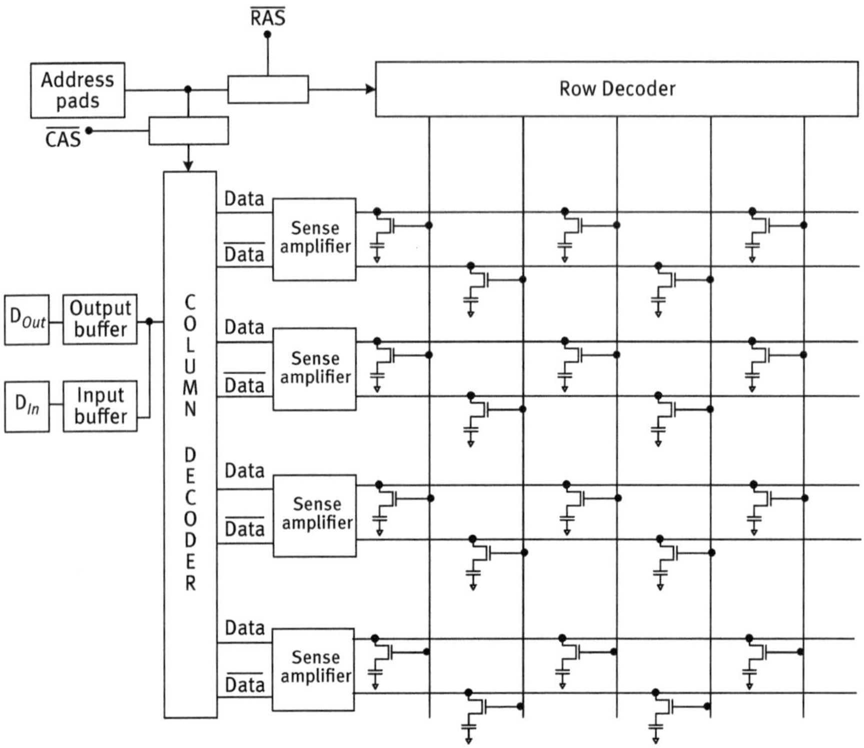

Figure 6.7: Simplified DRAM memory chip block diagram.

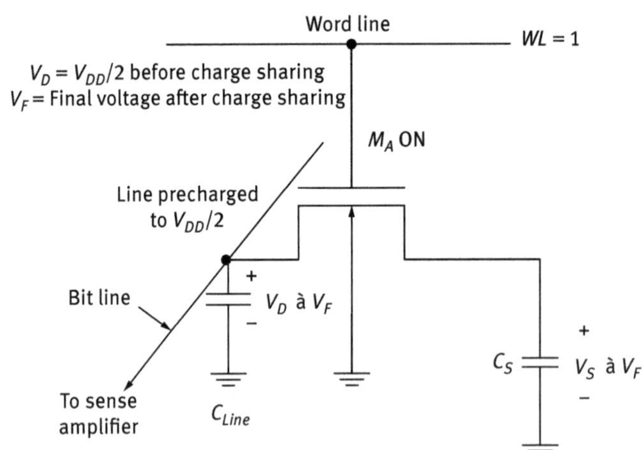

Figure 6.8: Read operation cycle in a DRAM cell.

the cell capacitance is typically smaller than bit line capacitance. According to charge sharing [18, 19] the voltage swing during readout is

$$\Delta V = \frac{V_{DD}}{2} \frac{C_S}{C_S + C_{Line}}$$

Notice that a large C_S is an important requirement to provide a reasonable voltage swing and also to retain the content of the cell for an arbitrary long time, i.e. to increase their charge retention time.

Still DRAM manufacturers have made a major effort to develop the technology to achieve a large capacitance cell with minimized silicon area. Different design solutions are proposed to reduce the cell area in size, principally the dimension of the DRAM capacitor responsible for storing charges. The most important technologies to shrink the capacitor area without altering its value are (1) to utilize new capacitor shapes to fit into a minimum surface area but having larger effective capacitor area and (2) to increase the dielectric constant of the trench oxide [20] using high-*k* dielectric. The memory cell capacitance has been improved by using advanced cell structures and a high permittivity (ε) dielectric material such as Ta_2O_5.

6.5 Capacitor in DRAM

The two established structures for C_S are the trench and the stack capacitors. The former is an adaptation of trench oxide isolation and is illustrated in its basic form in Figure 6.9.

The polysilicon fill-in is the storage node; the part of the substrate surrounding the trench oxide is the capacitor plate. There are various structures for such capacitors. The large sidewall area determines the capacitance and makes little demand on the

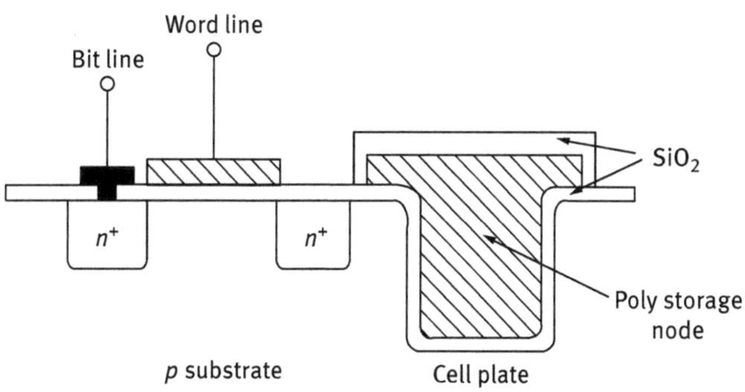

Figure 6.9: A vertical trench capacitor DRAM cell.

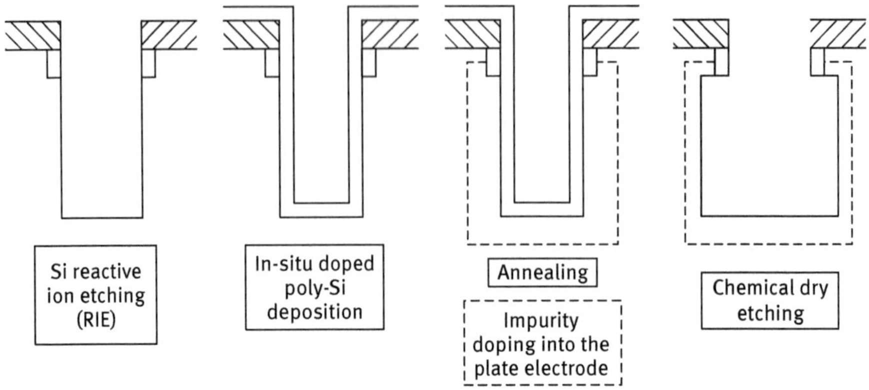

Figure 6.10: Evolution and varieties of trench capacitors highlighting their respective fabrication process.

silicon estate. The major trench capacitor designs and their evolutions are shown in Figure 6.10.

The fabrication process steps for the 1-T DRAM 32 nm MOS transistor device are reported by Wang et al. [21] and are specified as the following:

- Shallow trench isolation (STI) formation on silicon substrate (etching and thermal oxidation with initial deposition in two dimensions)
- Cylindrical trench formation followed by an oxide liner and a polysilicon fill
- Implantations for threshold-voltage adjustment
- Gate oxide and polysilicon gate formation
- Halo implantation followed by low doped drain (LDD) implantation
- Sidewall spacer formation
- Highly doped drain (HDD) implantation
- Rapid thermal annealing (RTA) at 1000°C
- Contact formation

On the other hand, the stack capacitor, whose structure is shown in Figure 6.11, benefits from the well-established technology of interconnects above the gate oxide. The plates are separated by thin oxide and can rest on the various interconnect layers. Trench and stack capacitors achieve similar capacitances for the same silicon area. The trench capacitance is readily increased by deepening the depth of the trench. For fabrication of this kind of structure, separate process steps are required, resulting in increased cost. The limitations of the trench capacitor are leakage current between trenches. These interfere with the stored charges residing in the substrate. This phenomenon becomes more serious for deep submicron device structures.

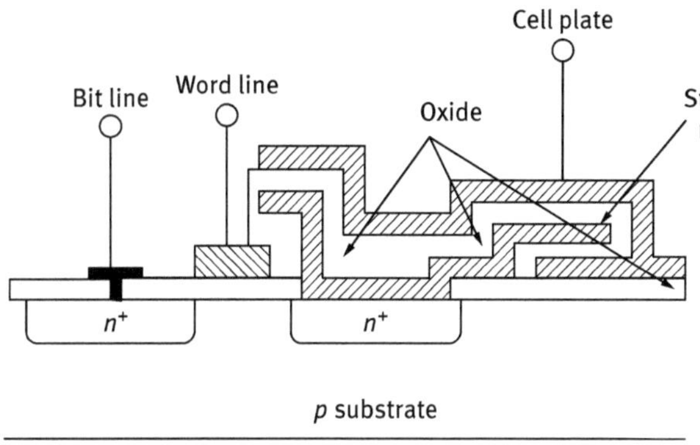

Figure 6.11: Basic stack capacitor DRAM cell.

6.6 Refresh Operation of DRAM

Memory units must be able to hold data so long as the power is applied to maintain the data integrity. To overcome the charge leakage problem, DRAM employs a refresh operation in every cycle where the data is periodically read from every cell, amplified and rewritten as shown in Figure 6.12. Refresh circuitry is included in the overhead logic that surrounds the cell array. The refresh cycle is designed to operate in the background and is transparent to the user. As mentioned earlier, the sense amplifier has dual roles: (1) it is responsible to send out the data to the output buffer for chosen memory cells in read operation and (2) it is also responsible to rewrite the data known for refreshing the memory cell. The burst or distributed refresh methods can be employed to accomplish this task as shown in Figure 6.13. A series of refresh clock cycles are passed in a burst technique until the completion of the accession of all rows, unlike a distributed refresh where periodic cycles are applied.

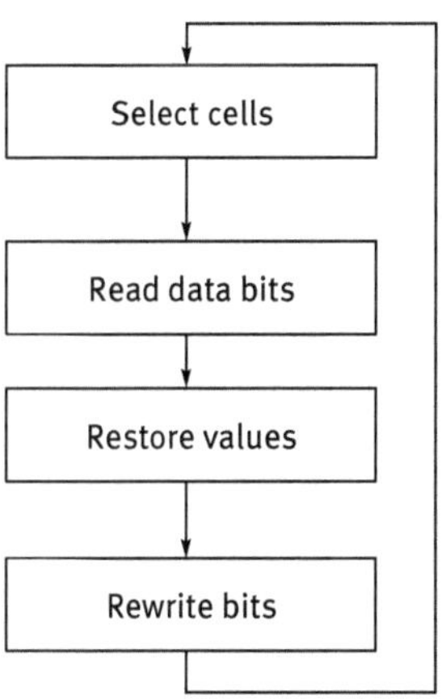

Figure 6.12: Refresh operation cycle.

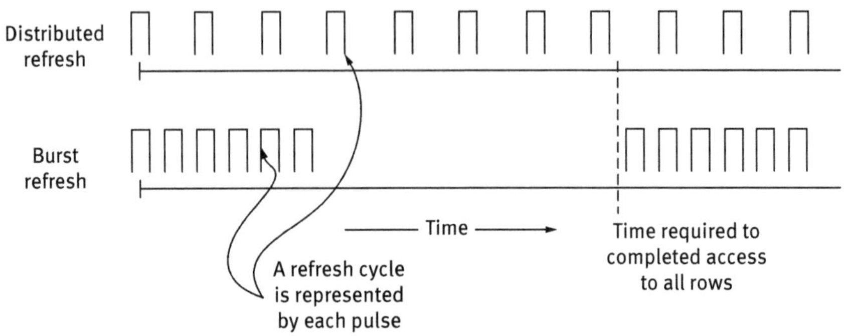

Figure 6.13: Distributed and burst refresh of DRAM.

6.7 DRAM Types

DRAMs have branched into many categories in order to meet the power and speed discrepancies. Figure 6.14 shows the current DRAM varieties.

6.7.1 FPM DRAMs

The addresses of a DRAM are multiplexed. A row address is supplied and then a column address is supplied in order to access a DRAM data. If the data demand is present in the same row address as the earlier applied row address, then simply changing the column address will permit the access to the recent data. In summary, with first page mode (FPM) DRAMs, the data stored on the same row can be accessed by changing the column address in order to achieve a faster access time. All standard DRAMs use this mode.

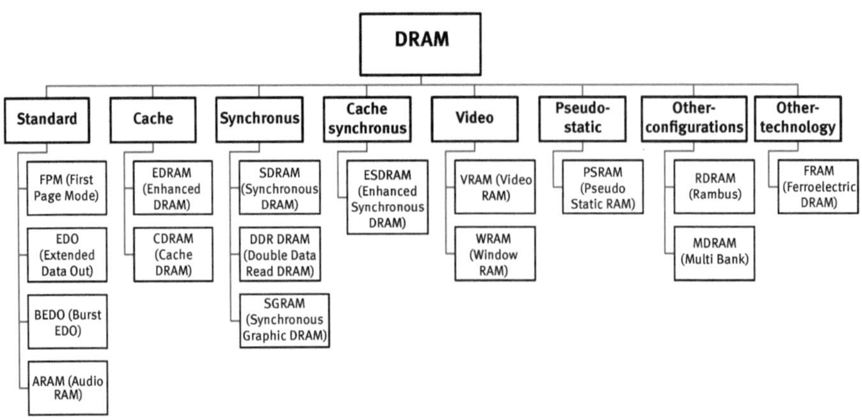

Figure 6.14: Current DRAM varieties.

6.7.2 Extended Data Out DRAMs

The design of the output buffer of the FPM DRAMs is changed in extended data out (EDO) technique to further reduce the access time by extending the valid data output time. In this technique, the accession of the new data and latching of old data are taking place simultaneously.

6.7.3 Burst EDO DRAMs

In this technique, a burst of n numbers of sequenced read and write cycles are applied to the memory chip. A design change in the output latch is provided to accomplish the "burst". A counter having m bit, where $m = \log_2(n)$ counting capability with a pipeline stage, is provided. It offers higher bandwidth as higher latency is obtained from the special pipeline stages. Moreover, burst EDO DRAM has the provision for an internal address counter. It requires the supply of only initial address in a burst of n number to complete the process.

6.7.4 ARAM

Due to fabrication process variability, some DRAMs are manufactured with some failed storage cells and are not passing through the testing phase. They are not good enough to be used as PC DRAMs. However, they are not thrown away but are used in audio applications (such as telephone, etc.) with a cheaper price where tolerance for error margin is less, known as Audio DRAM (ARAM).

6.7.5 Cache DRAM

In this technique, a DRAM and a six-transistor SRAM cache memory are integrated together in the same chip. A buffer is used to complete the data transmission between the SRAM and the DRAM. Usually it takes one complete clock cycle using the buffer in order to finish the operation. Figure 6.15 shows the chip organization of a cache DRAM.

6.7.6 Enhanced DRAM (EDRAM)

Enhanced DRAM (EDRAM) is a variant of cache DRAM, where read cycles take place from the cache all the time. Every time the comparator senses a "hit", it provides the data within a short time. In contrast, on every occasion when a "miss" is sensed, the whole cache is updated and the time taken to make the data available at the output is

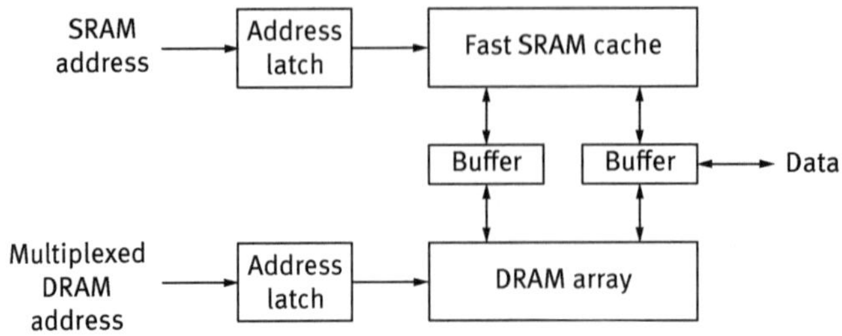

Figure 6.15: Cache DRAM organization.

higher. The major architectural difference between the cache DRAM and EDRAM is the presence of a fast page mode DRAM instead of a separate SRAM. The sense amplifiers located within the FPM DRAM act as a SRAM for reading and accessing data.

6.7.7 Synchronous DRAM

In synchronous DRAM (SDRAM) technology, the term *synchronous* refers to the synchronization of read and write cycles harmonized with processor clock. The provision of two separate banks offers each bank to have active rows simultaneously, leading to a concurrent access/refresh and recharge operation. These DRAMs offer programmability through the use of a programmable mode register to include different burst lengths, wrap sequence and CAS latency. In order to change the programmable features, the mode registers need to be modified.

6.7.8 Double Data Read DRAMs

Double data read DRAMs (DDR DRAMs) offer twice effective bandwidth at a given clock frequency by delivering the data for both the leading and falling edges of the processor clock supplied.

6.7.9 Synchronous Graphic RAM

Synchronous graphic RAM (SGRAM) provides graphic-specific precise features, for example, block write mode and masked write mode to target video applications.

6.7.10 Enhanced Synchronous DRAMs

Enhanced synchronous DRAM (ESDRAM) is a combination of SDRAM and cache DRAM fabricated on the same chip.

6.7.11 Video DRAMs

In video DRAM (VRAM) technology, to target video applications, which are inherently serial, in contrast to inherent parallel nature of DRAMs, a parallel-to-serial shift register is provided. These DRAMs are also known as dual-port DRAMs because of the provision for separate parallel and serial interface. VRAMs can be used as a DRAM or as a serial access memory (SAM). The typical organization of VRAM is shown in Figure 6.16. The parallel-to-serial shift register is separated into two halves. One-half is busy with reading the SAM, while the other half is responsible to load the data from the memory array. Nevertheless, for a nonreal-time operations such as loading during Cathode Ray Tube (CRT) retrace periods, VRAM is operated with a full register operational mode.

6.7.12 Window DRAMs

Window DRAM (WRAM) is a dual-ported VRAM with additional features like EDO and fast page mode access.

6.7.13 Pseudo-static RAMs

In pseudo-static RAM (PSRAM) technology, storage mechanism of DRAM technology with additional circuitry makes it behave like a SRAM. However, it is worth mentioning that the market of PSRAM has never grown strongly in spite of its presence for more than two decades.

6.7.14 Rambus DRAMs

The DRAMs designed by Rambus Inc. provides very high speed chip-to-chip interface due to a new DRAM architecture. In this architecture, an on-chip controller/processor is incorporated with a Rambus channel to send out information at a towering rate.

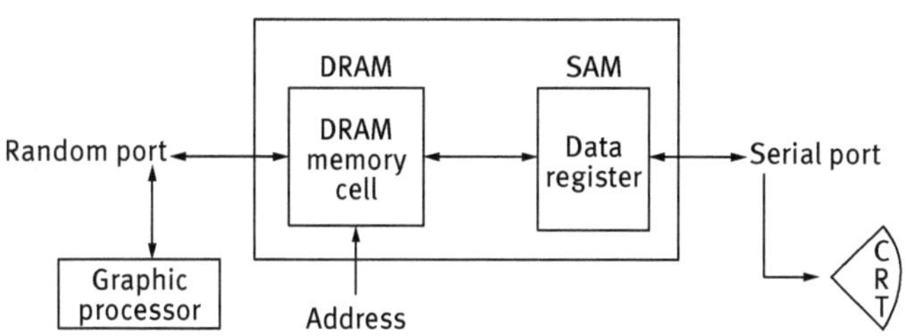

Figure 6.16: A typical organization of a VRAM.

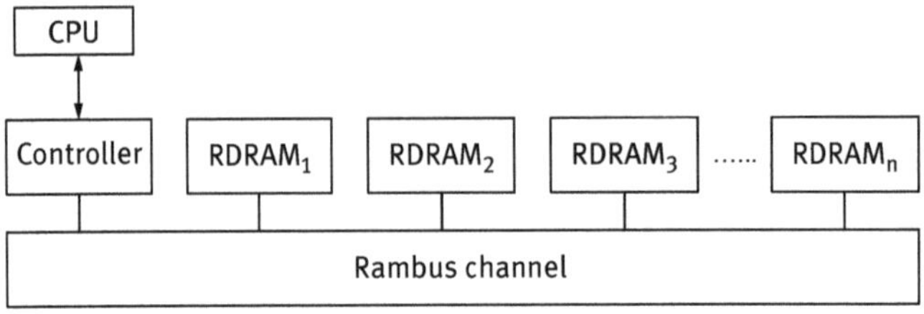

Figure 6.17: Typical Rambus configuration.

Figure 6.17 shows the typical Rambus configuration. Typical DRAM process and manufacturing technology are followed in Rambus DRAMs (RDRAMs). They can be divided into two parts: (a) DRAM core and (b) interface logic. The controller is designed to act as the interface between the Rambus channel and the RDRAMs.

6.7.15 Multibank DRAM

In multibank DRAM (MDRAM) technology, the idea is to use multiple bank of DRAM to reduce latency between the bursts. The next-generation DRAM is all multibank DRAMs. In contrast to a conventional DRAM, in a MDRAM, full access is given to its entire bank of memories, where each bank operates independently to offer a higher degree of memory granularity compared to other DRAM architecture alternatives.

6.7.16 Ferroelectric DRAM

What makes ferroelectric DRAM (FRAM) different from previous DRAMs is the presence of ferroelectric material in the capacitor. A ceramic film of PZT (made from lead zirconium titanite) is used to construct the film of the capacitor present in the DRAM used to tender nonvolatility in order to retain data, when power is switched off. The electrical schematic of a ferroelectric cell is shown in Figure 6.18.

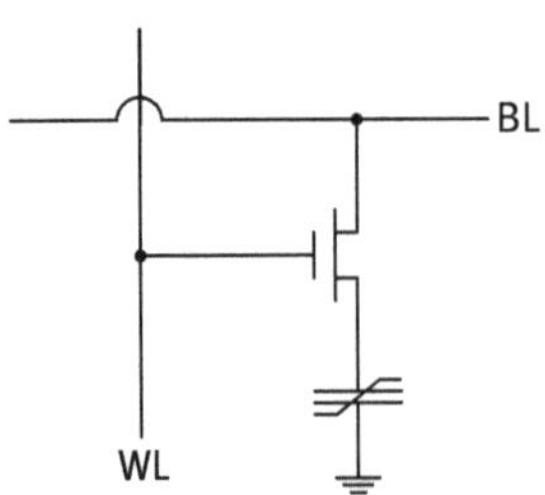

Figure 6.18: Ferroelectric cell.

6.8 SOI DRAM

One of the key challenges to achieve gigabit memories is the integration of very high density storage capacitors characterized with a shorter access time. In order to achieve this, several capacitor-less DRAMs have been proposed [22–25] in contrast to the conventional DRAMs where a vertical trench capacitor with a transistor must be involved. The floating body in SOI can be used as a storage capacitor [26] where the DRAM consists of a single transistor and three signal lines as shown in Figure 6.19. In this technique, the floating body is considered as the node in order to store the charge. Usually, the charges present in the floating body are considered as troublemaker to designers as it leads to shifting of thresholds introducing kink effect in the device characteristics and hysteresis in the circuit operation. However, they can be considered as an advantage since it can rule out the need of a separate storage capacitor, leading to a very high density memory.

6.8.1 Operating Principle

6.8.1.1 Write

The operation required for write cycle is shown in Figure 6.20. By creating positive charges using impact ionization method write "1" operation is accomplished after raising both word line (WL) and bit line (BL) to logic "1". The positive charges created from impact ionization method are trapped within the body when the WL is low. In contrast in order to write "0", a high voltage at gate and a low voltage at drain are applied to remove holes from the body owing to the forward-biased body-to-drain *p*-*n* junction.

6.8.1.2 Read

The read operation is started by activating the WL first. A small drain voltage is then applied to keep the stored charges intact so that the transistor operates in linear regime. As a result, the drain current appears through precharged BL on the sense amplifiers. Now, sense amplifiers are activated to sense the current in comparison to a reference current, and decision is made to output data.

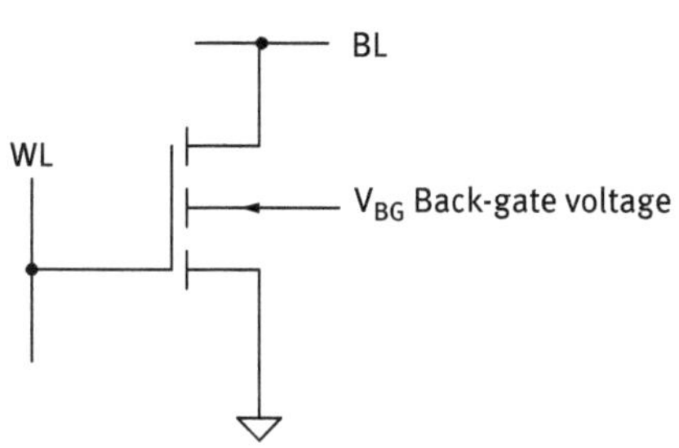

Figure 6.19: A single cell of the fully depleted (FD) SOI floating body 1-T DRAM with floating body cell.

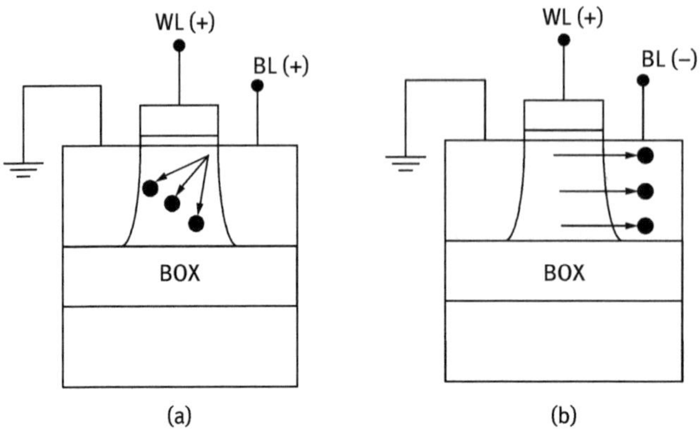

Figure 6.20: Write operation for (a) "1" (b) "0".

6.8.2 Design Considerations of SOI DRAM

The charges stored in the SOI transistor's body appear in the device characteristics due to kink effects. This floating body effect is much more pronounced in partially depleted (PD) SOI MOSFETs as compared to FD SOI MOSFETs. As the performance of 1-T SOI DRAM cell depends on the ability to store charges in its body, most DRAM cells use PD SOI technology. However, Refs [22, 23] use FD SOI technology to realize floating body DRAM cells use a negative voltage to the rear side (back gate) of the MOSFET to accumulate charges on the back side of the wafer by creating a potential well to store the holes injected into the body created by impact ionization method.

6.9 Introduction to SRAM

SRAM is used to store the data in a static fashion. The data remains stored as long as power is available. SRAM finds its usage as embedded memory such as cache memories in the microprocessors or data buffers in various Digital Signal Processors (DSP) chips.

The search for high speed, high density SRAMs is in demand in order to achieve enhanced system performance of computing systems. SRAMs require more transistors (usually 6) as compared to DRAMs. Hence, it is more expensive and requires more space to fabricate. The significant improvement in SRAM can be accomplished owing to scaled device dimension aspects, by optimizing the cell layout [27] resulting in fast, compact scalable SRAM.

Recently, SOI CMOS technology for SRAM and DRAM technologies has been used for achieving substantial on-chip RAM in computing machines. It is consistent with the trend of superseding of SOI technology over other technologies for emerging as chief contender of future-generation CMOS technology. However, design issue in SOI MOSFETs – floating body effects may upset the charge stored in a memory cell – needs to addressed properly.

6.10 SRAM Cell and Its Operation

Conventional 6-transistor (6-T) SRAM cells are used in various electronic applications in the industry as well as personal computing machines to store data. Since it is a static memory, it does not require periodic refreshment of stored charges.

Figure 6.21 illustrates the schematic representation of a 6-T SRAM memory cell to store one bit of information. Throughout the read and write operation, two nMOS access transistors offer access to the SRAM cell. Two bit lines (BL_A and BL_B) are employed to transfer the data during the read and write operations working in a differential manner. In order to achieve a better noise margin, the data and its complement are applied to BL_A and BL_B, respectively. The data is stored as two stable states, at storage nodes V_R and V_L, denoted as "0" and "1". SRAM works in mainly three modes:

(1) **Read operation cycle:** In this cycle, the data stored in SRAM cell is read by precharging both bit lines to logic "1" and then asserting a high on the WL to turn on the access transistors. If the cell content is a logic "0", the BL_A will get discharged through the right access transistor (AX_R) and right pull-down transistor (N_R) to logic "0", while BL_B will remain precharged to transfer the voltages of the storage nodes to the corresponding bit lines. In contrast, BL_B will get discharged and BL_A will remain precharged, if the cell stores a logic "1". To maintain the read-ability the cell-drive transistors should be wider than the access transistors in order to avoid the flipping of data during read operation.

(2) **Write operation cycle:** In this cycle, data ("0" or "1") which is to be written is applied to BL_A and its complement is applied to BL_B and then a logic high is asserted to WL. If the cell content is same as with the data to be written in it, then no charging/discharging will take place. Otherwise if the cell contains a "0" then discharge to "0" through AX_R and simultaneous pull-up of V_L to "1" through P_L will take place to write a "1". To write a "0" together with the initial cell content of "1", discharge to "0" through AX_L and simultaneous pull-up of V_R to "1" through P_R will take place to write a "0". For proper write operation, the input driver transistors should be stronger (wider) than the corresponding pull-up transistors so that the data stored in the cross-coupled inverters could easily be overwritten.

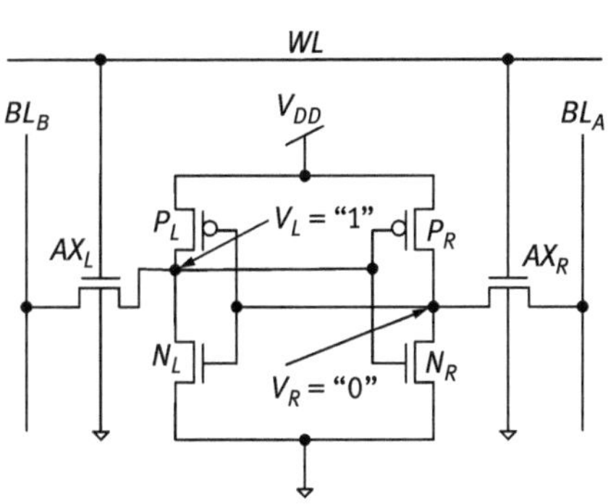

Figure 6.21: Schematic diagram of a 6-T SRAM cell.

(3) **Standby:** In this mode, the cell remains in idle state. The word line remains off, so the access transistor remains disconnected with bit lines, and hence cross-coupled inverters continuously reinforce each other to hold the stored data.

6.11 SRAM Cell Failures

The various failure mechanisms for a SRAM can be grouped under four types:

(1) **Read Failure:** The voltage at the node *R* increases and becomes greater than zero ($V_R > 0$), at some point in the read procedure cycle, as a result of the resistor divider act among access transistor AX_R and pull-down transistor N_R. It causes the pull-down transistor N_L to turn on weakly. The node *L* discharges due to the sub-threshold leakage through N_L, which in turn results in increase of V_R. As a result of this cyclic positive feedback action, V_R rises above the threshold position of the inverter located at left side (P_L–N_L) and turns over (reverses) the information stored of the cell [28].

(2) **Write Failure:** Consider logic "0" is stored at node *R* and a new data that equals to logic "1" has to be written next. Correspondingly, BL_A and BL_B are connected to logic "1" and logic "0", respectively. The node *L* would be discharged to lower voltage ($V_L < 1$) as a result of the resistor divider act among the pull-up transistor P_L and access transistor AX_L. If it is not possible to reduce V_L then the threshold position of the inverter located at the right side (P_R–N_R) during the time V_L is "1", and then a write failure occurs caused by the slow discharge of "1" due to the weakness of the access transistors.

(3) **Access Time Failure:** The time required to produce a voltage difference at the bit lines, so that sense amplifiers can read the state of the cell, is known as the access time. If discharging of bit lines is slow due to the weakness of pull-down and access transistors, then the voltage difference on the bit lines will not be sufficient to produce a significant voltage difference on the sense amplifiers to make the proper decisions, causing an access time failure.

(4) **Hold Failure**: In order to hold the data for a long time, the SRAM chip is placed in standby mode aiming to reduce the leakage current by decreasing the power supply. Nevertheless, the voltage level of the node storing logic "1" could possibly get lowered than the threshold voltage level of the inverter located at the right side (P_R–N_R) owing to large leakage of the pull-down nMOS transistor, causing hold failure by flipping of the cell state in standby mode.

6.12 Performance Metrics of SRAM

6.12.1 Static Noise Margin

It determines the stability of a SRAM cell in order to hold its data against the presence of noise [29]. To calculate static noise margin (SNM) of a SRAM, the voltage transfer characteristics (VTC) curves of two-cross-coupled inverter pairs present in the unit SRAM cell are plotted first. From that figure, a maximum square is drawn using the two VTCs. Then, the length of that square determines the SNM. It is broadly classified into two types: hold SNM and read SNM. Hold SNM is for an SRAM cell working in standby mode (W_L is disabled). However, read SNM is regarded as more crucial than hold SNM due to the susceptibility of the SRAM cell although the read operation cycle when WL is high. The reduction of SNM of the cell below the predefined threshold SNM of the cell causes the occurrence of a read failure. Other than the graphical method of finding the SNM by drawing the maximum possible square between the VTCs of inverter characteristics, the approach given in [29] is quick and reliable. Figure 6.22 shows the graphical representation of SNM in the two coordinate systems, which are rotated by 45°. This method can be implemented in commercial SPICE systems to calculate SNM.

6.12.2 Reliability Issues of 6-T SRAM

For the last few decades as a result of the development of microelectronics fabrication technology, the dimensions of MOS devices are shrinking. Therefore, reliability becomes a serious concern, especially at nanoscale dimensions. One of the

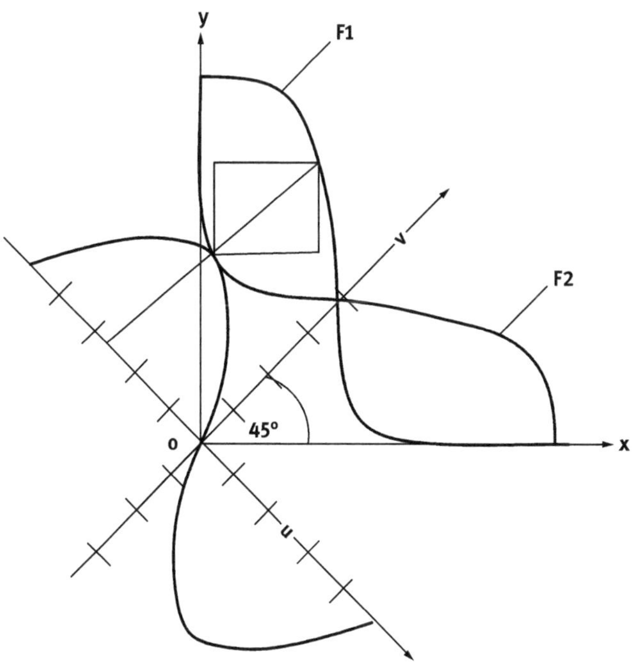

Figure 6.22: SNM calculation based on the maximum squares in a 45° rotated coordinate system.

major rising reliability issues is the aging effects, causing degradation of the device performance with time. NBTI (negative biased temperature instability) and PBTI (positive biased temperature instability) are the two most known aging effects that hinder the future downscaling of devices. NBTI and PBTI result in the threshold voltage degradation of pMOS and nMOS, respectively, owing to the presence of trapped charges.

Moreover, it is worth mentioning that NBTI and PBTI are also dependent on temperature and threshold voltage. It was observed that SRAM having low threshold voltage working at high temperature suffers severe performance degradation. As a result of the degradation of the SRAM performance, number of faulty cells increases as time progresses.

The researchers in industry and in academic institutes are continuously working to provide effective solutions for these reliability problems. There are some solutions such as replacement of SiO_2 by high-k dielectrics presenting benefits in terms of reduced gate-leakage current [30], providing relief to the reliability issues. However, there are many other reliability issues that appear as an obstacle to future downscaling of devices. NBTI results from interface-trapped charges from the broken Si-H bonds at the interface [31] whereas PBTI results from oxide-trapped charges [32]. It is worth mentioning that NBTI and PBTI are not dependent on supply voltage and temperature. Instead, they depend on threshold voltage and other technological parameters of the MOSFET. The different aspect of NBTI aging on SRAM reliability has been explored in [33, 34] for different MOSFET technologies. The impact of aging (NBTI and PBTI) on SRAM performance neglecting the variations in its threshold voltage and temperature is also studied [35–37].

Other than these, process variability and temperature variability are also considered as crucial reliability concerns. Process variability mainly emerges from the statistical variations in the length, width and doping of the channel, oxide thickness, line-edge roughness and random dopant fluctuations due to the imperfection of manufacturing process. It results in variation in threshold voltage from chip to chip (inter-die variation) and transistor to transistor (intra-die variation). As a result of this undesirable process variability, the transistors fabricated on the same chip might not behave identically, resulting in an undesirable output. On the other hand, the temperature might be different on different points within an IC or from IC to IC, depending on the load and operating conditions resulting in temperature variability.

6.13 Read-only Memory

A read-only memory (ROM) is essentially an AND plane of a Programmable Logic Array (PLA). ROM cells use only one transistor per bit of storage. Storing binary information at a particular address location can be achieved by the presence or absence of a

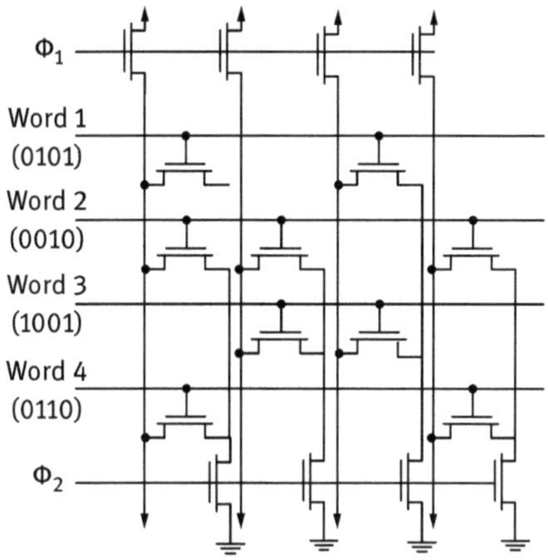

Figure 6.23: The structure of an nMOS ROM.

data path from the selected row (word line) to the selected column (bit line), which is equivalent to the presence or absence of a device at that particular location. The ROM memory contents are programmed by selectively placing transistors within the memory array. Therefore, the contents of a ROM memory are fixed as the part is manufactured and cannot be changed at a later time. It produces an output of n "words" or product terms each of m "variables" by an array of $n \times m$ nMOS switch matrices, as shown in Figure 6.23. The m bit lines are precharged to logic high for the duration of Φ_1 and a word line is selected. During Φ_2, the bit lines that have a ground path via a transistor controlled by the selected word line will be discharged to 0 V, signifying a 0 output. If a transistor is not present, a 1 output is produced. Read-only memories have been widely used to store microprograms, code conversion tables, tables of mathematical functions and general combinational functions.

The structure of a CMOS ROM is shown in Figure 6.24. CMOS is usually chosen for newer ROMs to reduce static power dissipation of the peripheral circuitry. It uses a precharge principle as in nMOS, but a single phase clock is sufficient (i.e. Φ_1 is replaced by $\overline{\Phi_1}$, and Φ_2 is replaced by Φ_1, and the circuit will still work).

Since the pattern of transistors in the switch matrix determines the stored information of the ROM, this pattern can be set during manufacture. Such a ROM is called a mask-programmable ROM. There are also field-programmable ROMs (PROMs), whose switch matrix can be set by the user. There are also erasable and reprogrammable ROMs (EPROMs), whose information can be erased by ultraviolet rays and reprogrammed with new information. Another MOSFET programmable ROM structure is shown in Figure 6.25 that contains data 0100,1111,1010,0001,1011,0111,1110,1001 in the address location 0–7 consecutively.

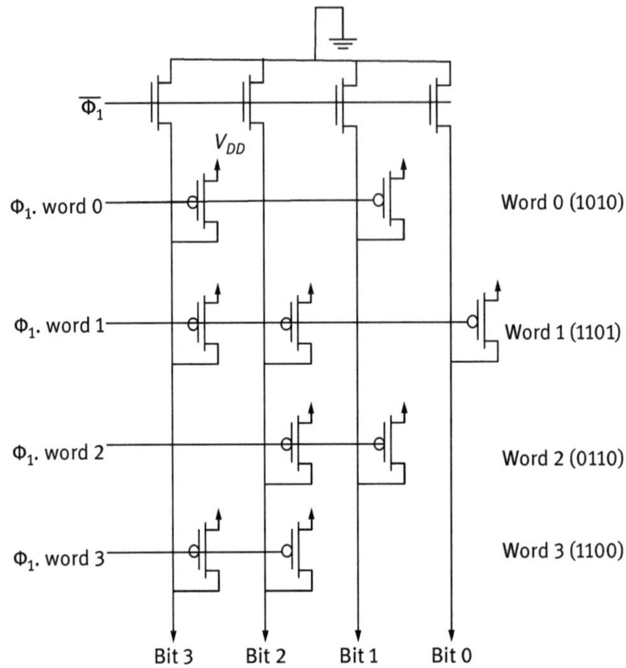

Figure 6.24: Structure of a CMOS ROM.

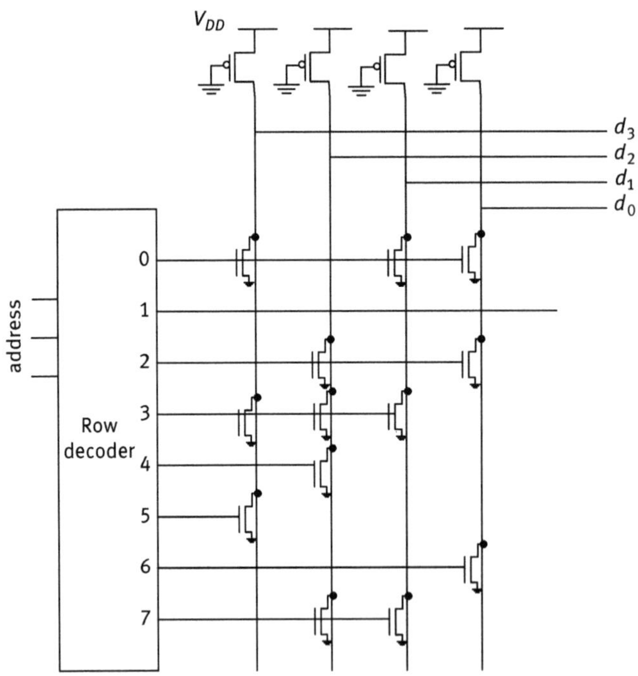

Figure 6.25: Structure of a MOSFET programmable ROM.

6.14 EPROM

Many applications require semiconductor memory that is nonvolatile like ROM, yet can be reprogrammed to correct unintentional errors in the contents of the memory or to change program-based characteristics of a system. The memory should have the nonvolatileness of ROM as well as the reprogram ability of RAM. As a solution to this, the EPROM (erasable programmable ROM) was developed. An EPROM provides dense, nonvolatile storage yet can be reprogrammed as necessary. The profile of the MOSFET for such a memory is developed by INTEL and was called FAMOS (floating gate avalanche transistor) technology and is illustrated in Figure 6.26. The cell comprises of two gates called floating (unconnected) gate and a control gate [38]. The oxide layer located below the floating gate and the oxide layer present in between the two gates possess identical thickness, which is equal to the thickness of the gate oxide (thin oxide) in an ordinary MOSFET. The programming consists of the application of large positive voltage (about 25 V) to control gate terminal and drain terminal followed by the grounding of the source terminal. The high electric field causes avalanche breakdown in the channel and produces an abundance of hot electrons, some of which have sufficient energy to penetrate the thin oxide and to be attracted to the floating gate by its positive potential, surmounting the energy barrier located between the substrate terminal and the gate tunneling oxide in order to do so [39]. The floating gate acquires a negative charge which it retains on removal of the programming voltage with the result that the transistor is nonconducting (due to charges present in the floating gate, threshold voltage enhanced) even when the control gate is taken to V_{DD}. If this programming step is omitted, the transistor behaves like a standard MOSFET. This is a one-transistor storage cell. The charge retention time is usually 10 years. The data storage mechanism of a unit memory cell depends upon the presence/absence of the charge in the floating gate terminal [40].

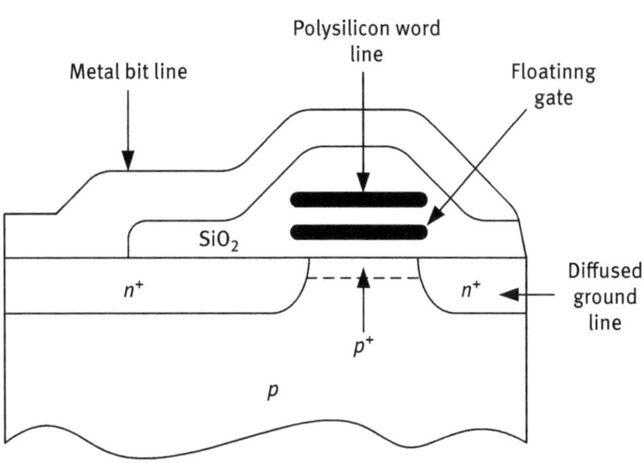

Figure 6.26: EPROM storage cell.

Memory is erased by removing undesired charge from the floating gate in the following manner. To wipe out the charges, the floating gate is exposed to intense light of the proper wavelength for a period of time; enough energy is imparted to the stored charge to remove it from the floating gate. A quartz window is incorporated into the memory chip package for light entry; thus, all memory cells are exposed and erased simultaneously. EPROMs normally present erase times in the range of 20 to 30 minutes.

6.15 Electrically Erasable Programmable Read-only Memory (E^2PROM)

UV erasure is slow and the next move in field programming therefore is the E^2PROM, in which both the programming and erasure are electrical. The basic memory cell for an E^2PROM comprises of a memory transistor and transistor for selection, as depicted in Figure 6.27. The vital difference between the structure of the FAMOS and the E^2PROM transistors is the extension of the floating gate over the drain with a protrusion which gives a ultrathin tunnel oxide region of 5–10 nm The floating gate is charged and discharged by Fowler –Nordheim (FN) tunneling [41] of cold electrons between the drain and the gate. These electrons, unlike FAMOS, tunnel through the oxide, which provided is thin enough, without surmounting the 3.2 eV barrier overcome by hot electrons.

After the assertion of a positive voltage at the control gate and grounding the drain the transistor is programmed. FN tunneling now invites electrons to the floating gate, putting the transistor in high ("1") state with high threshold voltage or in normal off state. The control gate is grounded and drain is connected to a high positive voltage to erase the transistor. Electrons tunnel from the floating gate terminal

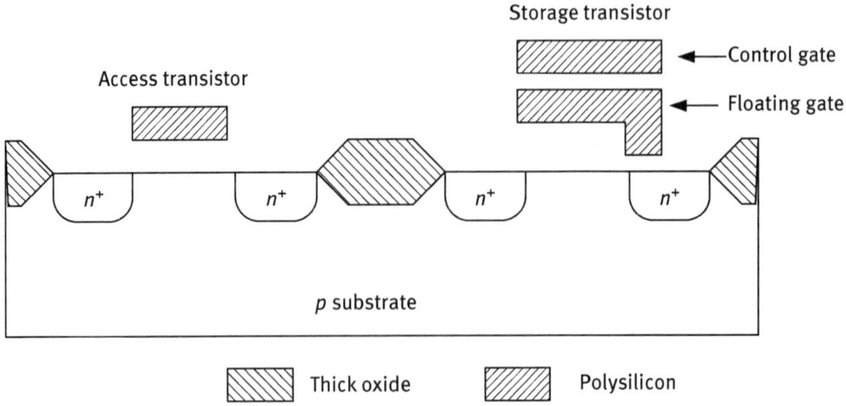

Figure 6.27: E^2PROM storage cell.

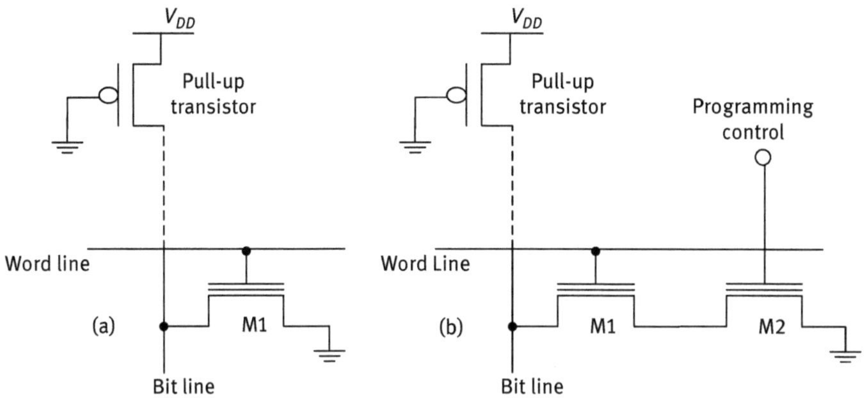

Figure 6.28: EPROM and E^2PROM cell placement.

to the drain terminal, leaving transistor in '0' state with low threshold voltage or in normal on condition. The control gate serves only for programming and erasure; it has no function for readout.

The placement of EPROM and E^2PROM cells within an array is shown in Figure 6.28. The critical step in the fabrication of E^2PROM transistor cell is the formation of the high quality tunnel oxide having an oxide thickness in the range of 5 to 10 nm.

The memory cell (transistor) can have two threshold voltages (two states) corresponding to the existence of charges (electrons) at the floating gate as shown in Figure 6.29. When electrons are accumulated at the floating gate, the threshold voltage of the memory cell raises, and as a result the memory cell is considered to be in a "1" state as a convention. This is because the memory cell is not turned on with the application of the read signal voltage (e.g. 5 V) to the control gate, and the bit line precharge level (e.g. V_{DD}) is maintained. It is possible to lower the threshold voltage of

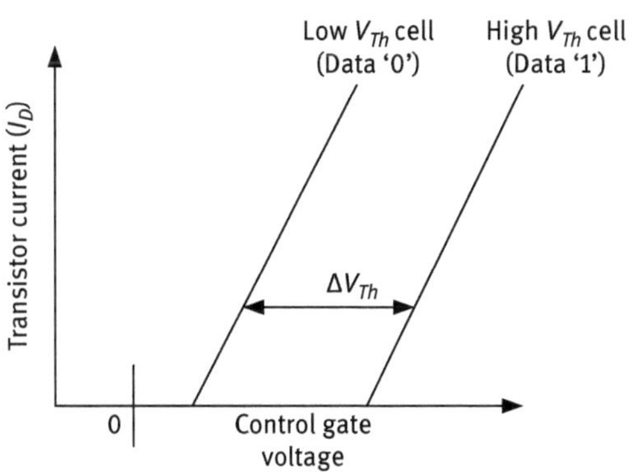

Figure 6.29: I–V characteristic curves of the flash memory cells with low and high threshold voltages for the control gate voltage.

the memory cell by taking away the electrons from the floating gate, resulting for the memory cell to be considered as to be in a “0” state. In this case, with the application of voltage, the cell transistor turns on with the simultaneous discharge of the bit line to the ground. Floating gate cells can be arranged in mainly two architectures: NOR and NAND. The NOR structure utilizes channel hot-electron (CHE) injection mechanism [42] for write and Fowler–Nordheim (FN) tunneling mechanism for erase. The NAND structures utilize Fowler–Nordheim (FN) tunneling mechanisms for both write and erase operations. The reading speed accession is lower for NAND-based structures due to their higher resistance of series connection. In contrast, lower program speed is offered by NOR-based structures as each cell must be written individually or in bytes due to high current involved in CHE process. As a result, writing a page or a block of data requires large time (in the range of ms). The cell size is larger for NOR-based structure as each NOR cell has its own connection to bit lines (requires larger metal-layer contacts and hence larger die area).

6.16 Flash Memory

It is formed by an array of several floating gate transistors. It has two gates: one is known as control gate (CG) like ordinary MOSFETs, and the other one is a floating gate surrounded by insulating oxide layer. An electrically isolated floating gate differentiated floating gate MOSFETs and normal MOSFETs. Quantum mechanical tunneling principle is used to transfer charges from the floating gate to the substrate through a thin insulating tunneling barrier. As the floating gate is isolated owing to the insulating oxide coating, every charge placed on it is trapped and thus stores the information. The two-transistor E^2PROM cell is complemented by the single transistor cell for the flash E^2PROM as shown in Figure 6.30. The split gate has a dual role. It functions like a conventional gate over the channel region and controls the erasure of the floating gate. The cell behaves like two transistors. Hot-electron injection is used for programming

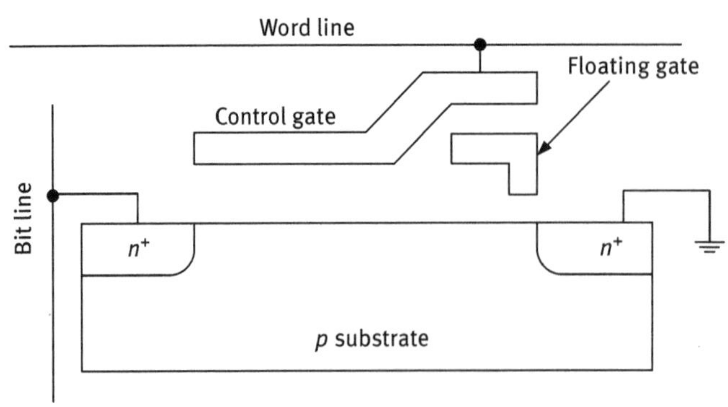

Figure 6.30: Split gate flash E^2PROM cell.

and cold-electron tunneling for erase is employed in this memory to operate. The reduction from two transistors to one transistor per bit increases the storage capability of flash E^2PROMs. The absence of a control gate makes this cell only suitable for flash erasure of the entire memory of large blocks thereof. Ten years' data retention time can be obtained if a tunneling oxide thickness larger than 8 nm is employed in the chip [43].

Programming has been simplified by on-chip generation of the programming voltage. Instead of requiring an external connection to a high positive voltage around 25 V, a circuit called a charge pump is provided to produce the essential programming voltage from the standard 5 V supply. It is worth mentioning that even though the flash memory has turned out to be the choice of the majority of current nonvolatile memory, it does exhibit some weaknesses: (1) slow operating speed as compared with volatile memory and (2) the duration of typical programming timings in μs and erase times in ms.

Currently flash memory is in demand for the reasons given subsequently:

(1) It can offer the highest chip density. In contrast to ferroelectric RAM (FeRAM) comprising of one transistor and one capacitor, MRAM has the need of one transistor and a magnetic tunnel junction [44]. Flash memory requires only one transistor.
(2) It can achieve the multi-bit per cell storage property [40]. Multi-bit storage capacity can be achieved with four dissimilar threshold voltage (V_{Th}) conditions attained in a flash memory cell by controlling the amount of charge stored in its floating gate. Two-bits/cell (with four V_{Th} states) flash memory cells have already been commercialized. SONOS flash memory offers multilayer integration for increased memory density [45].
(3) The fabrication process is compatible with regular commercial CMOS fabrication process steps and hence can provide suitable solution for embedded memory applications.

The main disadvantages related to conventional floating gate flash memory are the following:

(1) Program/erase speed is lengthy.
(2) It necessitates a high voltage to program/erase the stored data; operating voltage needs to be reduced below 1 V.
(3) Endurance is less (10^5 program/erase cycle).
(4) Scaling of flash memory technology is lagging after the CMOS logic device scaling technology due to the suffering from excessive SCEs and high leakage current resulting in higher power consumption.
(5) High voltage transistor located on the peripherals of the flash memory chip occupies a large area and thus hinders further downscaling.

In summary sending a short (several ms), high voltage pulse to the row line while grounding the drain causes hot-electron injection under high electric field from the

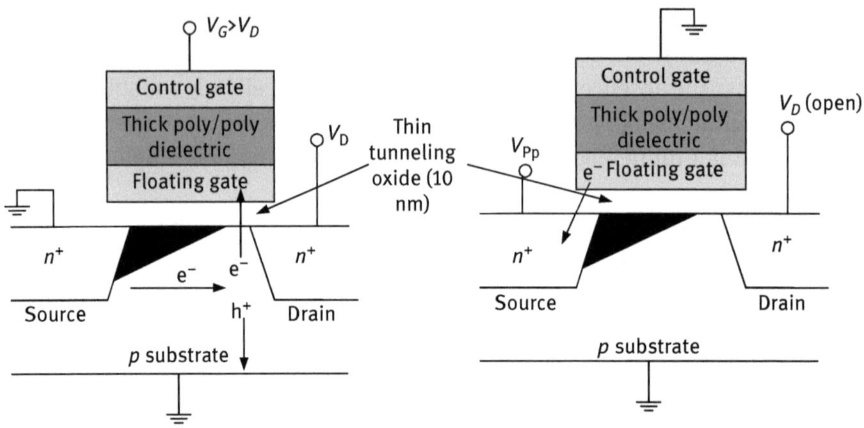

Figure 6.31: Data programming and erasing methods in the flash memory (a) hot-electron injection mechanism (b) Fowler–Nordheim tunneling mechanism.

drain terminal to the floating gate (program) terminal. A similar pulse to the drain with the row line grounded causes tunneling of electrons from the floating gate terminal to the drain terminal (erase), as shown in Fig. 6.31.

In brief, EPROM is programmed electrically, but it is possible to erase the data stored in it by exposing it to UV light. On the other hand, both the data stored in E^2PROM and flash can be wiped out electrically without being removed from the system. Flash is erased in bulk, so flash has become the most economical non-volatile storage. For example, flash cards are widely used in digital camera, mobile phone, etc.

6.17 Summary

The semiconductor memories exhibited an explosive growth of the flash memory market and its dominant position in market, motivated by the popularity of cellular phones and other of electronic gadgets such as tablet, PC, digital camera, MP3 player and so on.

Kahng and Sze in 1967 at Bell Labs [46] gave the picture of the first model for flash memory. Later Wegener et al. in 1967 used Si_3N_4 to trap charges [47]. In 1990s, polysilicon was used to trap charges. In 1995, Tiwari et al. examined the property of the gate having Si nano-crystals embedded on it and opened up a new horizon for research in this topic [48]. One of the extensions to the floating gate and charge-trapping technologies engrosses the use of nano-crystalline floating gates. Due to the discontinuous nature of the charge storage layer, the material is impervious to isolated defect paths in the primary tunnel oxide. Therefore, the tunnel oxide below the nano-crystal area could be scaled down and thus it is possible to reduce the operating voltage. In 2003, first 4 MB memory using Si nano-crystal technology using 90 nm technology was produced. Later, 24 MB memory array was produced by Freescale^{TM} based on silicon nano-crystals exhibiting excellent uniformity. Meanwhile, as the device dimensions reduce below 90 nm, it has become cumbersome to embed floating gate-based flash memories in a cost-effective way. The chip requires a high voltage (9–12 V) to write and erase the conventional floating gate memories. As a result of this high voltage, reliability issues such as memory failure and loss of data have become another major concern. Thus, nano-crystal memories are advanced class of next-generation memory technology. They are scalable; i.e. it is possible to scale the tunneling oxide thickness without having a serious impact on the data retention. The ultimate aim to scale nano-crystal-based memories is to use a single electron to represent a bit known as single electron transistor (SET) [49].

References

[1] H. Partovi, "Clocked storage elements," in *Design of High-performance Microprocessor Circuits*, A. Chandakasan, et al. (ed.), pp. 207–233, 2001.

[2] J.M. Rabaey, *Digital Integrated Circuits*. Prentice Hall, Upper Saddle River, NJ, 1996.

[3] Neil Weste and Kamran Eshraghian, *Principles of CMOS VLSI Design*. Addison-Wesley Publishing Company, New York, USA, 1985.

[4] John P. Uyemura. *Circuit Design for CMOS VLSI*. Kluwer Academic Publishers, 1992, Berlin, Germany.

[5] A. Hodges and H.J. Jackson, *Analysis and Design of Digital Integrated Circuits*, 2nd edn., McGraw-Hill, New York, 1988.

[6] John P. Hayes. *Computer Architecture and Organization*. McGraw-Hill, Inc., New York, USA, 1988.

[7] The International Technology Roadmap for Semiconductor, Overview, 2012 Update.

[8] F. Gamiz, A 20 nm low-power triple-gate multibody 1T-DRAM cell, International Symposium on VLSI Technology, Systems, and Applications (VLSI-TSA), 23–25 April 2012, Taiwan.

[9] Kurinec, K. Santosh, and Krzysztof Iniewski, eds. Nanoscale Semiconductor Memories: Technology and Applications. CRC Press, Boca Raton, USA, 2013.

[10] Tseung-Yuen Tseng and Simon M. Sze, *Nonvolatile Memories–Materials, Devices and Applications*. Wiley-IEEE Press, New York, USA, 2012.

[11] C. Shin, B. Nikolić, T.-J. King Liu, C.H. Tsai, M.H. Wu, C.F. Chang, Y.R. Liu, C.Y. Kao, G.S. Lin, K.L. Chiu, C.-S. Fu, C.-T. Tsai, C.W. Liang, "Tri-gate bulk CMOS technology for improved DRAM scalability," Proceedings of European Solid-State Device Research Conference (ESSDERC), September 2010.

[12] Thomas Vogelsang, Understanding the energy consumption of dynamic random access memories, MICRO '43 Proceedings of the 2010 43rd Annual IEEE/ACM International Symposium on Microarchitecture, 2010, pp. 363–374.

[13] Ningde Xie, Tong Zhang and Erich F. Haratsch, "Using embedded dynamic random access memory to reduce energy consumption of magnetic recording read channel." IEEE Transactions on Magnetics, 46.1 (2010): 87–91.

[14] Simon Deleonibus, *Electronic Device Architectures for the Nano-CMOS Era from Ultimate CMOS Scaling to Beyond CMOS Devices*. Pan Stanford Publishing Pvt. Ltd., Singapore, 2009.

[15] C. Galup-Montoro, and M. C. Schneider, (2007). MOSFET modeling for circuit analysis and design. World scientific Publishing Pte., Ltd., Singapore.

[16] Jean-Pierre Colinge (ed.), *FinFETs and Other Multi-Gate Transistors*. Springer US, Berlin, Germany, pp. 1–48, 2008.
[17] Sung-Mo Kang and Yusuf Leblebici, *CMOS Digital Integrated Circuits*. McGraw-Hill, New York, 1996.
[18] Randall L. Geiger, Phillip E. Allen and Noel R. Strader, *VLSI Design Techniques for Analog and Digital Circuits*. McGraw-Hill Inc., New York, USA, 1990–.
[19] L.J. Herbst, *Integrated Circuit Engineering*. Oxford University Press, New York, USA, 1996.
[20] C. Cho, S. Song, S. Kim, S. Jang, S. Lee, H. Kim, ... and K. Kim, (2004, June). Integrated device and process technology for sub-70nm low power DRAM. In VLSI Technology, 2004. Digest of Technical Papers. 2004 Symposium on (pp. 32–33). IEEE.
[21] G. Wang, D. Anand, N. Butt, A. Cestero, M. Chudzik, J. Ervin, ... and B. Kim, (2009, December). Scaling deep trench based eDRAM on SOI to 32nm and Beyond. In Electron Devices Meeting (IEDM), 2009 IEEE International (pp. 1–4). IEEE.
[22] P. C. Fazan, S. Okhonin, M. Nagoga, and J. M. Sallese, (2002). A simple 1-transistor capacitor-less memory cell for high performance embedded DRAMs. In Custom Integrated Circuits Conference, 2002. Proceedings of the IEEE 2002 (pp. 99–102). IEEE.
[23] R. Ranica, A. Villaret, C. Fenouillet-Beranger, P. Malinge, P. Mazoyer, P. Masson, ... and T. Skotnicki, (2004, December). A capacitor-less DRAM cell on 75nm gate length, 16nm thin fully depleted SOI device for high density embedded memories. In Electron Devices Meeting, 2004. IEDM Technical Digest. IEEE International (pp. 277–280). IEEE.
[24] T. Shino, T. Higashi, N. Kusunoki, K. Fujita, T. Ohsawa, N. Aoki, ... and H. Nakajima, (2004). Fully-depleted FBC (floating body cell) with enlarged signal window and excellent logic process compatibility. Tech. Dig. IEDM, 281–284.
[25] K. Hatsuda, K. Fujita, and T. Ohsawa, (2005, September). A 333MHz random cycle DRAM using the floating body cell. In Custom Integrated Circuits Conference, 2005. Proceedings of the IEEE 2005 (pp. 259–262). IEEE.
[26] S. Okhonin, M. Nagoga, J.M. Sallese and P. Fazan, "SOI capacitorless 1TDRAM concept." Proceedings of IEEE International SOI Conference, October 2001, pp. 153–154.
[27] K. Sasaki, K. Ueda, K. Takasugi, H. Toyoshima, K. Ishibashi, T. Yamanaka, N. Hashimoto and N. Ohki, "A 16-Mb CMOS SRAM with a 2.3-pm2 single-bit-line memory cell." IEEE Journal of Solid-State Circuits, 28 (1993): 1125.
[28] S. Mukhopadhyay, H. Mahmoodi, and K. Roy, (2005). Modeling of failure probability and statistical design of SRAM array for yield enhancement in nanoscaled CMOS. Computer-Aided Design of Integrated Circuits and Systems, IEEE Transactions on, 24(12), pp. 1859–1880.
[29] E. Seevinck, F. J. List, and J. Lohstroh (1987). Static-noise margin analysis of MOS SRAM cells. Solid-State Circuits, IEEE Journal of, 22(5), pp. 748–754.
[30] J. Hicks, D. Bergstrom, M. Hattendorf, J. Jopling, J. Maiz, S. Pae, ... and J. Wiedemer, (2008). 45nm Transistor Reliability. Intel Technology Journal, 12(2), pp. 131–144.
[31] Rakesh Vattikonda, Wenping Wang, Yu Cao, "Modeling and minimization of PMOS NBTI effect for robust nanometer design." Design Automation Conference, July 2006.
[32] Sufi Zafar, Alessandro Callegari, Evgeni Gusev and Massimo V. Fischetti, "Charge trapping related threshold voltage instabilities in high permittivity gate dielectric stacks." Journal of Applied Physics, 93 (2003): 9298–9303.
[33] K. Kang, H. Kufluoglu, K. Roy, and M. A. Alam, (2007). Impact of negative-bias temperature instability in nanoscale SRAM array: modeling and analysis. Computer-Aided Design of Integrated Circuits and Systems, IEEE Transactions on, 26(10), 1770–1781.
[34] Kunhyuk Kang, Muhammad Ashraful Alam and Kaushik Roy "Characterization of NBTI induced temporal performance degradation in nanoscale SRAM array using IDDQ." IEEE International Test Conference (TC) in Proceeding of 2007 IEEE International Test Conference, ITC 2007, Santa Clara, California, USA, October 21–26, 2007.

[35] Aditya Bansal, Rahul Rao, Jae-Joon Kim, Sufi Zafar, James H. Stathis and Ching-Te Chuang, "Impacts of NBTI and PBTI on SRAM static/dynamic noise margins and cell failure probability." Microelectronics Reliability, 49.6 (2009): 642–649.
[36] A. Bansal, R. Rao, Jae-Joon Kim, Sufi Zafar, J.H. Stathis and Ching-Te Chuang, "Impact of NBTI and PBTI in SRAM bit-cells: Relative sensitivities and guidelines for application-specific target stability/performance." IEEE International Reliability Physics Symposium, 2009, pp. 745, 749, 26–30 April 2009.
[37] J.C. Lin A.S. Oates, and C.H. Yu, "Time Dependent Vccmin Degradation of SRAM Fabricated with High-k Gate Dielectrics." Proceedings of 45th Annual. IEEE International Reliability Physics Symposium, 2007, pp. 439, 444, 15–19 April 2007.
[38] R. Bez, E. Camerlenghim A. Modelli and A. Visconti, "Introduction on flash memory." Proceedings of IEEE, 91, pp. 489–502, April 2003.
[39] R. Bez, A. Pirovano, "Non-volatile memory technologies: Emerging concepts and new materials." Material Science in Semiconductor Processing, 7 (2004): 349–355.
[40] P. Cappelletti, C. Golla, P. Olivo and E. Zanoni, *Flash Memories*. Kluwer Academic Publishers, Berlin, Germany, 1999.
[41] R. Fowler and L. Nordheim, "Electron emission in intense electric fields." Proceedings of the Royal Society of London, A119 (1928): pp. 173–181.
[42] H.A.R. Wegener, A.J. Lincoln, H.C. Pao, M.R. O' Connell and R.E. Oleksiok, "The variable threshold transistor. A new electrically alterable. Non-destructive read-only storage device." IEEE IEDM Technical Digest, 1967.
[43] Seiichi Aritome, "Advanced flash memory technology and trends for files storage application." IEDM, 2002, p. 763.
[44] "Advanced Memory Technology and Architecture", Short course, IEDM, 2001.
[45] A. J. Walker, S. Nallamothu, E. H. Chen, M. Mahajani, S. B. Herner, M. Clark, ... and S. Hu, (2003, June). 3D TFT-SONOS memory cell for ultra-high density file storage applications. In VLSI Technology, 2003. Digest of Technical Papers. 2003 Symposium on (pp. 29–30). IEEE.
[46] D. Kahng and S.M. Sze, "A floating gate and its application to memory devices." The Bell System Technical Journal, 46.4 (1967): 1288–1295.
[47] H.A.R. Wegener, A.J. Lincoln, H.C. Pao, M.R. O' Connell, and R.E. Oleksiok, "Metal insulator semiconductor transistors as a nonvolatile storage." IEEE IEDM, Abstract 59, Washington, DC.
[48] S. Tiwari, F. Rana, K. Chan, H. Hanafi, W. Chan and D. Buchanan, "Volatile and non-volatile memories in silicon with nano-crystal storage." IEDM Technical Digest, 1995, p. 521.
[49] H. Grabert and Michel H. Devoret, *Single Charge Tunneling: Coulomb Blockade Phenomena in Nanostructures*. Plenum Press, New York, 1992.

7 Analog Low Power VLSI Circuit Design

7.1 Analog Low Power Design: Problems with Transistor Mismatch

The quest for high speed, low power and accurate design of analog integrated circuit (IC) design has always been supported by never-ending dimension scaling of the transistors for the last few decades. It is well known that accuracy of high speed, low power circuits depends on the amount of matching between the transistors. For example, in a classical A/D or D/A converter, bit accuracy varies which depends on transistor matching performance. Hence, in this section, impact of mismatch in analog IC design will be discussed.

To model mismatch between two closely spaced identical transistors, random variation of the difference in their threshold voltage V_{Th} and current factor β is considered, with zero substrate bias. Considering a popular and widely accepted model with normal distribution, zero mean with variance depending upon channel length L and width W is given by [1]

$$\sigma(\Delta V_{Th}) = -\frac{A_{V_{Th}}}{\sqrt{W.L}}, \tag{7.1}$$

$$\sigma\left(\frac{\Delta\beta}{\beta}\right) = -\frac{A_{\beta}}{\sqrt{W.L}}. \tag{7.2}$$

Table 7.1 shows $A_{V_{Th}}$ and A_{β} for commercial CMOS technologies [2].

Now let us consider the current value to be mirrored in a current mirror circuit as a design parameter, and using the above equations, it is given by

$$\left(\frac{\sigma_I}{I}\right) = \frac{KP}{2IL^2}\left[4.AV_{Th}^2 + A_{\beta}^2\left(V_{GS} - V_{Th}\right)^2\right].$$

The above equation is plotted as a function of input current in Figure 7.1 for overdrive voltage 200 mV and 500 mV.

Figure 7.1 indicates that if high accuracy together with high speed (minimal length of transistor) is critical, then large current values resulting in a high power drain is necessary. For example, for achieving 1% current mismatch with 1 μm gate length, a biasing current of 500 μA is required.

It is clear that in order to obtain higher accuracy, larger chip area (proportional to W.L) is required. However, it will increase the capacitive loading at each node, which affects the speed. Therefore, it may be concluded that analog circuit design is a complex task involving multiple trade-offs between speed, power consumption and accuracy with several degrees of freedom like drain current and transistor dimensions. In deep-submicron technologies with sub-100 nm MOSFETs, the situation is getting more complex in presence of short channel effects (SCEs).

Table 7.1: A_{VT} and A_{β} for different commercial CMOS technologies.

Technology (µm)	Type	$A_{V_{Th}}$ (mV/µm)	A_{β} (%µm)
2.5	nMOS	30	2.3
1.2	nMOS	21	1.8
0.7	nMOS	13	1.9

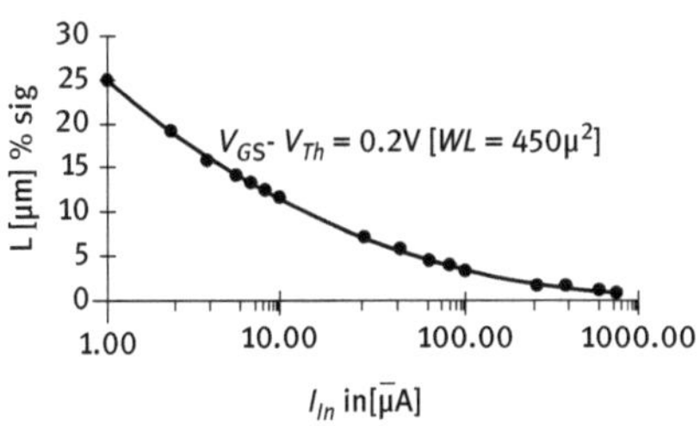

Figure 7.1: Minimum transistor length plotted as a function of biasing current for an error $\sigma = 1\%$, with a transistor length that equals to 1 µm for overdrive voltages 200 mV and 500 mV.

7.2 Mixed-signal Design with Sub-100 nm Technology

A true system-on-chip (SoC) integration requires the inclusion of digital logic along with mixed-signal (MS) processing that allows the IC to connect with the outside world. In the world of communication for cost reduction, an SoC contains high frequency Radio Frequency (RF) blocks, processing based on digital logic and high performance analog blocks [3, 4]. Other than cost savings, SoC offers some added advantages like reduced jitter and phase noise due to on-chip processing, and great power savings can be achieved due to the removal of differential signaling (Low-Voltage Differential Signaling (LVDS), an interconnect standard) interfaces for high performance ADCs, to name a few. However, the SoC design still relies on the meticulous execution of time-consuming custom-design process. The popular digital logic design techniques like design reuse, logic synthesis and soft macros are still not available in the design of MS modules. Until now, design of MS module is done by dropping of hard macros into the SoC. In future, to deal with the power and reliability issues, scaling of supply voltage is mandatory for tomorrow's downscaled devices with reduced feature size, which is sufficient to make the life of an analog MS designer gradually more difficult. It is worth mentioning that the reduction of power supply voltage with feature size is expected to saturate due to the likely death of Moore's prediction in near future. However, in sub-100 nm space, the design of MS module is further complicated by the issue of process variability. As a result, due to the deficiency of a standard design methodology, the increasing intricacies associated with MS design unavoidably make the nonrecurring engineering cost to go sky-high, thus making the design impractical. This is the reason why some researchers suggested discarding the SoC concept for integration of analog and RF modules by using advanced packaging techniques to throwing out some of the advantages obtained from SoC [5]. On the other hand, there are people who believe that by investigating in different ways MS design will go hand in hand with continuously scaled digital world [6].

7.3 Challenges in MS Design in Sub-100 nm Space

In this section, the challenges and difficulties in sub-100 nm MS design are presented.

7.3.1 Lack of Convergence of Technology

The main motto of technology scaling is to make ICs cheaper, faster and smaller. In a similar manner, cost reduction is considered as the predominant motivation behind SoC development to integrate total solutions within a single chip [7]. Additionally, along with reduced cost, SoC also offers better performance and reduction in power.

This plan does not hold well in the current scaling scenario. To understand this in a better way, let us introduce the concept of optimal technology. To characterize a technology, a set of parameters like minimum device feature size (channel length), threshold voltage, supply voltage and the reliability-related voltage limits is considered. The technology that is perfectly suitable for one component (like SRAM, DRAM, analog, digital, RF, etc.) is considered as optimal technology. For an SoC, the optimal technology must go well with all the different units that are integrated within the chip. Normally, the technology that best fits the majority component is chosen. There is no price to be paid if the technology has enough headroom. For example, in the past, when the supply voltage is adequately high, the technology that is used to realize nearly all analog functions is also utilized to implement digital and memory functions on the same die by paying almost zero or little price.

In recent years, it has been a common practice to consider power (energy) as important as the overall performance of the product. Therefore, power is considered as the most influential factor to determine the downscaled process parameters [8]. Because of this, downscaling may not offer significant improvements in performance (normally in terms of speed) and/or power. One more crucial property of this type of scaling is that the optimal technologies are diverging for different components of the chip.

To exemplify this point, consider optimal supply voltage, one of the most pertinent characteristics of scaling. As per constant field scaling [9], the supply voltage is to be scaled in proportion to the minimum device feature size. In general, this was followed in the past, especially when supply voltage was adequately high (Figure 7.2). However, for sub-100 nm technologies, International Technology Roadmap for Semiconductors (ITRS) [10] envisaged a deceleration in the supply

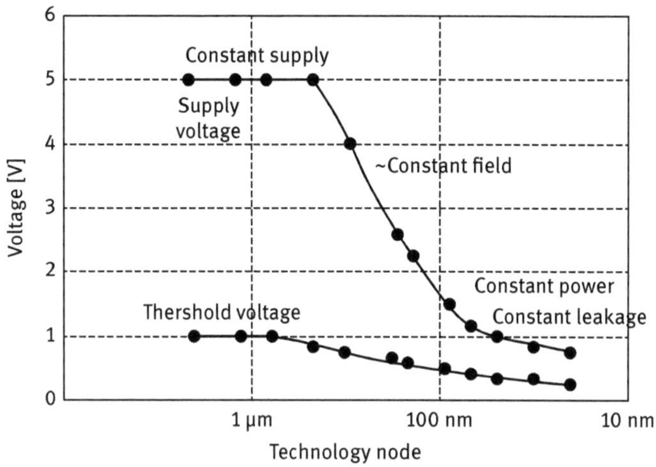

Figure 7.2: Trend in the supply voltage scaling.

voltage scaling. This slowdown can be primarily ascribed to the lack of ability of scaling of threshold voltage.

According to the ITRS, approximately 1 V supply is to be used by 45 nm technology node devices. For a 1 V supply, it becomes extremely difficult for the analog designers to design analog circuits with high dynamic range. For low power digital operations with supply voltage less than 1 V, analog functionality will be seriously affected. Therefore, it is possible to conclude that while supply voltage scaling is profitable for digital voltage scaling, it becomes ineffective for analog component design.

Recent ITRS prediction is also reflecting this divergence in scaling, where different scaling methodologies have been prescribed for almost every type of application (Figure 7.3). Even the digital scaling domain has been divided into several types like high performance (HP), low operating power (LOP) and low standby power (LSTP). Moreover, in the MS scaling realm, several options like high bandwidth, high resolution, RF and MEMS categories have been identified. This divergence in technology has serious ramifications in the SoC design due to the requirement of extra processing steps with the requirement of complex technologies, which ultimately undermines the cost-effectiveness of SoC as compared to the system-in-package (SiP) methodologies. For example, as the higher operating voltage is an essential requirement for I/O blocks, they require thicker oxide thickness as compared to digital logic transistors with thinner oxide thickness, thus necessitating extra processing steps.

7.3.2 Digital Scaling

Digital logic corresponds to the majority of the total integration within an SoC. Most digital logic transistor uses minimum channel length and the latest technology node

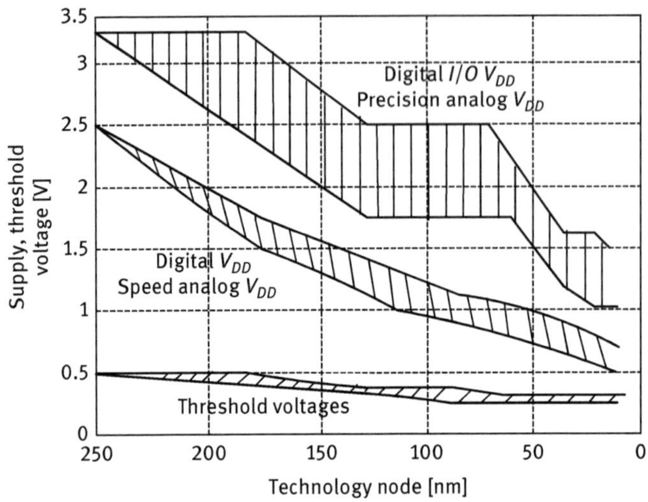

Figure 7.3: The trends of divergence in supply voltage scaling, as per ITRS.

available. Blocks identified as high activity and high performance are preferred to be designed with low supply and low threshold transistors. Whereas, blocks identified as low activity and low performance (applications in mobile phone) are preferred to be designed with high threshold transistors. Therefore, it may be concluded that there exists an optimal supply and threshold voltage for each of the digital logic applications, depending upon their activities.

7.3.3 Memory Scaling

Memory is an integral part of an SoC. It is of paramount interest to have high bandwidth and low latency on-chip memory within an SoC. However, its scaling provides a suite of challenges, which is dissimilar to digital scaling. Unlike conventional CMOS process, DRAM requires use of processes like multiple polysilicon layers and formation of deep trenches to fulfill its requirement of large storage capacitors. This exemplifies the technology divergence from mainstream CMOS process and DRAM fabrication. Therefore, to embed DRAM within an SoC, extra processing steps are mandatory. It is worth mentioning that considering the cost-effectiveness, embedding DRAM within an SoC has occasionally been made, whereas SiP or board-level integration approaches have been given preference.

On the other hand, SRAM constitutes a large amount of SoC area. Future memory scaling is being hindered mainly by the challenges like leakage, extra processing steps [11] and process variations. Moreover, reduction in supply voltage used to address the power consumption [12] affects the cell stability measured by static noise margin [13].

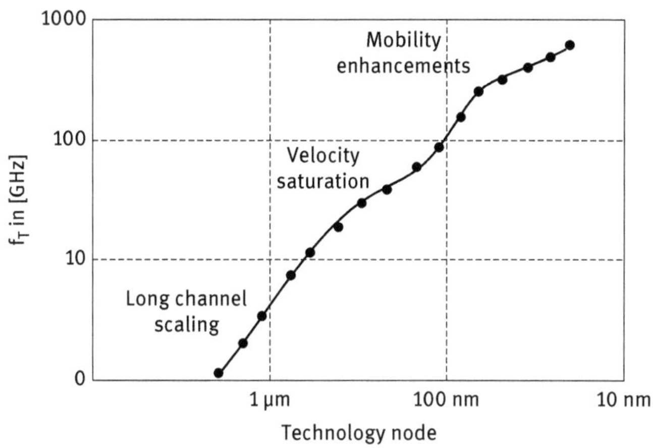

Figure 7.4: Trend of variation of unity-gain frequency with scaling.

7.3.4 Analog Scaling

In contrast to debatable existence of optimal technologies in scaling of digital blocks, there exist few optimal technologies in less standardized scaling of analog blocks. One of them is unity-gate frequency also known as cut-off frequency f_T of devices (Figure 7.4). With scaling, f_T increases but is degrading all other things related to analog circuit design [14].

Therefore, a high speed analog block can make use of the high f_T if it is possible to manage the other difficulties that may come along. In contrast, scaling of supply voltage with minimum channel-length device is not always going to provide improvement in analog design. For example, a high dynamic-range Nyquist converter requires higher supply voltage.

7.3.5 Degraded SNR

Reduction in supply voltage causes reduction in available voltage excursions, resulting in degradation in signal-to-noise ratio (SNR). In addition, due to scaling the transistor noise increases gradually. The thermal noise component of the drain current and flicker noise (1/f) component appearing in the 2nd term of (1) exhibits significant increase in short channel device [15].

$$\overline{i_d^2} = 4kTyg_{DS0}\Delta f + K\frac{I_D^\alpha}{f}\Delta f \tag{7.3}$$

Moreover, parameter y also exhibits a steady increase with channel-length scaling. In order to increase the SNR in the sampled-data circuits suffering with kT/C

noise component, capacitor size needs to be increased, resulting in increased power consumption [16].

7.3.6 Degradation in Intrinsic Gain

The intrinsic gain ($g_m R_O$) of a device is considered as the product of g_m and R_O, where g_m is the transconductance and R_O denoted the output resistance. The intrinsic gain decreases with scaling due to the increased drain-induced barrier lowering (DIBL) resulting from reduced gate-electrostatic control, increased channel-length modulation (CLM) and impact ionization dominated hot-carrier effects (HCEs). This degradation causes serious limitations in designing analog circuits [17]. For example, if intrinsic gain is small, then the requirement of precise charge manipulation by sampled-voltage circuits like high precision data converters cannot be fulfilled. This is the reason why a constant $g_m R_O$ has been predicted by ITRS for future generation MS components design.

7.3.7 Device Leakage

One of the most striking features of MOSFET technology is that it offers good switches to provide precise manipulation/control over the stored charges. However, with scaling, due to the increase in the gate and drain leakage current components, it is hard to maintain the fidelity of the charges stored in the node connected to the leaky transistor [18].

7.3.8 Mismatch due to Reduced Matching

For low precision application, size reduction is possible as the matching between two close transistors is proportional to device area, and it increases with reduction in oxide thickness [19] as shown in Figure 7.5. More on this topic is discussed in Section 7.1 of this chapter.

7.3.9 Availability of Models

Analog design requires a stable and mature device models. However, due to unavailability of mature models in deep submicron devices, the analog circuit designer experiences additional challenges.

7.3.10 Passives

Scaling technology, largely aiming at digital technologies, shows deficiency in high-Q, high density linear capacitors as required in sampled-data circuits. The gate and

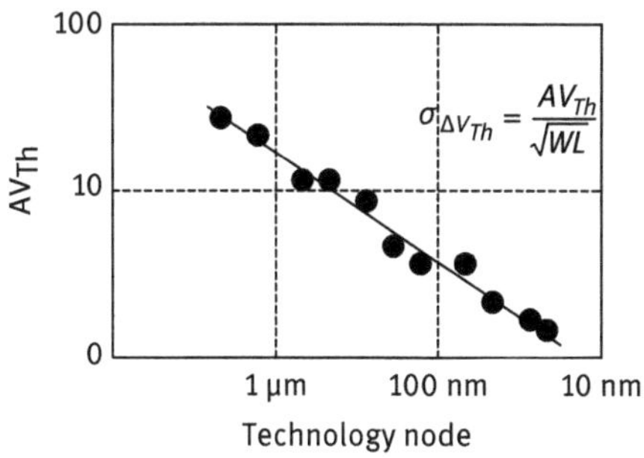

Figure 7.5: Trend of variation of transistor matching (shown in terms of AV_{Th}, an industry standard) as a function of scaling.

diffusion capacitances cannot be used because of their linearity and parasitic resistances. Hence double poly or metal-insulator-metal (MiM) capacitor finds their usage with a higher cost.

7.3.11 RF Scaling

CMOS has been the technology of choice for the mainstream digital logic and memory ICs over the years. On the other hand, RF and analog IC technologies depend on many different materials and device structures to provide the optimal solutions. In the past, RF and analog ICs have been mainly developed using the bipolar and the compound semiconductor technologies due to their better performance. In the recent years, the advancement made in CMOS technology allowed the analog and the RF circuits to be built with the same digital CMOS technology for the analog and RF circuits applications leading to the following threefold advantages:
(a) Performance improvement
(b) Cost advantage
(c) Ease of integration

The deep submicron MOSFETs scaled for digital operations are also capable of operating in the GHz regime because of their very high unit-gain frequencies in the range of GHz and increase in NF_{Min} [20]. Therefore, in addition to the high levels of integration for digital circuit design offered by the advanced CMOS processes, new opportunities for the RF circuit applications and new markets in the microwave and millimeter wave region have been created as the high frequency capabilities of the MOSFETs have reached well into the GHz range [21]. There is a significant difference in approach between the microwave designer and RF IC designer. A microwave designer would have concentrated more on metrics like G_{Max}, stability, power gain, circular plots in s-domain in order to find the optimal transistor size, bias and impedance matching network. However, in contrast, an analog Radio Frequency Integrated Circuits (RFIC) designer concentrates more on voltage gain [22], without studying impedance matching.

Previously GaAs was used mainly for microwave devices (s-parameter community) and Si was used for analog IC design ($g_m R_O$ centric community). However, today with scaling and advancement in fabrication process of III-V-based semiconductors, this type of community division has been come under dispute [23].

7.3.12 Issues Related with Power Devices

Power devices require voltages around 3.6 to 5 V and therefore are not compatible with deep submicron mainstream CMOS. Furthermore, large capacitor or inductor is often required. To incorporate these power devices, SiP approach clearly gives more advantage than SoC approach.

7.4 Basics of Switched-capacitor Circuits

This section of the book introduces switched-capacitor (SC) circuits. To start with, the resistor emulation using SC has been introduced initially. Later, filter using SC is shown. The basic components of an SC circuits is MOS switches and MOS capacitors. A typical MOS switch can provide high off resistance and low on resistance and thus is almost ideal to be used in SC circuits. A typical MOS switch can provide off resistance in the range of GΩ and on resistance in the range of KΩ, depending upon their sizes. If one increases the width of the MOSFET, then both these resistances will decrease. However, it is worth mentioning that increasing the size of the MOS transistor also increases the nonlinear voltage and size-dependent parasitic capacitances associated with the MOS switch as shown in Figure 7.6 which may influence the transfer function of the filter designed with SC circuits. In an SC circuit, the switch can be implemented by an nMOS, pMOS or a parallel combination of both. A parallel combination of nMOS and pMOS not only decreases the resistance but also improves the linearity.

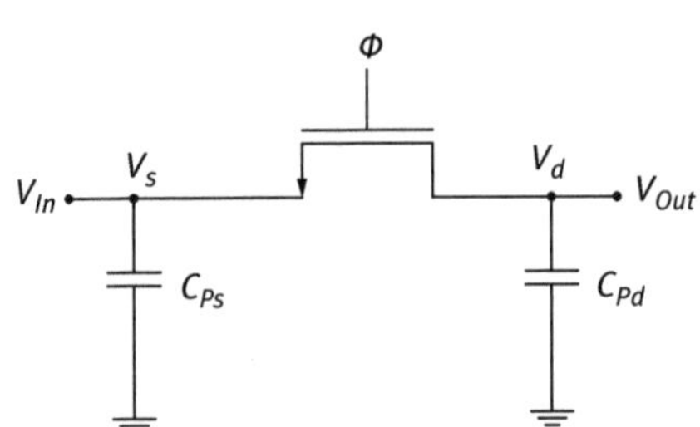

Figure 7.6: MOS switch including source and drain parasitic capacitances.

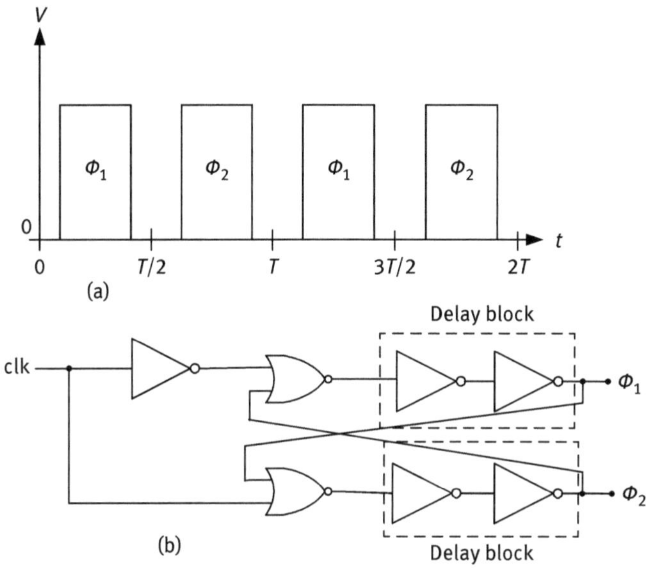

Figure 7.7: (a) Nonoverlapping clock signals (b) simple nonoverlapping clock signal generator.

The switches present in the SC circuit require at least two nonoverlapping clock signals as shown in Figure 7.7 in order to transfer the charge as described below. It is essential to ensure that the two clocks must not overlap at any instant so that two switches should not be closed at the same time. Otherwise, charges may be accidentally lost leading to a failure in SC circuit operation.

7.4.1 Resistor Emulation Using SC Network

The purpose of a resistor (R) is to transfer charge from one end (V_{In}) to the other end (V_{Out}). The average current that passes through the resistance is given by

$$i_{Rms} = \frac{V_{In} - V_{Out}}{R}. \tag{7.4}$$

Let us consider a simple parallel SC network with two nonoverlapping clocks applied as shown in Figure 7.8.

During the $0 \leq t \leq T/2$ when Φ_1 is high and Φ_2 is low, the switch located at the left closes and the switch located at the right is open; thus capacitor C charges due to the current flowing from the input given by

$$i_{Rms} = \frac{1}{T} \int_{0}^{T/2} i_C(t)\, dt. \tag{7.5}$$

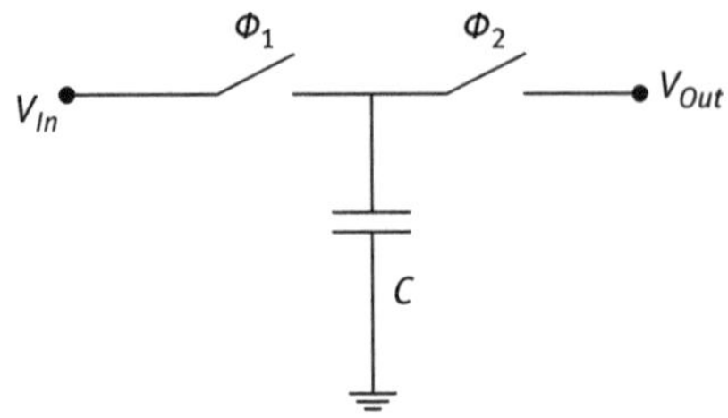

Figure 7.8: Parallel SC network to emulate a resistor (R).

Since

$$i(t) = \frac{dq(t)}{dt}, \tag{7.6}$$

$$i_{Rms} = \frac{1}{T}\int_0^{T/2} dq_C(t) = \frac{Q_C(T/2) - Q_C(0)}{T} \tag{7.7}$$

During $T/2 \leq t \leq T$ when Φ_1 is low and Φ_2 is high, the switch located at the right closes and the switch located at the left opens. Thus capacitor C discharges due to the current flowing from the capacitor to the output. Thus in one clock cycle T, overall charge transfer takes place, and the average current is given by

$$i_{Rms} = \frac{(V_{In} - V_{Out})\,C}{T}. \tag{7.8}$$

Comparing eqs (7.4) and (7.8), we obtain

$$R_{Eq} = \frac{T}{C}. \tag{7.9}$$

The advantages of this type of equivalent resistance in IC design are as follows:

(a) It is possible to fabricate high value resistors consuming very little silicon area. For example, a 1-MΩ resistor can be realized with a 10-pF capacitor switched at a clock rate of 100 kHz.

(b) Very accurate time constants can be realized because the time constant is proportional to the ratio of capacitances and inversely proportional to the clock frequency.

Table 7.2 shows some examples of SC network with their equivalent resistances and the capacitor charges in each phase [24].

7.4.2 Integrator Using SC Circuits

Figure 7.9 shows a standard OP-AMP integrator and Figure 7.10 shows its SC version.

Table 7.2: Popular SC network to emulate resistor.

Circuit	Schematic	R_{Eq}	$Q(\Phi_1)$	$Q(\Phi_2)$
Parallel	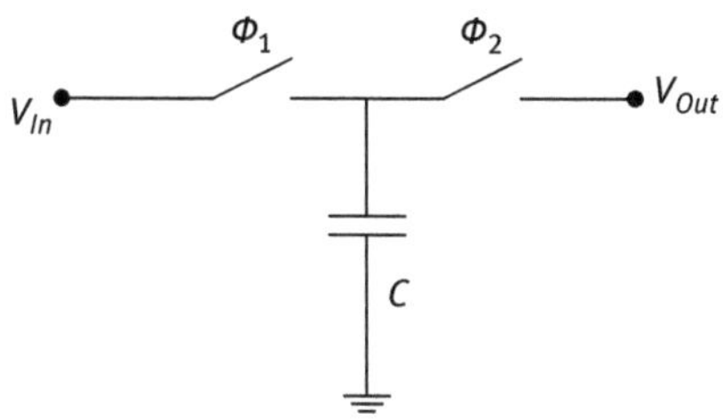	T$\frac{T}{C}$	$V_{In}C$	$V_{Out}C$
Series	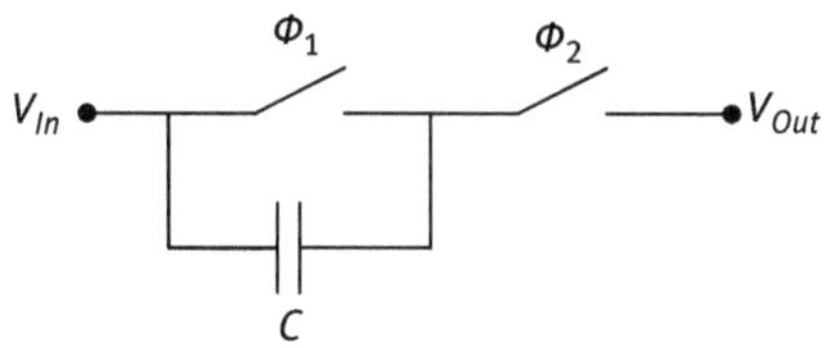	$\frac{T}{C}$	0	$(V_{In}-V_{Out})C$
Series-Parallel	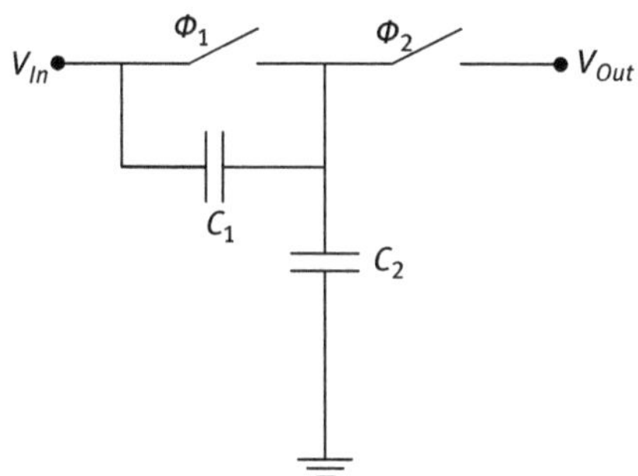	$\frac{T}{C_1+C_2}$	0 $V_{In}C_2$	$(V_{In}-V_{Out})C_1$ $V_{Out}C_2$
Bilinear	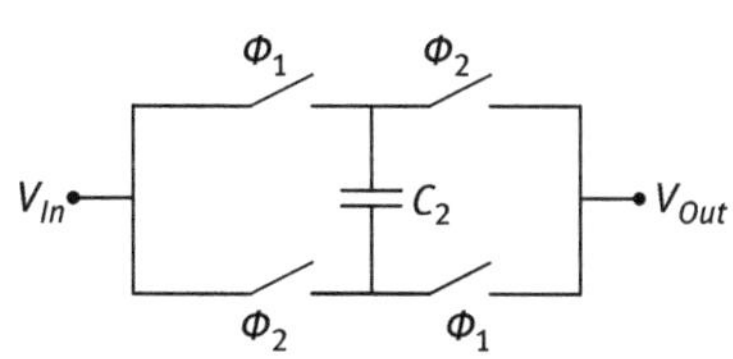	$\frac{T}{4C}$	$(V_{In}-V_{Out})C$	$(V_{Out}-V_{In})C$

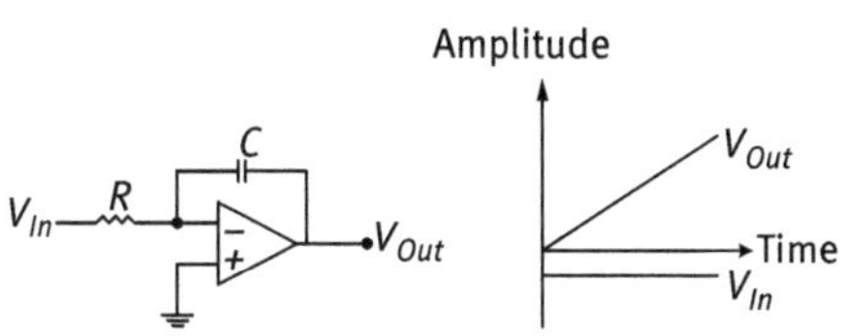

Figure 7.9: A standard OP-AMP integrator with ideal step input and integrated ideal ramp output.

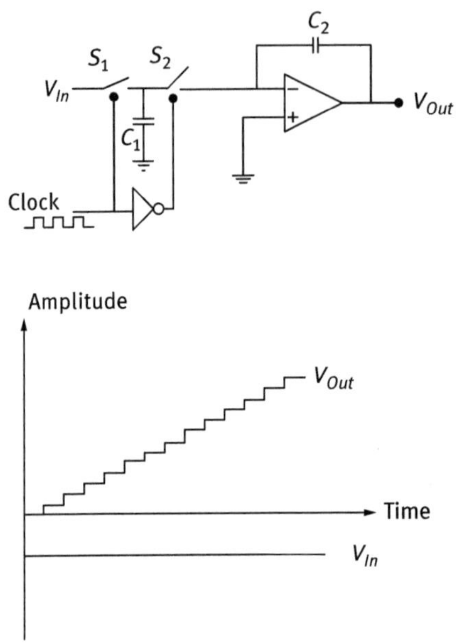

Figure 7.10: An ideal SC OP-AMP integrator with ideal continuous time step input, and for a continuous time step input, the output is a staircase ramp function.

The transfer function of the OP-AMP integrator shown in Figure 7.10 is given by

$$H(f) = \frac{V_{Out}(f)}{V_{In}(f)} = \frac{-1}{j\left(\frac{f}{f_0}\right)} \tag{7.10}$$

with

$$f_0 = \frac{1}{2\pi Rc}. \tag{7.11}$$

Now after replacing the resistance with SC equivalent network as shown in Figure 7.11, it is possible to express

$$f_0 = \frac{1}{2\pi}\frac{C_1}{C_2}f_{Clk}. \tag{7.12}$$

Now, the integrator does not require resistance to consume excessive silicon die area. Moreover, the −3dB frequency now depends on the ratio of capacitances, not on the RC product. Furthermore, the tolerance of a ratio is easier to control than the product of two tolerances. However, several constraints are to be maintained during the design of an SC filters:

(a) Since an SC circuit is a sampled-data system, the clock frequency must be much higher than the critical frequency set by the RC products in the circuit.

(b) SC equivalent cannot be used to close the negative feedback path in an OP-AMP all by itself. It is due to the fact that in order to ensure stability, the negative feedback path must be closed continuously. However, SC equivalents are sample data construction equivalence of resistor, not continuous.

(c) The bottom plate of the MOS capacitor needs to be grounded or connected to a voltage source. The top and bottom plate of an MOS capacitor is associated with nonlinear voltage-dependent parasitic capacitances with 5% and 20% of the capacitance value [25]. Therefore, to ensure that these nonlinear capacitances do not affect the overall response of the system, they are connected to the AC signal ground.

7.4.3 SC Integrator Sensitive to Parasitic

Figure 7.11 shows the SC integrator shown earlier in Figure 7.10 with parasitic elements.

C_{p1} corresponds to the top plate capacitances of C_2 and the switches. C_{p2} denotes the bottom plate capacitance of C_2. C_{p3} represents the top plate parasitic capacitance of C_1, and OP-AMP input capacitance and parasitic capacitance of switch No. 2. C_{p4} characterizes the bottom plate capacitance of C_1 and the next stage input capacitance. Table 7.3 represents the charges stored in the capacitances in each phase.

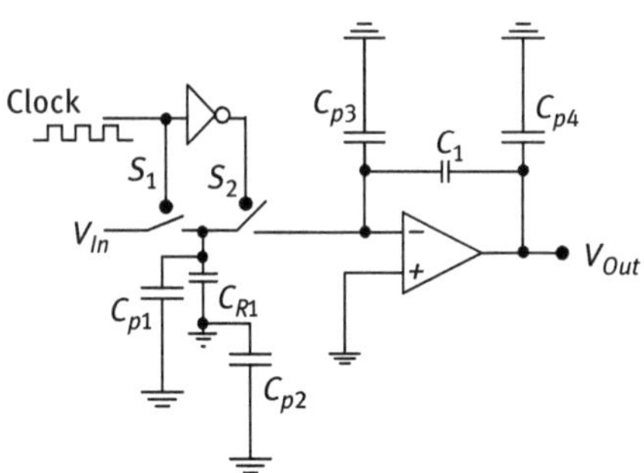

Figure 7.11: SC integrator shown in Figure 7.10 with parasitic elements.

Table 7.3: Stored charges in the capacitances during the two non-overlapped clock phases.

	$(n-1)T$	$(n-0.5)T$	a) nT
$Q_{C_{R1}}$	$V_{In}\,[(n-1)T]C_{R1}$	0	$V_{In}\,[nT\,]C_{R1}$
Q_{C1}	$-V_{Out}\,[(n-1)T]C_1$	$-V_{Out}\,[(n-0.5)T]C_1$	$-V_{Out}\,[nT]C_1$
$Q_{C_{P1}}$	$V_{In}\,[(n-1)T]C_{p1}$	0	$V_{In}[nT\,]C_{p1}$
$Q_{C_{P2}}$	0	0	0
$Q_{C_{P3}}$	0	0	0
$Q_{C_{P4}}$	$V_{Out}\,[(n-1)T]C_{p4}$	$V_{Out}\,[(n-0.5)T]C_{p4}$	$V_{Out}\,[nT]C_{p4}$

By considering one single clock period $(n-1)T$ to nT, and also by considering capacitances are connected to virtual ground at the end of the transition, it is possible to write [26]

$$[(n-1)\,T]\,\grave{a}\,[(n-0.5)\,T] : V_{In}\,[(n-1)\,T]\,(C_{R1} + C_{p1}) - V_{Out}\,[(n-1)\,T]\,C_1 = - V_{Out}\,[(n-0.5)\,T]\,C_1, \tag{7.13}$$

$$[(n-0.5)\,T]\,\grave{a}\,[nT] : -V_{Out}\,[(n-0.5)\,T]\,C_1 = -V_{Out}\,[nT]\,C_1. \tag{7.14}$$

Taking Z-transform, the transfer function is obtained as

$$H(z) = \frac{V_{Out}}{V_{In}} = -\left(\frac{C_{R1} + C_{p1}}{C_1}\right)\left[\frac{z^{-1}}{1-z^{-1}}\right]. \tag{7.15}$$

It is observed that $H(z)$ depends on parasitic capacitance C_{p1}.

To overcome this limitations, i.e. in order to design an SC integrator insensitive to parasitic capacitances, Martin et al. [27] developed a new structure as shown in Figure 7.12. The effects of stray capacitance have been reduced greatly by dual-switch configurations.

7.4.4 Low Power Switched-capacitor Circuit

The continuous downscaling of MOS transistors has introduced nanoscale (sub-100 nm) CMOS technologies. Consequently, the power supply voltage is also scaled down. However, threshold voltage of MOS transistor is almost maintained at a constant value. Due to this low supply voltage, the implementation of some well-known analog design techniques like SC becomes problematic. In a typical ΣΔ modulator designed

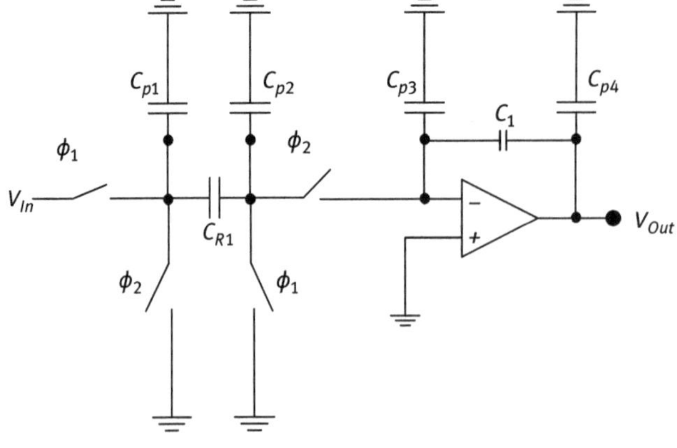

Figure 7.12: Parasitic-independent SC integrator.

with SC circuits, the switch driving problem becomes a major limitation with downscaled low power supply voltage. The low supply voltage does not provide enough overdrive to turn on the transistor to act as a switch anymore.

A classic SC circuit employs a complementary switch. For the input voltage ranges from V_{Ss} (0 V) to V_{Tn} (threshold voltage of nMOS transistor), the nMOS transistor conducts. On the other hand, for input voltage V_{Tp} (threshold voltage of pMOS transistor) to V_{DD}, the pMOS transistor conducts. Figure 7.13 shows the simulated switch conductance versus input voltage characteristics for two different supply voltages (5 V and 1.5 V).

For very low value of supply voltage ($V_{DD} < V_{Tn}+|V_{Tp}|$), the complementary switch behavior is lost. Figure 7.13(b) indicates that for V_{DD} = 1.5 V, there is a range of input signal for which capacitor cannot sample the signal. Thus, the advantage of rail-to-rail operation of the SC circuit is lost. However, a small value of input signal can be sampled, if it lies within the range where at least one switch is conducting, with the other switch remaining off, and thus can be excluded. It indicates that a single transistor SC circuit with a reduced power supply voltage imposes severe restriction on the input signal swing.

It is known that the RC time constant of SC circuit depends on the on resistance of the switch together with the sampling capacitor. After the sampling of the input signal by the sampling capacitor, the settling of the sampling process totally depends on this RC time constant. The on resistance of an nMOS transistor is given by [28]

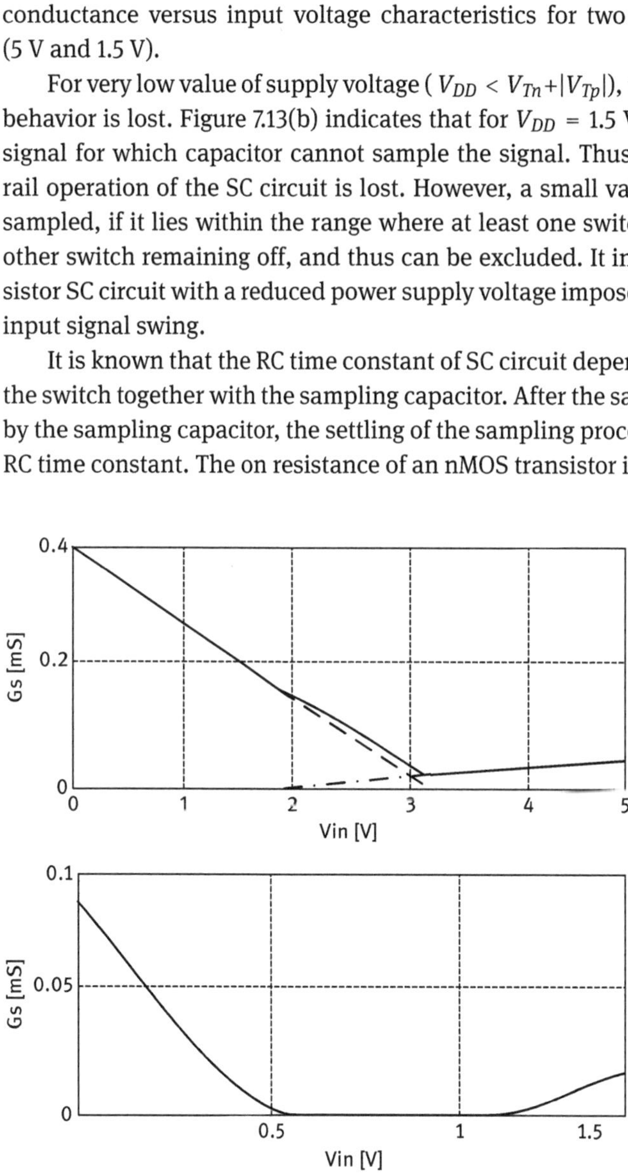

Figure 7.13: Complementary switch conductance versus input voltage for (a) V_{DD} = 5 V (b) V_{DD} = 1.5 V.

$$R_{sw,n} = \left[KP\frac{W}{L}\left(V_{DD} - V_S - V_{Th,n}\right) \right]^{-1}, \tag{7.16}$$

where

$$V_{Th,n} = V_{TO,n} + y_n\left(\sqrt{|2\varnothing_F - V_S|} - \sqrt{|2\varnothing_F|}\right), \tag{7.17}$$

where V_S is the applied signal. The above expression indicates that downscaling of device increases the on resistance, and thus resulting in higher settling time. On the other hand, upscaling results in lower settling time. However, it is important to remember that clock feedthrough and charge injection impose a limit on the improvement caused by upscaling. Due to this parasitic effect, a trade-off between clock feedthrough and high speed is needed to be managed for design of high speed SC circuits. The problem becomes more severe with the decrease in the supply voltage. Therefore, it may be concluded that low voltage SC circuits generally result in low speed of operation.

To overcome this problem, severe techniques can be employed as listed below:

(a) To use transistor with low threshold voltage: The problem with this technique is that a transistor with low threshold voltage results in a shift in the intersection of the subthreshold slope (Log (I_D) axis with V_{GS}) with the vertical axis that causes an increase in the subthershold leakage current. It indicates a higher switch-off leakage current. This increase in the leakage current affects the integrator capacitor to leak away the stored charge. This signal-dependent leakage produces harmonic distortion.

(b) Another technique, using voltage multiplier to produce higher supply voltage on chip to generate clock signal, is also used [29, 30]. Other than the clock signals, the rest of the circuits like the operational transconductance amplifiers (OTAs) and digital circuitry are designed to be operated with low supply voltages. For example, in Ref. [29], the OTA is operated with 1.8 V supply, whereas this voltage is multiplied to generate 3.4 V to drive the switches. It has the advantage of reduced power dissipation as the OTA is powered with low supply voltage. However, the main disadvantage of this technique is the area and power consumed by the voltage multiplier. Moreover, in nanoscale MOS transistors with extremely scaled power supplies, it is difficult to produce higher supply voltage.

(c) Switched OP-AMP technique allows the design of true low voltage SC technique as reported in Refs [31, 32]. With the requirement of extra active elements, they benefit from large signal overdrive with reduced power supply voltage.

7.5 Current Source/Sink

A *current sink* and *current source* are two terminal components whose current at any instant of time is independent of the voltage across their terminals. The current of a

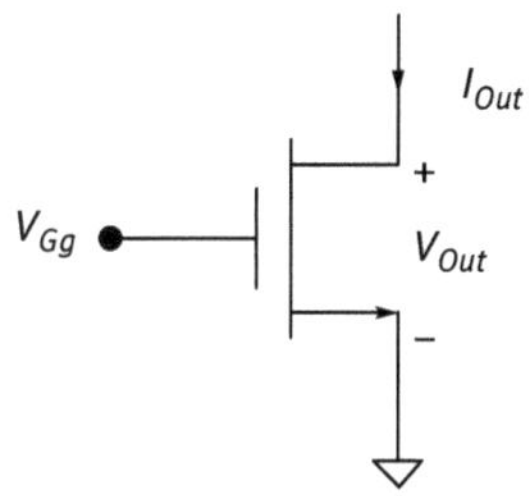

Figure 7.14: Current sink circuit.

current sink or source flows from the positive node, through the sink or source to the negative node. A current sink typically has the negative node at V_{SS} and the current source has the positive node at V_{DD}.

Figure 7.14 shows the MOS implementation of a current sink and the I-V Characteristics is shown in Fig. 7.15. The gate is taken to whatever voltage is necessary to create the desired value of current.

When a MOSFET is used as a linear amplifying device or switching device, it must be biased properly. The biasing can be done using a constant current source/sink. An nMOS can act as current sink, where negative node is connected to V_{SS}. A pMOS can act as a current source, where positive node is connected to V_{DD}. The gate is taken to whatever voltage is necessary to create the desired value of current. Typically, a voltage divider using MOS resistor is used to create the bias voltage of the gate (Figure 7.15).

Non-saturation region of a MOSFET is not a good choice for current source application.

MOSFET can act as a current source/sink in the saturation region, where the output current is constant subjected to no CLM ($\lambda = 0$). However, due to finite CLM, resulting in finite output resistance, the drain-to-source current increases with increase in drain-to-source voltage. However, for a better current source/sink, the current must be constant over a wide output voltage range. Therefore, in order to obtain a better MOS current sink/source we need to increase the output resistance.

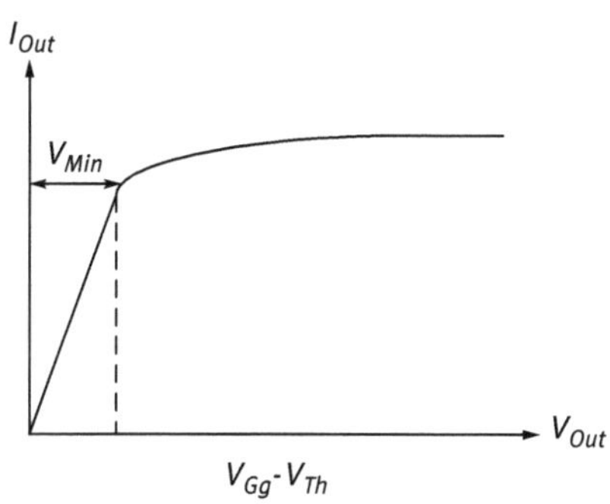

Figure 7.15: I–V characteristics of the current sink.

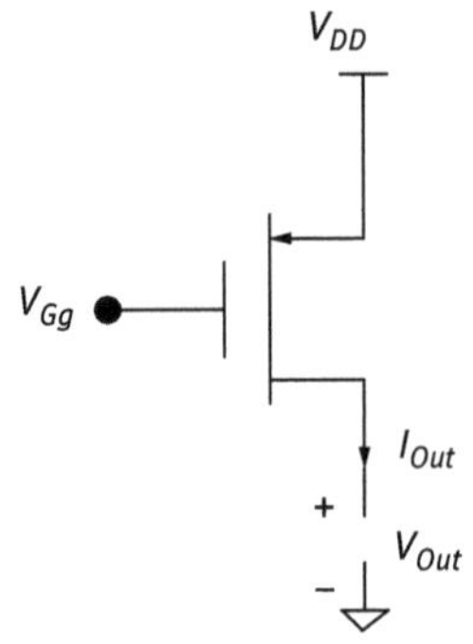

Figure 7.16: Current source circuit.

For proper working of the current sink, the condition is

$$V_{Out} \geq (V_{Gg} - V_{Th}).$$

Figure 7.16 shows an implementation of a current source using a *p*-channel transistor.

Again, the gate is taken to a constant potential as is the source. With the definition of V_{Out} and I_{Out} of the source as shown in Figure 7.16, the large-signal I_{Out} vs. V_{Out} characteristics are shown in Figure 7.17. This current source only works for $V_{Out} \leq V_{Gg} - |V_{Th}|$.

The advantage of the current sink and source of Figures 7.14 and 7.16 is their simplicity. However, there are two areas in which their performance may need to be improved for certain applications. One improvement is to increase the small signal output resistance resulting in a more constant current over the range of V_{Out} values. The second is to reduce the value of V_{Min}.

The small signal output resistance can be increased using the principle illustrated in Figure 7.18.

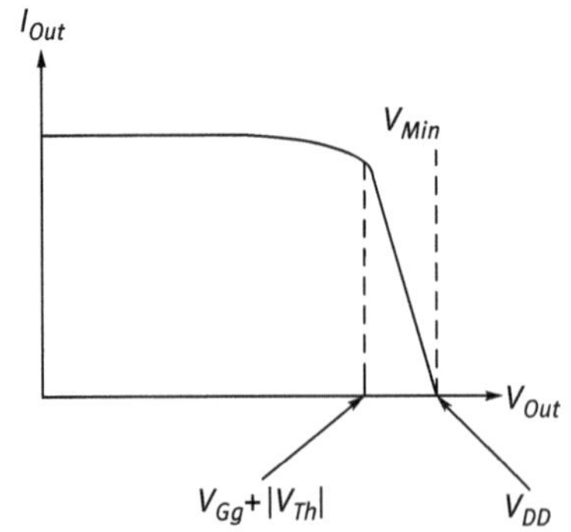

Figure 7.17: $I-V$ characteristics of the current source.

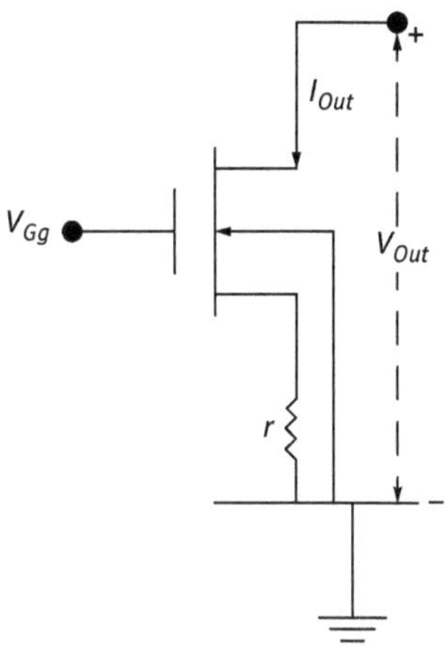

Figure 7.18: Circuit for increasing the output resistance of resistor *r*.

7.5.1 Technique to Increase Output Resistance

The principle uses the common-gate configuration to multiply the source resistance *r* by the approximate voltage gain of the common-gate configuration with an infinite load resistance.

The exact small signal output resistance rout can be calculated from the small signal model of Figure 7.19 as

$$r_{Out} = \frac{v_{Out}}{I_{Out}} = r + r_{Ds} + [(g_m + g_{Mbs})\, r_{Ds}]\, r$$
$$\approx (g_m r_{Ds})\, r, \tag{7.18}$$

where $(g_m r_{Ds}) \gg 1$ and $g_m > g_{Mbs}$.

Now, to implement the above principle, in place of a resistor *r* in Figure 7.19, a MOSFET M_1 is provided in Figure 7.20 with small signal equivalent circuit shown in

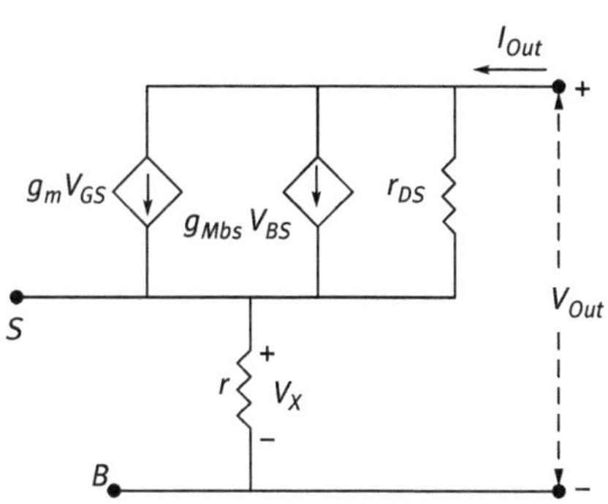

Figure 7.19: Small signal model for the circuit for increasing the output resistance of resistor *r*.

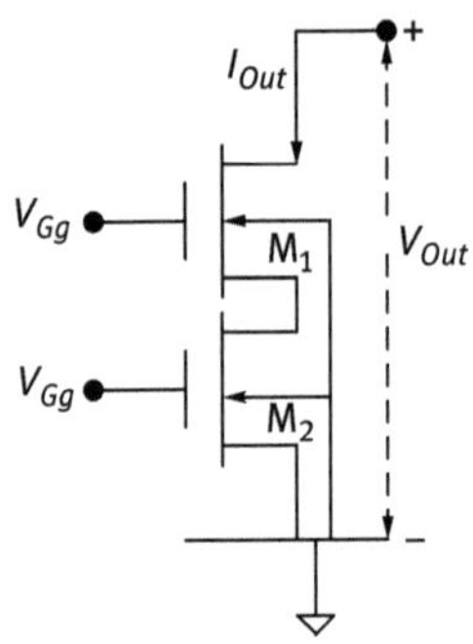

Figure 7.20: Circuit for increasing the output resistance of a current sink.

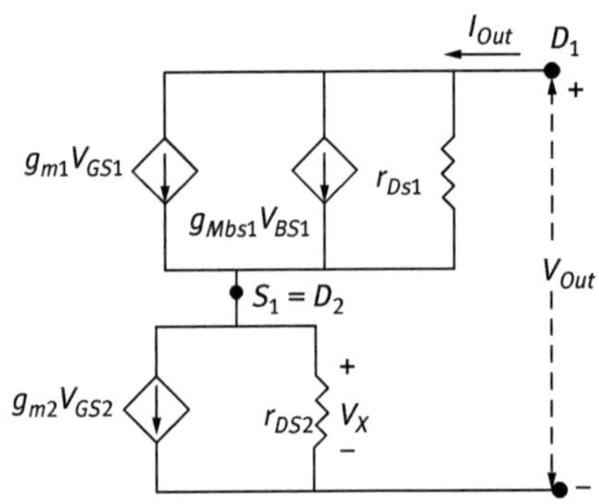

Figure 7.21: Small signal model for the circuit for increasing the output resistance of a current sink.

Figure 7.21. Applying KCL, it is possible to obtain the output resistance r_{Ds1} of the current sink of Figure 7.14 should be increased by the common-gate voltage gain of M_2.

$$v_{Gs2} = 0, v_X = r_{Ds2}I_{Out}, v_{Bs1} = -v_X, v_{Gs1} = -v_X.$$

Applying KCL at $S_1 = D_2$, we obtain

$$g_{m1}v_{Gs1} + \frac{v_{Out}-v_X}{r_{Ds1}} + g_{Mbs1}v_{Bs1} - \frac{v_X}{r_{Ds2}} = 0$$

$$-g_{m1}r_{Ds2}I_{Out} + \frac{v_{Out}+r_{Ds2}I_{Out}}{r_{Ds1}} - g_{Mbs1}r_{Ds2}I_{Out} + \frac{r_{Ds2}I_{Out}}{r_{Ds2}} = 0$$

$$r_{Out} = \frac{v_{Out}}{I_{Out}} = r_{Ds1}\left[g_{m1}r_{Ds2} + \frac{r_{Ds2}}{r_{Ds1}} + g_{Mbs1}r_{Ds2} + 1\right]$$

for $g_{m1} \gg g_{Mbs1}$ and $g_{m1}r_{Ds2} \gg 1$

$$r_{Out} = (g_{m1}r_{Ds1})\, r_{Ds2}. \tag{7.19}$$

We see that the small signal output resistance of the current sink is increased by the factor of $g_{m1}r_{Ds1}$. This principle of increasing output resistance in order to obtain a constant current over a wide range of V_{Out} is known as cascode current sink technique. In a similar manner higher output resistance can be obtained by using cascode current source.

7.6 Low Power Current Mirror

The current mirror is used to generate a replica of a given reference current. Electrical function wise, it is a current-controlled current source (CCCS). Figure 7.22 (iii) shows an ideal current mirror. However, it is worth mentioning that in a real circuit, current mirror cannot perform the task of a CCCS exactly.

7.6.1 Use of Current Mirrors in IC

The biasing technique of transistors with discrete components is different from that of transistors fabricated within an IC. It is due to the fact that it is expensive to fabricate precise resistors within ICs and they occupy a large area within an IC. However, this is not a problem for discrete components circuits. Therefore, common techniques like voltage divider bias and self-bias using resistor and DC voltage sources are employed or biasing analog circuits with discrete components. In contrast, DC current sources are usually used for biasing transistors within an IC. These current sources within an IC are typically designed using MOSFETs. This current source biasing allows the designer to take advantage of matched devices, which are manufactured simultaneously with the same device parameters (parameters from chip to chip will vary). Now the question is, how should a MOSFET be biased to be used as a stable current source? A simple resistor divider can be used as shown in Figure 7.22 (i). Assuming MOSFET in saturation, the current expression is

$$I_{Out} = \frac{1}{2}\mu_n C_{Ox}\frac{W}{L}(V_{GS} - V_{Th})^2 . \qquad (7.20)$$

neglecting CLM and other second-order effects. In this case, V_{Th} may vary from wafer to wafer due to process variability. Also μ_n and V_{Th} can vary due to temperature variation. This temperature and process dependence exists even if the gate voltage is not a function of supply voltage. Thus we must seek other method of biasing the MOSFET current source. Biasing in the analog IC design is based on the use of constant current sources. The design of current sources in analog circuits is based on "copying" currents from a reference. Typically, a complex circuit with external adjustment is used to generate a stable reference current, I_{Ref}, which is copied to many current sources in the system. The process is known a current steering.

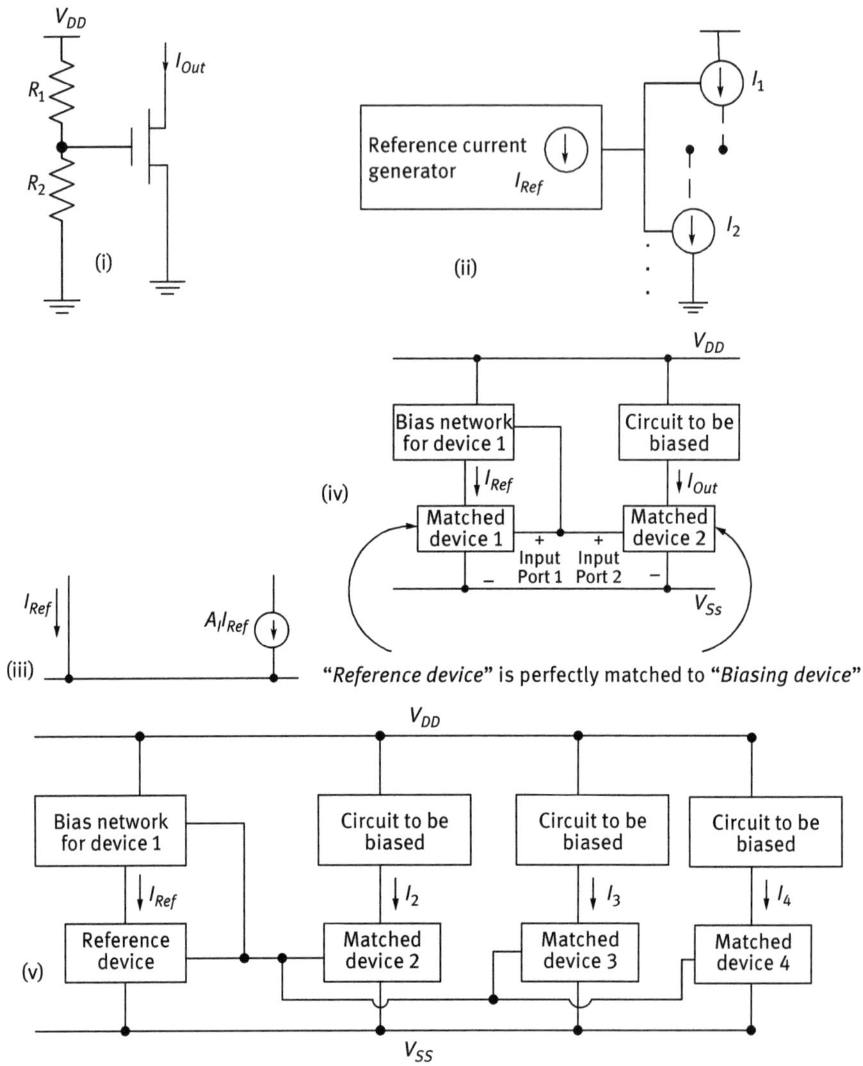

Figure 7.22: (i) Definition of current source by resistor divider; (ii) reference generator; (iii) basic action of current mirror; (iv) basic topology of current mirror circuits; (v) three circuits biased by a common current mirror reference.

Figure 7.22(i) shows the use of a reference to generate various currents. Let us consider for a MOSFET, I_D= f(V_{GS}), where f(.) denotes the functionality or dependency of I_D over V_{GS}. Now, V_{GS} = $f^{-1}(I_D)$. That is, if a transistor is biased at I_{Ref}, it produces V_{GS}= $f^{-1}(I_{Ref})$. Thus, if this voltage is applied to the gate and source of a 2nd transistor, then the resulting current will be $I_{Out} = f\,f^{-1}(I_{Ref}) = I_{Ref}$. In another point of view, two identical MOSFETs having equal gate-to-source voltage operating in saturation region carry equal currents. This structure using two transistors to copy current from one portion of the circuit to other is known as simple current mirror. Current mirror finds wide applicability in analog circuits. The main advantage of this topology is the independence of temperature and process variation. The ratio of I_{Ref} and I_{Out} can be controlled by their relative size and aspect ratio (W/L ratio); thus any value of current can be obtained for the design analog ICs.

Moreover, the current to be copied is not measured ideally as it would be necessary to show a short circuit. Instead, to measure the reference current, a diode-connected MOS transistor is normally used. We can design a number of circuits which accomplish the current mirror function. The ones mostly used (and studied below) are

- simple current mirror,
- Wilson current mirror and
- cascode current mirror.

7.6.2 Simple Current Mirror

The implementation of the current mirror shown in Figure 7.23 is the simplest form of the required function: it is composed of two transistors, of which one, M_1, is diode connected. M_1 receives the reference current I_{Ref} and measures it by developing at its gate the voltage V_{GS1}; this voltage biases the gate of M_2. We assume that both transistors operate in the saturation region; therefore, the currents are

$$I_{Ref} = I_1 = \mu_n \frac{C_{Ox}}{2} \left(\frac{W}{L}\right)_1 (V_{GS1} - V_{Th})^2 (1 + \lambda V_{DS1}) \tag{7.21}$$

$$I_{Out} = \mu_n \frac{C_{Ox}}{2} \left(\frac{W}{L}\right)_2 (V_{GS1} - V_{Th})^2 (1 + \lambda V_{DS2}), \tag{7.22}$$

which enable us to express the output current I_{Out} as a function of I_{Ref}, V_{Out} and V_{DS2}. To simplify algebraic calculations, we assume the term $\lambda V_{DS1} = 0$.

Thus, V_{GS1} is directly derived from the former equation, and after substitution in the latter, gives

$$I_{Out} = I_{Ref} \frac{(W/L)_2}{(W/L)_1} (1 + \lambda V_{Out}). \tag{7.23}$$

Apart from the term $(1 + \lambda V_{Out})$ the output current is a replica of I_{Ref} multiplied by the aspect ratios, W/L of the transistors M_2 and M_1. The term $(1+\lambda V_{Out})$ takes into account the finite output resistance of M_2 that, for small signal, is $r_{Out} = \frac{1}{\lambda I_{Out}}$.

Unfortunately, the value of output resistance which can be achieved with the technologies and medium value currents used is not large enough for a number of applications. Assuming $\lambda = 1/30V^{-1}$, r_{Out} is as low as $300k\Omega$ for $I_{Out} = 100\mu A$. Thus,

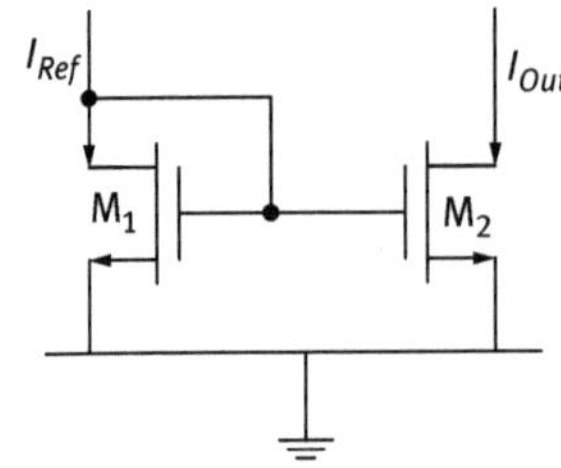

Figure 7.23: Simple current mirror.

as we will study shortly, other solutions must be used when output resistance is a key design issue. However, for low voltage applications the simple scheme in Figure 7.23 may be preferred because of its excellent output dynamic range: the output node (the drain of M_2) can swing down to the saturation voltage of M_2, whose value, for common designs, is as low as a few hundred mV. More complex solutions allow us to increase the output impedance but we normally have to pay for this benefit with a reduced dynamic range. Both the simple current mirror and the other schemes that we are going to study may deviate from ideal behavior. This is due to

- imperfect geometrical matching,
- technological parameter mismatch and
- parasitic resistances.

Let us again consider eqs (7.21) and (7.22). We assumed equal values of technological parameters and properly ratioed geometrical dimensions of transistors to obtain eq. (7.24). This is not attained in a real circuit and certain mismatches in the geometrical dimension and the technological parameters will always exist. Any mismatches will produce an error in the generated current.

If all the parameters in eqs (7.21) and (7.22) are statistically independent we obtain

$$\left(\frac{\partial I_{Out}}{I_{Out}}\right)^2 = \left(\frac{\partial W}{W}\right)^2 + \left(\frac{\partial L}{L}\right)^2 + \left(\frac{\partial \mu_n}{\mu_n}\right)^2 + \left(\frac{\partial C_{Ox}}{C_{Ox}}\right)^2 + 2\left(\frac{\partial V_{Th}}{V_{GS} - V_{Th}}\right)^2 + 2\left(\frac{\partial V_{GS}}{V_{GS} - V_{Th}}\right)^2 . \tag{7.24}$$

Hence, current incorrectness derives from the quadratic superposing of relative geometrical and technological mismatches. Inaccuracy in geometrical dimensions comes from photolithographic processes and etching. To limit these, the layout should take into account undercut and boundary-dependent effects. The simple current mirror is superior to all the other architecture studied here when the output dynamic range is the key target. Errors due to mobility and oxide thickness mainly come from unavoidable technology gradients along the surface of the chip. We can reduce this effect by using an inter-digitized or a common centroid structures which minimize the distance between the transistors.

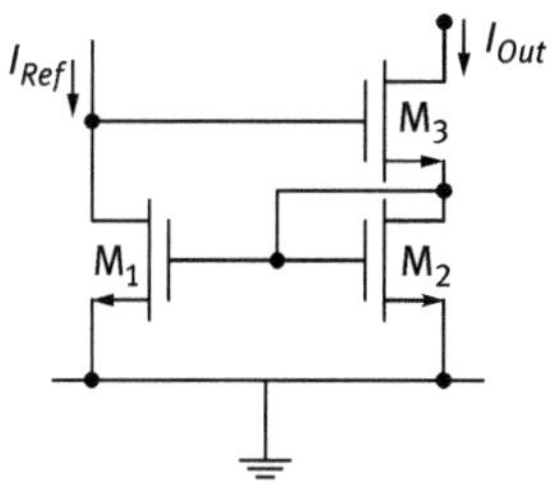

Figure 7.24: Wilson current mirror.

7.6.3 Wilson Current Mirror

The relatively low value of the output resistance of the simple current mirror can be improved with the Wilson scheme shown in Figure 7.24. The gate-to-source voltages of M_1 and M_2 are equal, therefore ensuring similar operation to the circuit in Figure 7.23. However, as we shall see, the addition of M_3 and the established local feedback allows us to increase the output resistance. The small signal equivalent circuit of the stage is shown in Figure 7.25.

Resistance R_L represents the external load seen from the reference current connection. Analysis of the circuit provides

$$\begin{aligned} v_{g2} &= v_{s3} = i_x/g_{m2}, \\ v_{g3} &= -g_{m1}v_{g2}r_T, \\ v_X &= \frac{i_X}{g_{m2}} + (i_X - g_{m3}v_{Gs3})r_{Ds3}, \end{aligned} \tag{7.25}$$

where r_T denotes the parallel connection of R_L with r_{Ds1}. The Wilson current mirror earns a high output resistance only if the reference current comes from a high impedance source for which small signal resistance must be higher or comparable to r_{Ds1}.

From (7.24) the output resistance is given by

$$r_{Out} = \frac{1}{g_{m2}} + r_{Ds3}\left[1 + \frac{g_{m3}}{g_{m2}}(1 + g_{m1}r_T)\right]. \tag{7.26}$$

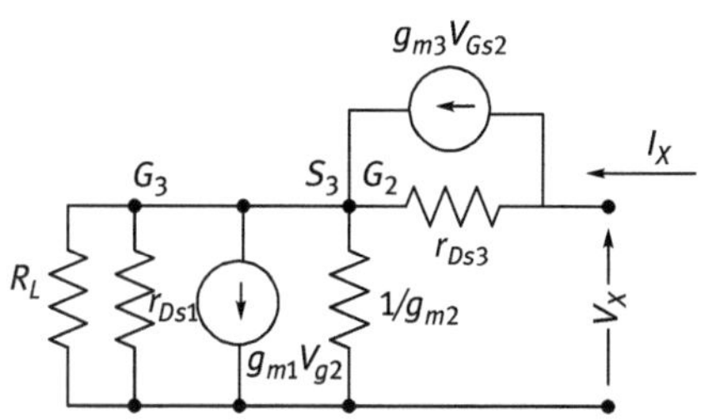

Figure 7.25: Small signal equivalent circuit of Wilson current mirror.

Transconductances gm_2 and gm_3 have same order of magnitude because M_2 and M_3 are carrying the same current. Therefore, the output resistance of the circuit is approximately determined by the output resistance of M_3 (r_{DS3}) amplified by the factor $r_T\,gm_1$. It is large if r_T is large and, in turn, if R_L is sufficiently large. This condition is naturally met when the reference current comes from a current source. The circuit in Figure 7.23 suffers from a shortcoming. The drain-to-source voltages of M_1 and M_2 are systematically different, in fact

$$V_{DS1} = V_{GS3} + V_{DS2}.$$

This, because of the CLM effect (or the finite small signal output conductance), reveals a systematic mismatch between reference current and output current.

Approximately, the reference current is larger than the generated one by the amount V_{GS3}/r_{DS1}, remembering that $1/r_{DS1} = \lambda_1 I_{D1}$ the current difference becomes $\lambda_1 I_{D1} V_{GS3}$.

7.6.4 Cascode Current Mirror

By increasing the output resistance, CLM can be avoided. In the current mirror circuit, V_{DS} may not be always equal, leading to imperfection in mirror action. An alternative way to increase the output resistance is to use a cascode configuration since it is known that a cascode amplifier offers higher output resistance. High output impedance arises from the fact that if the output node voltage is changed by ΔV, the resulting change at the source of the cascode device is much less, which is known as shielding property. In the circuit of simple current mirror shown in Figure 7.23, V_{DS2} may not be equal to $V_{GS2} = V_{GS1} = V_{DS1}$, because of the circuitry fed by M_2. Therefore, in order to suppress CLM and to make V_{DS} equal, a 3rd transistor is connected in cascode to M_2 as shown in Figure 7.26. The voltage V_{Bias} is chosen such that $V_A = V_B$. This ensures $I_{Out} = I_{Ref}$.

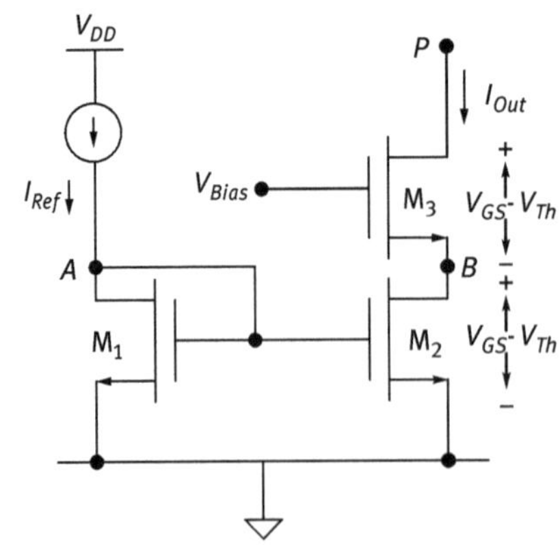

Figure 7.26: Adding a 3rd transistor to simple current mirror circuit in order to mitigate channel length modulation.

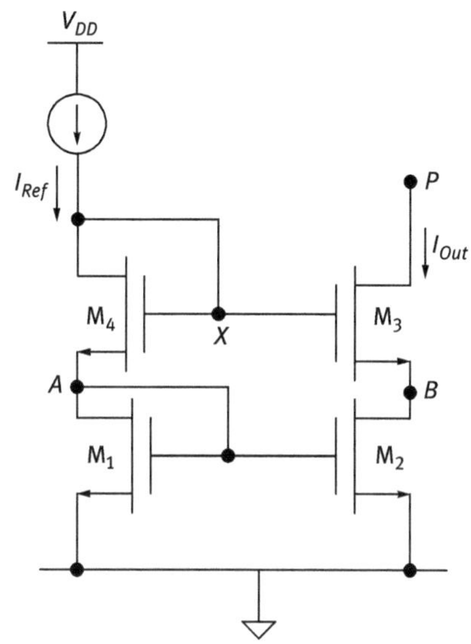

Figure 7.27: Cascode current mirror circuit.

The voltage V_{Bias} is given by

$$V_{Bias} = V_{GS3} + V_B.$$

For proper mirroring action, $V_B = V_A$. Therefore, $\mathrm{V_{Bias}} = \mathrm{V_{GS3}} + \mathrm{V_A}$. Therefore, it is clear that V_{Bias} can be obtained by adding V_A to gate-to-source voltage of M_3. Now consider the circuit shown in Figure 7.27 showing a possible scheme. Now,

$$V_X = V_{GS4} + V_X.$$

As node X is connected to the gate of M_3

$$V_{GS4} + V_A = V_{GS3} + V_B.$$

Therefore, if we can ensure $V_{GS3} = V_{GS4}$ by choosing $\frac{(W/L)_3}{(W/L)_4} = \frac{(W/L)_2}{(W/L)_1}$ then it yields $V_A = V_B$, which causes proper current mirror action by providing $I_{Ref} = I_{Out}$. It can be shown that this result is valid even if M_4 and M_3 suffer from substrate-bias effects. Minimum voltage at P to keep M_2 and M_3 in saturation is equal to $V_P = 2V_{Ov} = 2(V_{GS} - V_{Th})$.

The output stage consists of two transistors ($M_2 - M_3$) in the cascode arrangement. Their biases result from two other transistors ($M_1 - M_4$) which are diode connected. Again, as for the previously studied current mirrors, the V_{GS} voltages of M_1 and M_2 are set equal.

Therefore, a replica of the current in M_1 is generated by M_2. The output resistance increases because of the cascode arrangement which can be explained by examining the small signal analysis. However, the increased output resistance obtained from cascode configuration is traded off by a reduction of the output dynamic range. It consumes substantial voltage in the headroom. To ensure M_3 in saturation, the minimum allowable output voltage at node P is equal to

$$\begin{aligned} V_P &= V_{DS3} \geq V_{GS3} - V_{Th}, \\ &\geq V_X - V_{Th}, \\ &\geq (V_{GS4} + V_{GS1}) - V_{Th}, \\ &\geq (V_{GS4} - V_{Th}) + (V_{GS1} - V_{Th}) + V_{Th}, \\ &\geq 2V_{Ov} + V_{Th}. \end{aligned}$$

i.e. it is equal to two overdrive voltage plus one threshold voltage. Thus, compared to the circuit shown in Figure 7.26, this circuit shown in Figure 7.27 wastes one threshold voltage in the headroom and reduces the output swing.

7.6.5 Low Voltage Current Mirror

A current mirror is considered as one of the fundamental building blocks of analog IC designing and is used to copy the current through one MOSFET by controlling the current in other MOSFET of the circuit, keeping the output current independent of loading in the circuit. The main task of a current mirror is to perform current amplification and to provide biasing in analog IC. The efficiency of current mirror has a direct impact on the analog IC performance as current mirror is used in a large number of IC building blocks like OTA, current conveyors, CFOAs (current feedback operational amplifiers), filters etc. [33].

The relevant performance parameters of a current mirror for use in typical analog IC are [34] the following:

(1) Linearity in current transfer
(2) Power consumption
(3) Minimum supply voltage
(4) Input and output resistances

An ideal current mirror has zero input resistance and infinite output resistance. However, in practice, practical CM exhibits nonzero input and finite output resistance. The aim of the designer is to achieve smallest input resistance and largest output resistance so that error ($\Delta I = I_{In} - I_{Out}$) in the current transfer equals to $I_{In}(R_{In}/R_{Out})$.

In ideal situation, the output current is independent of the voltage at the output node. In practice, however, the minimum voltage at the output, which ensures the device is operating at the saturation region, is called the output compliance voltage.

In summary, the desirable characteristics of a current mirror are the following:

(1) Current transfer ratio must be adjusted by (W/L) ratio, not by temperature

(2) High output impedance (large R_{Out} and small C_{Out}) in order to ensure the independency of output current of output voltage
(3) Low input resistance
(4) Low input and output compliance voltage ensuring a higher voltage swing

Conventional current mirrors suffer with high input impedance and low output impedance. This drawback has been studied by many researchers [35].

It was observed that the insertion of a voltage amplifier between the drain and the gate of the input transistor causes a lowering of input impedance of the current mirror. Several possible implementation of the voltage amplifier has also been reported. However, the major drawback associated with this configuration is the poor stability. An improvement of performance can be obtained by replacing the voltage amplifier with differential amplifier [36].

However, differential amplifier suffers with the drawbacks of higher supply voltage and associated higher power consumption. Later, it was claimed that the introduction of an active input-regulated cascode (AIRC) in the differential amplifier in the input and output side of the current mirror causes a decrease in the input resistance and an increase in the output resistance [37].

On the other hand, the current mirrors using flipped voltage follower (FVF) scheme suffer with transient and bandwidth performance degradation, circuit complexity and a careful design of biasing network [38].

Alternatively current mirror based on bulk-driven, subthreshold and floating-gate technology has also been proposed. The bulk-driven current mirrors suffer with higher parasitic components and reduced swing. In contrast, the subthreshold current mirrors have the drawbacks of low transconductance resulting in poor bandwidth and low matching accuracy [39–42].

In a conventional current mirror, the input compliance voltage and output compliance voltage are given by

$$V_{In} = V_{Th} + \sqrt{\frac{2I_{In}}{\beta_i}} \tag{7.27}$$

$$V_{Out} = V_{DS(sat)}$$

$$\beta_i = \mu_n C_{Ox} \frac{W}{L},$$

where V_{Th} is the threshold voltage, I_{In} is the input current, μ_n is the carrier mobility and C_{Ox} is oxide capacitance. W and L refer to the width and length of the MOS transistors.

The lower limit of V_{In} in eq. (7.27) is at least the value of the threshold voltage V_{Th}. If the input current I_{In} is small, it forces to act M_1 in the subthreshold region where $V_{In} = V_{Ref}$ is sufficiently low (<100 mV).

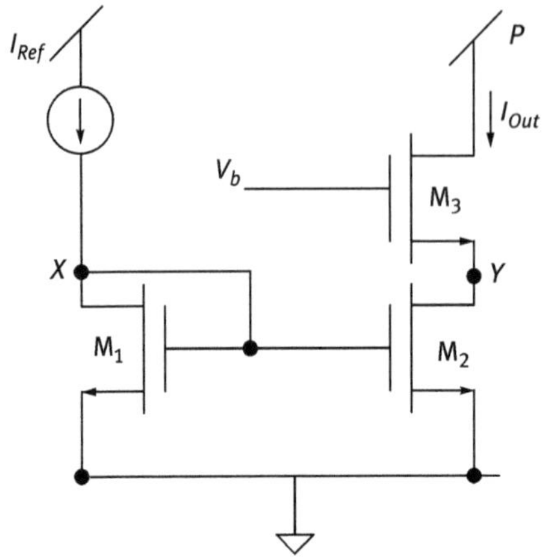

Figure 7.28: Current mirror circuit with single transistor at the input port.

The conventional CM has mainly two drawbacks:
(1) Input compliance voltage is always greater than V_{Th}.
(2) Low output compliance voltage.

Alternatively cascade structure can be used to increase the output resistance [43–45]. The input compliance voltage of cascode CM is given by

$$V_{In} = 2V_{Th} + 2\sqrt{\frac{2I_{In}}{\beta_i}}. \tag{7.28}$$

The input compliance voltage of cascade CM is double than that of a conventional CM and hence cannot be used in low voltage applications. The analogue designers

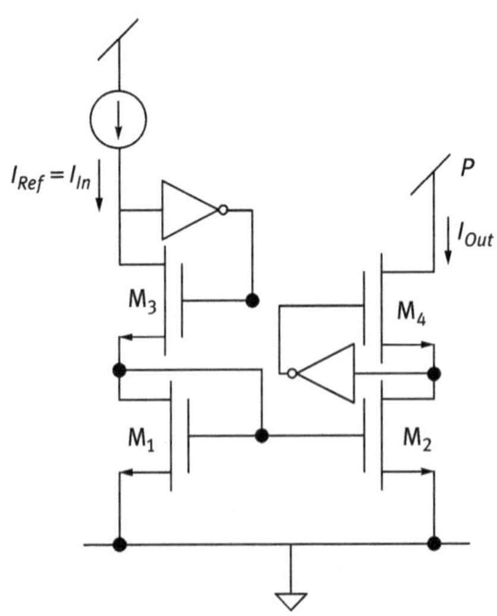

Figure 7.29: Current mirror circuit with amplifiers for bias voltage.

prefer single transistor at the input port while keeping the cascode structure at the output port as shown in Figure 7.28. Problem with this circuit is to provide suitable biasing voltage Vb for charging the gate of M_3. Recently, a current mirror with amplifiers for bias voltage has been proposed to improve the input and output resistance performance as shown in Figure 7.29 [46].

7.7 Fundamentals of Current/Voltage Reference

An ideal current or voltage reference is independent of power supply and temperature. The large signal current and voltage characteristics of an ideal current and voltage reference are shown in Figure 7.30.

These characteristics are identical to those of the ideal current and voltage source. The term "reference" is used when the current or voltage values have more precision and stability than ordinarily found in a source.

A reference is typically dependent on the load connected to it. In the discussion that follows, it will be assumed that an HP voltage reference can be used to implement an HP current reference and vice versa.

A very crude voltage reference can be made from a voltage divider between the power supplies as shown in Figures 7.31 and 7.32. Passive or active components can be used as the divider element.

Figures 7.31 and 7.32 show an example of each. Unfortunately the value of V_{Ref} is directly proportional to the power supply. Let us introduce the concept of sensitivity S. The sensitivity of V_{Ref} to V_{DD} can be expressed as

$$S_{DD}^{V_{Ref}} = \frac{(\partial V_{Ref}/V_{Ref})}{(\partial V_{DD}/V_{DD})} = \frac{V_{DD}}{V_{Ref}}\left(\frac{\partial V_{Ref}}{\partial V_{DD}}\right)$$

$$S_{DD}^{V_{Ref}} = \frac{(\partial V_{Ref}/V_{Ref})}{(\partial V_{DD}/V_{DD})} = \frac{V_{DD}}{V_{Ref}}\left(\frac{\partial V_{Ref}}{\partial V_{DD}}\right). \tag{7.29}$$

Equation (7.29) can be interpreted as follows: if the sensitivity is 1, then a 10% change in V_{DD} will result in a 10% change in V_{Ref} (which is undesirable for a voltage reference). A simple way of obtaining a better voltage reference is to use an active device as shown in Figure 7.33.

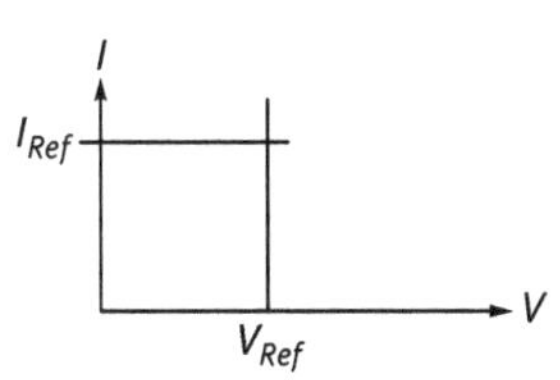

Figure 7.30: I–V characteristics of ideal current and voltage reference.

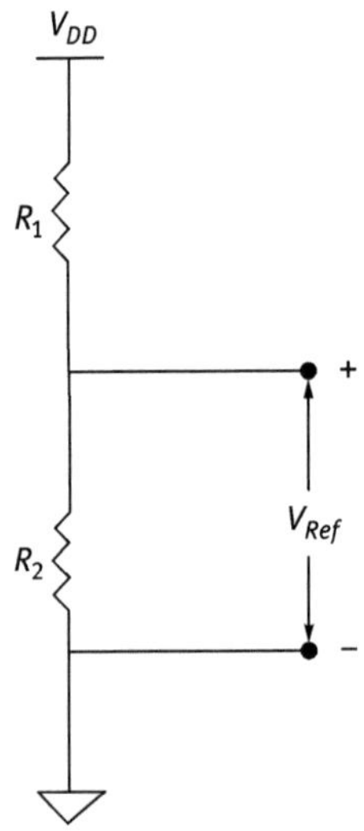

Figure 7.31: Resistor implementation of voltage references using voltage divider.

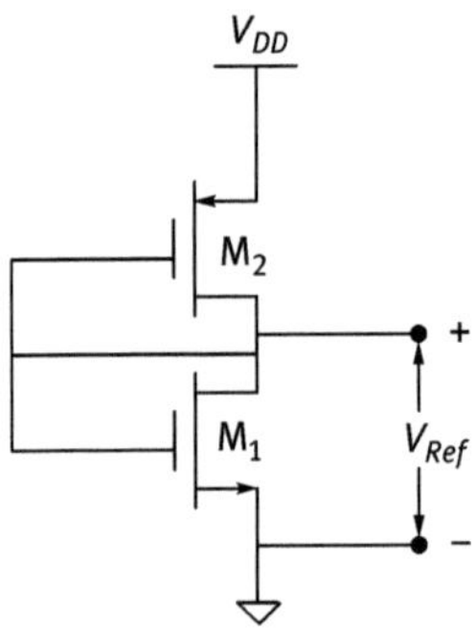

Figure 7.32: Active device implementation of voltage references using voltage divider.

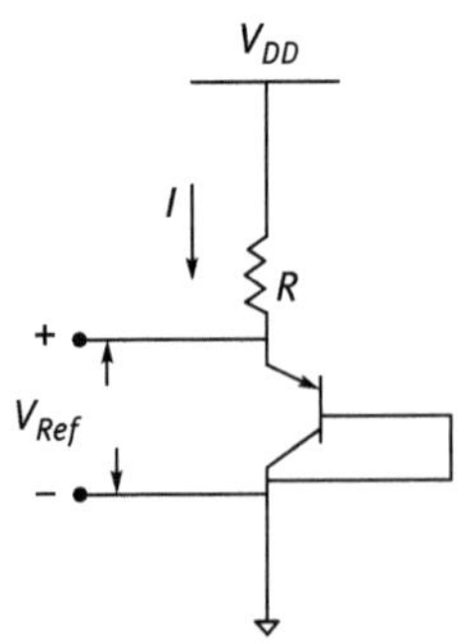

Figure 7.33: PN-junction voltage reference.

In Figure 7.33, the substrate Bipolar Junction Transistor (BJT) has been connected to the power supply through a resistance R. The voltage across the PN-junction is given as

$$V_{Ref} = V_{Eb} = \frac{kT}{q}\ln\left(\frac{I}{I_S}\right),$$

where I_S is the junction saturation current.

If $V_{DD} >> V_{Eb}$ then current I is given as

$$V_{DD} \gg V_{Eb}.$$

So the reference voltage of this circuit is given as

$$I = \frac{V_{DD} - V_{Eb}}{R} \approx \frac{V_{DD}}{R}$$

$$V_{Ref} \approx \frac{kT}{q}\ln\left(\frac{V_{DD}}{RI_S}\right).$$

The sensitivity of V_{Ref} to V_{DD} is given by

$$S_{DD}^{V_{Ref}} = \frac{(\partial V_{Ref}/V_{Ref})}{(\partial V_{DD}/V_{DD})} = \frac{V_{DD}}{V_{Ref}}\left(\frac{\partial V_{Ref}}{\partial V_{DD}}\right) = \frac{1}{\ln\frac{V_{DD}}{RI_S}} = \frac{1}{\ln\frac{I}{I_S}}. \tag{7.30}$$

Since I is normally greater than I_S, the sensitivity of V_{Ref} of Figure 7.33 is less than unity. The V_{Ref} obtained in the above manner is small. Figure 7.34 shows a method of increasing the value of V_{Ref}. The reference voltage of Figure 7.34 can be written as

$$V_{Ref} \approx V_{Eb}\left(\frac{R_1 + R_2}{R_1}\right). \tag{7.31}$$

If $(R_1 + R_2)$ is large, V_{Ref} will also be large.

The BJT of Figure 7.34 may be replaced with an MOS enhancement device to achieve a voltage that is less dependent on V_{DD} as shown in Figure 7.35.

$$V_{GS} = V_{Th} + \sqrt{2\frac{I}{\beta}} \tag{7.32}$$

Ignoring CLM, it is possible to write

$$V_{Ref} = V_{Th} - \frac{1}{\beta R} + \sqrt{\frac{2(V_{DD} - V_{Th})}{\beta R} + \frac{1}{\beta^2 R^2}}. \tag{7.33}$$

Sensitivity can be expressed by

$$S_{Dd}^{V_{Ref}} = \frac{V_{DD}}{V_{Ref}}\frac{1}{1 + \beta(V_{Ref} - V_{Th})R} \tag{7.34}$$

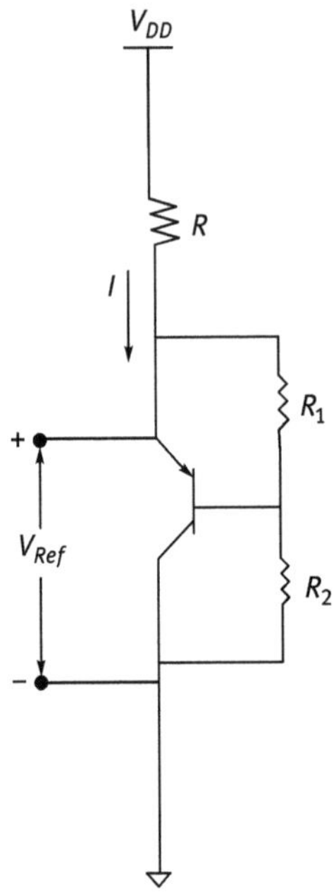

Figure 7.34: Circuit for increasing V_{Ref} of the PN-junction voltage reference.

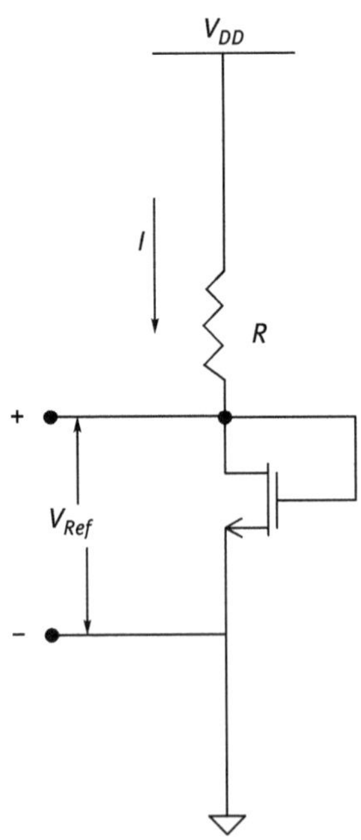

Figure 7.35: MOS equivalent of PN-junction voltage reference.

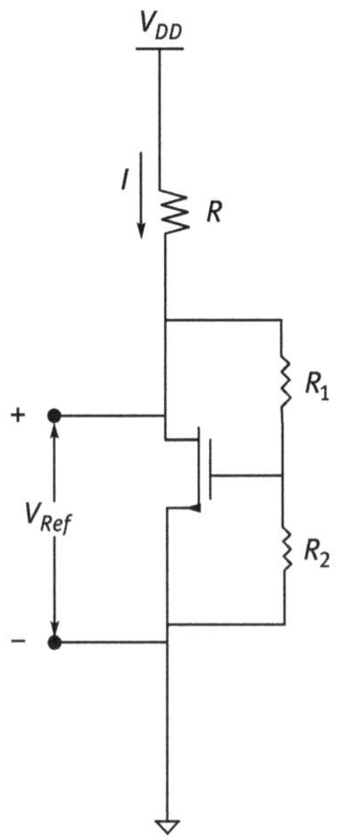

Figure 7.36: Circuit for increasing V_{Ref} of the MOS equivalent of the PN-junction voltage reference.

From Figure 7.36,

$$V_{Ref} = V_{GS}\left(1 + \frac{R_2}{R_1}\right). \tag{7.35}$$

In order to adjust the value of V_{Ref}, Geometry dependent parameters can be varied. In the BJT reference the geometry-dependent parameter is I_S and for the MOS reference it is W/L. The small signal output resistance of these references is a measure of how dependent the reference will be on the load.

We have seen that since I is normally greater than I_S, the sensitivity of V_{Ref} across an active device is less than unity. If the voltage across the active device is used to create a current and which in turn is used to give the original current through the device then we obtain a voltage or current independent of V_{DD}. This technique is named V_{Th} referenced source and is also called a bootstrap reference.

The circuit in Figure 7.37 is an example of this technique using MOS devices. Here M_3 and M_4 are doing current mirroring. This causes the current I_1 to be equal to I_2. Due to current I_1 flowing through M_1 a voltage drop V_{GS1} is developed. Also due to the current I_2 flowing through R a voltage drop I_2R is developed. Since these two voltages are connected with each other, an equilibrium point is established as shown in Figure 7.38. Here I_1 and I_2 are plotted as a function of the voltage. The equilibrium point Q is obtained by the intersection of these curves.

At equilibrium, $I_2R = V_{GS1}$. We know that the current I_1 is given by

$$I_1 = \frac{1}{2}K_n\frac{W_1}{L_1}(V_{GS1} - V_{Th})^2.$$

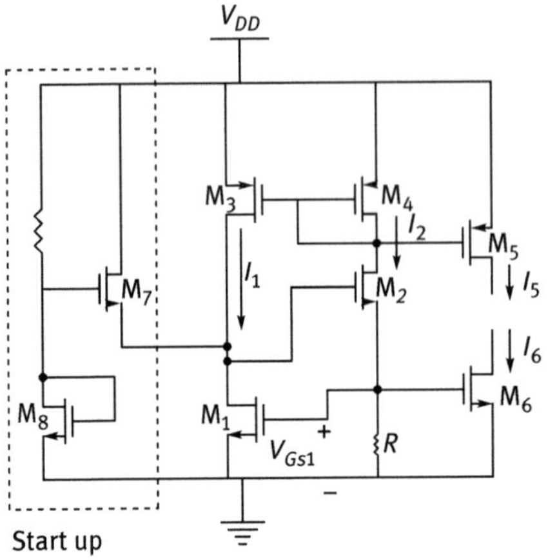

Figure 7.37: Threshold referenced circuit.

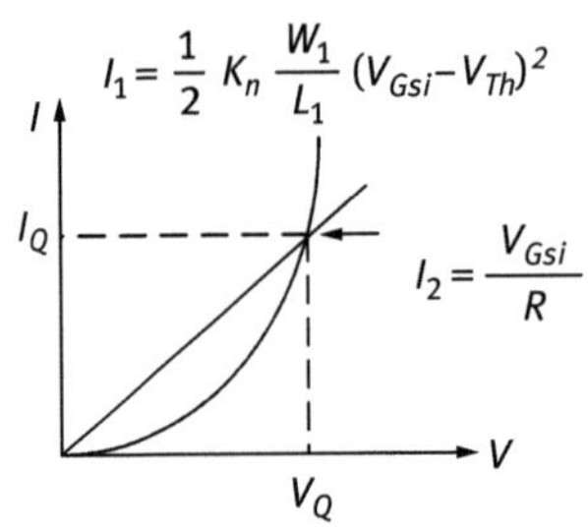

Figure 7.38: I–V characteristics of the threshold referenced circuit showing how the bias point is established.

This equation can be solved for $I_1 = I_2 = I_Q$.

$$V_{GS1} = V_{Th} + \sqrt{\frac{2L_1 I_1}{K_n W_1}} = I_2 R$$

$$I_1 = I_2 = I_Q$$

$$I_1 = I_2 = I_Q = \frac{V_{Th}}{R} + \frac{1}{\beta_1 R^2} + \frac{1}{R}\sqrt{\frac{2V_{Th}}{\beta_1 R} + \frac{1}{\beta_1^2 R^2}}. \tag{7.36}$$

So neither I_1 nor I_2 changes as a function of V_{DD}. Therefore, the sensitivity of I_Q to V_{DD} is zero which is the ideal one. By mirroring $I_2 = I_Q$ through M_5 or M_6 a voltage reference can be achieved using a resistor. However, there are two intersection points in the I versus Vcurves in Figure 7.38: one at the origin and the other at Q. So there are two equilibrium points and for the circuit to choose the correct point a start-up circuit is required. Both I_1 and I_2 are zero at the origin of the I versus V characteristics. But due to M_7 which provides a current through M_1 the circuit will move to the equilibrium point Q. As the Q point is approached, due to the increase in the source voltage of M_7 the current through it decreases. At Q the current through M_1 and M_3 are equal.

Alternately a circuit shown in Figure 7.39 uses V_{BE} to reference the voltage or current. This circuit also has two reference points and a start-up circuit is required to choose the correct one.

Both the circuits in Figures 7.38 and 7.39 can be operated in the weak-threshold inversion for developing a low power and supply voltage reference.

7.7.1 Another Way to Obtain Simple Bootstrap Voltage Reference Circuit with Start-up Circuit

Our objective is to obtain a supply (V_{DD})-independent bootstrap voltage reference; i.e. if the voltage across the active device is used to create a current and this current is somehow used to provide the original current through the device, then a current or voltage could be obtained that for all practical purposes becomes independent of supply voltage V_{DD}.

Let us consider the circuit shown in Figure 7.40.

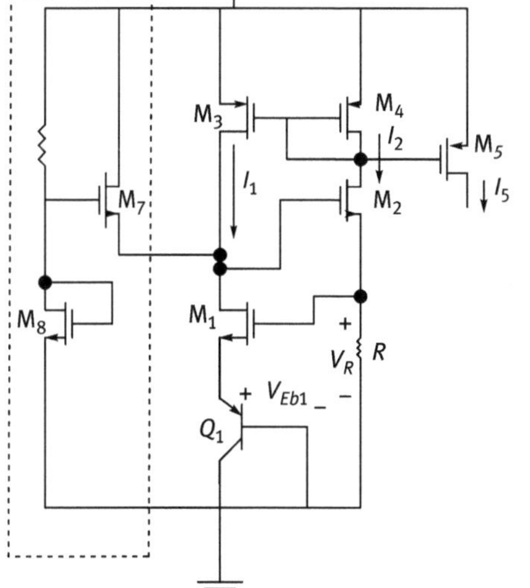

Figure 7.39: Base-emitter voltage referenced circuit.

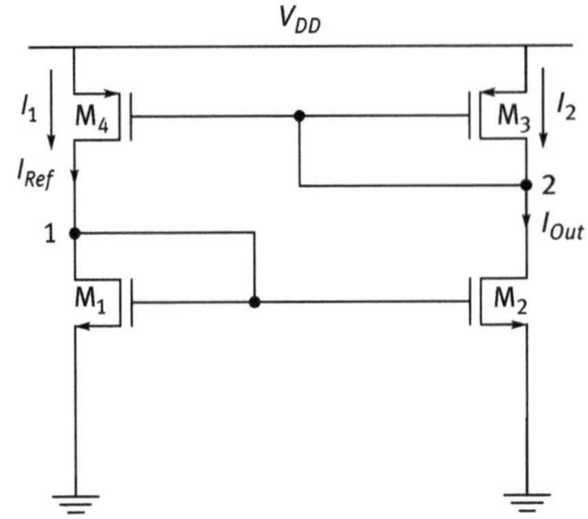

Figure 7.40: Crude circuit for bootstrap voltage reference.

If we set the ratio of size of M_4:M_3 = 1:1 and M_1:M_2 = 1:1 the current through the two paths becomes equal; i.e. the current I_1 through $M_4 - M_1$ is equal to the current I_2 through $M_2 - M_3$. Another important assumption to carry out is the voltage at node 2 is equal to the voltage at node 1; i.e. $V_2 = V_1$. To prove it, let us consider for example, V_1 is larger than V_2. Now if V_1 is higher than V_2 then I_1 will be less than I_2. Therefore, current through M_1 is less than current through M_2. Thus, higher current through M_2 causes higher voltage drop across M_2, which indicates that V_2 is larger than V_1. This is a contradiction and is just opposite to our starting assumption. Therefore, V_1 must equal V_2.

Therefore, in this circuit I_{Ref} is derived from I_{Out}. Thus, I_{Ref} is bootstrapped to I_{Out} as M_3– and M_4 copy I_{Out}, thereby defining I_{Ref}. If CLM is neglected, since each diode-connected device feeds from a current source, I_{Out} and I_{Ref} are relatively independent of V_{DD}. Now in order to obtain $I_{Out} = KI_{Ref}$ for all general applications, we make some changes in the circuit as shown in Figure 7.41.

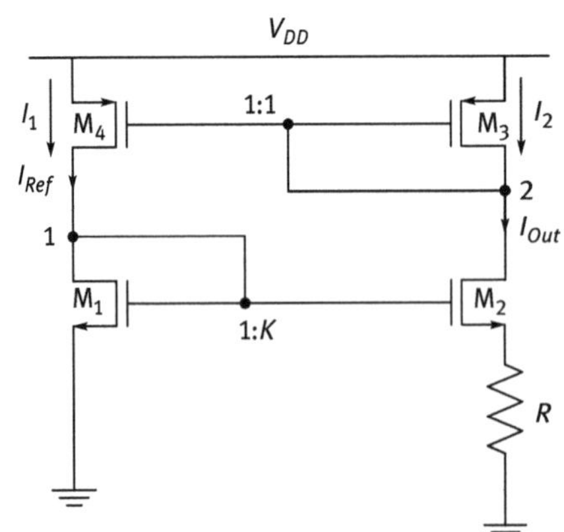

Figure 7.41: Modified version of bootstrap reference.

In this circuit, for example if we force $I_{Ref} = 10\mu A$, I_{Out} that equals K × 10 μA circulates around the loop, sustaining those current levels on the left-side and right-side branches indefinitely. To find the value of the current I_{Out}, it is possible to write

$$V_{GS1} = V_{GS2} + I_{D2}R$$
$$\sqrt{\frac{2I_{Out}}{\mu_n C_{Ox}(W/L)_N}} + V_{Th1} = \sqrt{\frac{2I_{Out}}{\mu_n C_{Ox} K (W/L)_N}} + V_{Th2} + I_{Out}R.$$

Assuming $V_{Th1} = V_{Th2}$

$$\sqrt{\frac{2I_{Out}}{\mu_n C_{Ox}(W/L)_N}}\left(1 - \frac{1}{\sqrt{K}}\right) = I_{Out}R$$
$$\frac{2I_{Out}}{\mu_n C_{Ox}(W/L)_N}\left(1 - \frac{1}{\sqrt{K}}\right)^2 = I_{Out}^2 R^2$$

$$I_{Out} = \frac{2}{\mu_n C_{Ox}(W/L)_N}\frac{1}{R^2}\left(1 - \frac{1}{\sqrt{K}}\right)^2. \tag{7.37}$$

Now, there is another question regarding the circuit: how does the circuit work when we just switch on the power? If the currents through the two branches equal to zero, then zero current will circulate through the loop. In the above expression of current, there is no term of supply voltage V_{DD}. Therefore, only switching on V_{DD} does not have any impact on the current flowing through the circuit. Since initially before switching on the V_{DD}, as the current is zero, after switching on V_{DD}, zero current will flow through the circuit using current mirror action. Therefore, when we switch on V_{DD}, we need to introduce artificially some amount of current in the circuit. If we inject some amount of initial current, the circuit itself balances itself. Therefore, some extra circuitry is required to inject some initial current, known as start-up circuit. Figure 7.42 shows the modified voltage reference with start-up circuit.

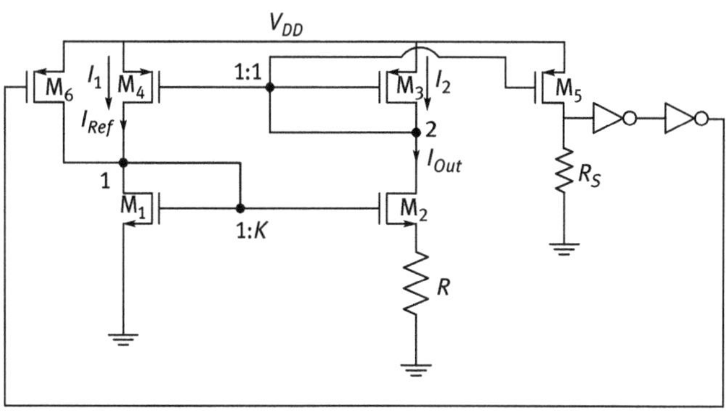

Figure 7.42: Bootstrap voltage reference with start-up circuit.

To describe its working, let us consider at the initial stage all the node voltages and currents are zero. Now as the power is switched on, the current through M_2 is zero or very small. Therefore, current through M_5 is also zero as M_5 and M_3 are mirror connected. Therefore, the drop across R_S is also small. Thus, the input to the first digital CMOS inverter is logic "0", i.e. below the threshold voltage of the inverter. Then output of the second inverter is logic "0", which turns on M_6, which is very much conductive. As M_6 is turned on, it introduces and injects current in the circuit from V_{DD} through M_6 and M_1. Because of mirror action, this current circulates through the loop. Therefore, current through M_2 and M_3 increases. This causes increase in current through M_5, which causes an increase in the voltage drop across R_S. As this built-up voltage drop across R_S crosses the threshold voltage of the inverter, the first inverter identifies its input as a logic "1". Then the logic "1" output of the second inverter causes M_6 to cease its conduction. Then the circuit balances itself. Therefore, introduction of this start-up circuit causes introduction of some current through the circuit; later the circuit stabilizes itself to its required I_{Out} value.

7.8 Bandgap Voltage Reference

Analog IC incorporates voltage and current references. As mentioned earlier, the objective of reference generation is to establish a DC voltage or current that is independent of the supply, process and temperature variation. The bootstrapped reference discussed in earlier section is not sensitive to supply variation; i.e. they are supply independent. However, in addition to supply variability, process and temperature variability may be critical as well. It is worth mentioning that reference voltage or currents that exhibit little dependence on temperature prove critical in many applications. It is interesting to note that, since most process parameters vary with temperature, if a reference is temperature-independent, then usually it is process independent as well.

Now the question is how we can generate a quantity that remains constant with temperature. We postulate that if two quantities having opposite temperature coefficients (TCs) are added with proper weighting, they result a zero TC. For example, consider two voltages V_1 and V_2 with opposite TCs. Now, if we choose a_1 and b_1 in such a way so that

$a_1 \frac{\partial V_1}{\partial T} + b_1 \frac{\partial V_2}{\partial T} = 0$, then we can obtain a reference voltage $V_{Ref} = a_1 V_1 + b_1 V_2$ with zero TC. We should now identify two voltages having positive and negative TCs.

7.8.1 Positive TC Voltage

It was observed that the base-emitter voltage (V_{BE}) difference between two bipolar transistors is proportional to absolute temperature (PTAT). The circuit shown in Figure 7.43 shows two identical transistors (with equal scale current $I_{S1} = I_{S2}$) are

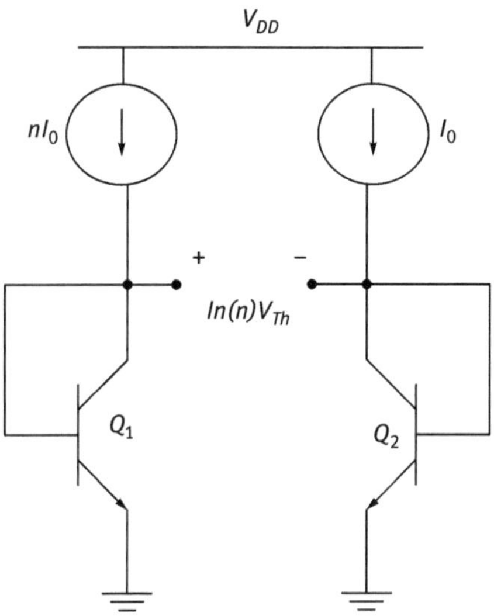

Figure 7.43: Generation of PTAT voltage.

biased at collector currents of nI_0 and I_0. Considering negligible base currents, it is possible to write

$$\Delta V_{BE} = V_{BE1} - V_{BE2} = V_{Th} \ln\left(\frac{nI_0}{I_{S1}}\right) - V_{Th} \ln\left(\frac{nI_0}{I_{S1}}\right) = \frac{KT}{q} \ln n = \ln(n) V_{Th},$$

where K is the Boltzman's constant, and V_{Th} is the thermal voltage.

Thus, the V_{BE} difference exhibits a positive TC since $\frac{\partial \Delta V_{BE}}{\partial T} = \frac{K}{q} \ln\ n.$

7.8.2 Negative TC Voltage

It is known that the V_{BE} of bipolar transistors exhibits a negative TC. For a bipolar device, we can write

$$I_C = I_S e^{\left(\frac{V_{BE}}{V_{Th}}\right)}. \tag{7.38}$$

Now

$$\begin{aligned}
&I_S\, \mu\, \mu K T n_i^2 \\
&\mu\, \mu\, \mu_0 T^m, \text{where } m \gg -3/2 \\
&n_i^2\, \mu\, T^3 \exp\left(-\frac{E_g}{KT}\right),
\end{aligned}$$

where $E_g \approx 1.12\text{eV}$ for silicon.

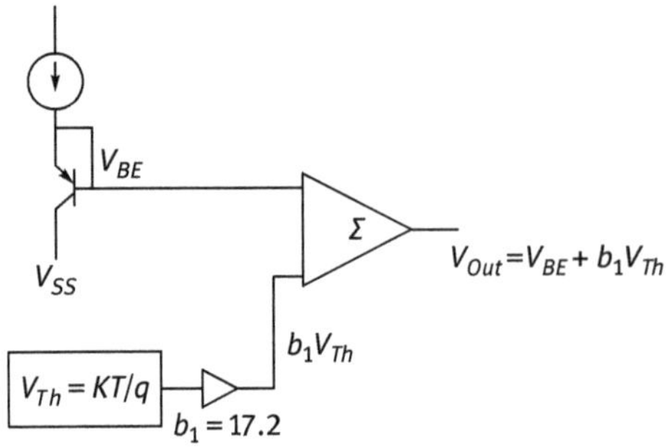

Figure 7.44: Conceptual generation of bandgap reference.

Thus, $I_S = bT^{4+m}exp\left(-\frac{E_g}{KT}\right)$, where b is proportionality constant.

$$\frac{\partial I_S}{\partial T} = b\,(4+m)\,T^{3+m}exp\left(-\frac{E_g}{KT}\right) + bT^{4+m}exp\left(-\frac{E_g}{KT}\right)\left(\frac{E_g}{KT^2}\right)$$

Now, $\frac{\partial V_{BE}}{\partial T} = \frac{\partial V_{Th}}{\partial T}\, ln\frac{I_C}{I_S} - \frac{V_{Th}}{I_S}\frac{\partial I_S}{\partial T}$

Substituting and simplifying, we obtain

$$\frac{\partial V_{BE}}{\partial T} = \frac{V_{BE} - (4+m)V_{Th} - \frac{E_g}{q}}{T}.$$

At T = 300°K, with a V_{BE} = 750 mV, $\frac{\partial V_{BE}}{\partial T}$ = −1.5 mV/°K, which is complementary to absolute temperature (CTAT—reduces with increasing temperature).

Conventional bandgap reference: with the availability of PTAT and CTAT, we can now develop a reference having zero TC. We write $V_{Ref} = a_1V_1 + b_1V_2$ such that $a_1\frac{\partial V_1}{\partial T} + b_1\frac{\partial V_2}{\partial T} = 0$. Therefore, $V_{Ref} = a_1V_{BE} + b_1\left(ln(n)V_{Th}\right)$.

Now $\frac{\partial V_{BE}}{\partial T}$ = −1.5 mV/°K and $\frac{\partial V_{Th}}{\partial T}$ = +0.087 mV/°K, then we choose $a_1 = 1$ and the reference voltage is given by

$$V_{Ref} = V_{BE} + \frac{1.5}{0.087}V_{Th} = V_{BE} + 17.2V_{Th}. \tag{7.39}$$

The whole principal is shown in Figure 7.44.

7.9 An Introduction to Analog Design Automation

IC manufacturing industry has shown a rapid advance in terms of complexity, functionality and productivity. This has been possible only by the prevalent use of computer-aided design (CAD) tools. All CAD tools are intended for use within a well-defined design flow. The design flow for digital IC design has become a well-defined standard. Such a standard design flow does not exist in analog IC design domain. Research on analog design automation (ADA) is very limited compared to digital design automation (DA). There is no such well-defined structured design methodology. Researchers have concentrated mostly either on specific tool designed for specific applications or on tools targeted toward specific aspects of a design like SPICE which is a tool used for circuit-level simulation. The major reason behind this lack of availability of ADA tool is the difficulty of the problem. In analog design the numbers of parameters are large with a huge interactivity among them. Due to the lack of a structured design methodology and automation tools, analog design remained more of an art than a science.

A short time to market is essential to make profit for an IC manufacturer. Due to availability of modern computers and DA software, digital designers are able to design multimillion transistor chips in less than a year, whereas a team of analog designers may require a similar time to design an analog chip (or an analog section of a chip) consisting of a few hundred transistors. The digital designers can update or change their designs in a few weeks, whereas the analog designers may require a time similar to the original design time. The situation described above is not continuing in the same direction. There is strong demand generating for ADA tools. Recently, the IC Design industry has been moving to SoC designs, where a single chip may contain both analog and digital circuits. An SoC requires various design blocks from various fields. A designer, or even a company, cannot be an expert on all these different fields and may choose to use the designs of others. This brings us to the concept of intellectual property (IP) reuse in IC design.

An **ADA** environment is applying IP reuse efficiently but in a different manner from digital domain. This leads to a general design flow methodology for ADA tools with well-defined inputs, outputs and duties. Analog blocks typically constitute only a small fraction of the components on MS ICs and emerging SoC designs. But due to the increasing levels of integration available in silicon technology and the growing requirement for digital systems to communicate with the continuous valued external world, there is a growing need for CAD tools that increase the design productivity and improve the quality of analog ICs. This chapter describes the motivation and evolution of these tools and outlines progress on the various design problems involved such as simulation and modeling, symbolic analysis, synthesis and optimization, layout generation, yield analysis and design centering.

In the coming section of the chapter, a brief survey on previous analog design flow is presented with related terminologies.

7.9.1 Survey of Previous Analog Design Flow

This survey [47–50] presents a complete and recent overview of recent advances in the state of the art for CAD tools for analog and MS ICs. In recent years, complete

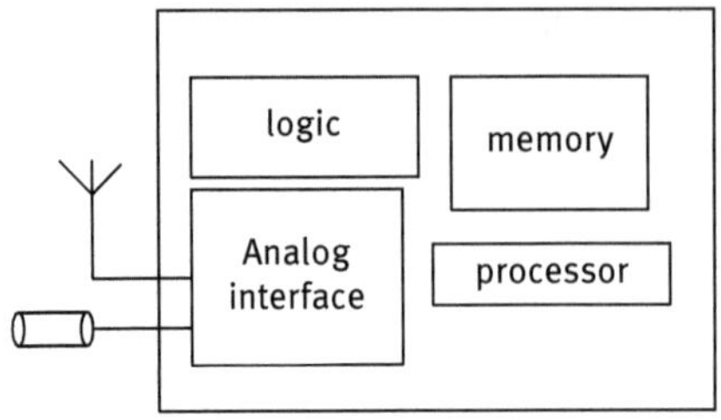

Figure 7.45: System-on-chip design.

systems that previously occupied one or more boards have been integrated on a few chips or even one single chip. Examples of such SoC are the single-chip TV or the single-chip camera or new generations of integrated telecommunication systems that include analog, digital and eventually radio frequency (RF) sections on one chip. Normally the technology chosen is CMOS but BiCMOS can be used for RF and analog circuits. Analog circuits are needed to make the interface between the real world and electronic system, although other major parts are digital in nature. A typical SoC looks like that shown in Figure 7.45. It contains memory, processor, logic and also some analog interface to communicate with the external world.

There is a trend to replace analog circuit functions with digital computations (e.g. digital signal processing in place of analog filtering), but still there are some typical functions that will always remain analog. Analog circuits are indispensable in all electronic applications that interface with the outside world as shown below:
- The input side of a system is a typical analog function: signals from a sensor, microphone, antenna, etc., must be sensed or received and then amplified and filtered up to a level that allows digitization with sufficient SNR ratio.
- The second typical analog function is on the output side of a system: the signal is reconverted from digital to analog form and it has to be strengthened so that it can drive the outside load (e.g. actuator, antenna, and loudspeaker) without too much distortion. Typical analog circuits used here are drivers and buffers, filters, oscillators and mixers (in case of up-conversion).
- The third type of blocks is the true MS circuits that interface the above analog circuits with the DSP part of the system. Typical circuits used here are the sample-and-hold circuits for signal sampling, analog-to-digital converters for amplitude discretization, digital-to-analog converters for signal reconstruction, and phase-locked loops and frequency synthesizers to generate a timing reference or perform timing synchronization.
- In addition, the above circuits need stable absolute references for their operation, which are generated by voltage and current reference circuits, crystal oscillators, etc.

When both analog (possibly RF) and digital circuits are needed in a system, it becomes obvious to integrate them together to reduce cost and improve performance, provided the technology allows us to do so. Therefore, with the advancement of VLSI technology there is a need for higher level of integration. As a result of which the market shares of mixed signal ICs increase. Recent developments in CMOS technology (shrinkage to submicrometer region) have offered the possibility to combine good and scalable digital performance with adequate analog performance on the same die.

The possibilities of higher levels of system integration offer analog MOS transistor performance that approaches the performance of a bipolar transistor. This explains why CMOS is the technology of choice today, and why other technologies like BiCMOS are only used when more aggressive bipolar device characteristics (e.g. power, noise or distortion) are really needed. The technology shift from bipolar to CMOS (or BiCMOS) has been apparent in most applications. Even fields like RF, where traditionally GaAs and bipolar were the dominant technologies, now show a trend toward BiCMOS (preferably with a SiGe option) and even plain CMOS for reasons of higher integration and cost reduction. These higher levels of MS integration, however, also introduce a whole new set of problems and design effects that need to be accounted for in the design process.

With the increasing complexity the following factors are popping up:

(1) More and more transistor-performing digital and analog operations must be considered together with embedded software.
(2) New signal processing algorithms must be developed to cater for the functionality and performance requirement.
(3) Due to the rapid evolution of process technologies, the expectation for changing process technology parameters needs to be accounted for in the design cycle.
(4) Most importantly, the time-to-market factor is very critical for Application Specific Integrated Circuits (ASICs) because if one misses the initial market window relative to the competition, prices and, therefore, profit can be seriously affected.

All the above-mentioned effects of increased design complexity can be managed with the help of CAD and verification tools. What is needed to expedite the analog and MS design process is a structured methodology and supporting CAD tools to manage the entire design process and design complexity.

CAD tools are also needed to assist or automate many of the routine and repetitive design tasks, taking away the tedium of manually designing these sections and providing the designer with more time to focus on the creative aspects of design.

Therefore, analog CAD and circuit DA are likely to play a key role in the design process of the next generation of MS ICs and ASICs.

Analog CAD effort started with mixed-signal ASIC but in recent years it has moved toward a complete SoC design.

The story of automation is different for analog and digital side. Due to the lack of well-matured CAD tool in analog domain, the analog part design in an MS IC is a bottleneck. There are not yet any robust commercial CAD tools to support or automate analog circuit design apart from circuit simulators like SPICE and layout editing environments and their accompanying tools (layout verification tools, extraction, layout versus schematic (LVS), etc.). Analog design in general is perceived as less systematic and more heuristic and knowledge-intensive in nature than digital design. That is why till now, it was not possible to establish a higher level of abstraction which shields the lower level process details from the higher level design. Analog design requires skill obtained through years of experience.

The number of conflicting requirements, variety of schematics, dependability on process or device sizes and non-idealities and parasitic effect in analog design explain why digital design flow cannot be adopted for ADA. Instead a separate specific analog solutions need to be developed that are targeted at the analog design paradigm and complexity.

Despite the lack of commercial analog CAD tools, analog CAD and DA over the past 15 years has been a field of profound academic and industrial research activity, although with not quite as many researchers as in the digital world, resulting in a slow but steady progress. Analog circuit and layout synthesis have recently shown promising results at the research level, but commercial solutions are only starting to appear in the marketplace.

The development of analog and MS hardware description languages like ***VHDL-AMS*** and ***Verilog-AMS*** is intended to provide a unifying trend that will link the various ADA tasks in a coherent framework that supports a more structured analog design methodology from the design conceptualization stage to the manufacturing stage. They also provide a link between the analog and the digital domains, as needed in designing mixed analog–digital ICs and the SoC of the future.

7.9.2 Analog and Mixed-signal Design Process

A hierarchical design methodology consists of a top-down synthesis and a bottom-up verification at every level of the hierarchy. The various stages that are traversed in the design process are as follows:

- ***Conceptual Design***: In this stage the product is in conceptual stage and usually described by mathematical tools such as Matlab/Simulink after gathering the specifications.
- ***System Design***: After system concept, system design is carried out followed by simulations and verification. This stage contains little implementation details. The whole system is portioned into software and hardware parts. Hardware parts are described using behavioral model. This stage includes decisions about implementation issues, such as package selection, choice of the target technology and general test strategy.
- ***Architectural Design***: The partitioning of the system into digital and analog blocks is done in this stage. The various blocks are described by appropriate hardware description language (e.g. VHDL and VHDL-AMS) using behavioral model.

- ***Cell Design***: The analog blocks with the given specifications, and in the selected technology process, result in a fully sized device-level circuit schematic. The resulting circuit design is then verified against the specifications using SPICE-type circuit simulations. Selection of proper process technology and decision about device sizing is required in this stage.
- ***Cell Layout***: The circuit in the form of schematic is transformed into geometric shapes (layout) in this stage. This stage involves area optimization to make the total area small. After the generation of the layout, parasitics are extracted to view their effects and to ensure functionality of the circuit even in the presence of parasitics.
- ***System Layout***: System-level layout is generated by block place and route. In addition, power-grid routing is done in this stage. Test structures are inserted in this stage. Detailed verification or timing analysis is done in this stage. Finally, the system is verified by co-simulating the hardware part with the embedded software.
- ***Fabrication and Testing:*** This is the processing stage where the masks are generated and the ICs fabricated. Testing is performed during and after fabrication in order to reject defective devices.

The system design is then given as input to the architectural design phase, which is the input to the cell design, cell layout and system layout phases. Each of these phases has a feedback loop consisting of simulation and verification as shown in Figure 7.46. The top-down direction is called ***forward progress***, whereas the bottom-up phase is called ***backtracking and redesign***.

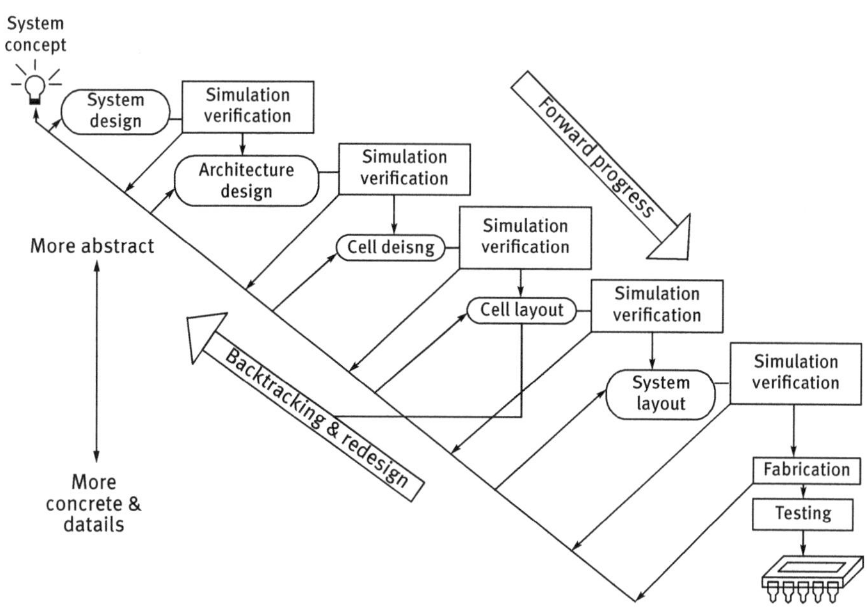

Figure 7.46: High level view of the analog or MS IC design process.

7.9.3 Hierarchical Analog Design Methodology

The hierarchical design methodology is prevalent in many of the emerging analog CAD systems.

As for example for the design of a phase-locked loop, the analog block is subdivided into smaller blocks (for example, loop filter or a comparator, etc.). The subblocks are designed after the derived specification from the original block or the subblock can be further divided into smaller subblocks. This process ensures that the top-level block fulfill its specification. The decomposition goes on until a level where a block finds its physical implementation (transistor level or by standard cells).

This type of decomposition is a ***top-down process***. It is followed by a ***bottom-up layout implementation and design verification process***. IC manufacturing is expensive; therefore verification is very important which ensures that the design is meeting its specification; otherwise redesign is done.

Each block in the top-down direction takes an architecture or topology from the preceding block as well as performance constraints. Its output is a more detailed circuit diagram (or finally layout) as well as new specifications for the next block. Thus, each block in the forward direction does a specification translation and topology selection (Figure 7.47). Most experimental analog CAD systems today use a ***performance-driven design strategy*** within such analog design hierarchy. This strategy consists of the alternation of the following steps in between any two levels and of the design hierarchy.

(1) **Top-down path**
   - Topology selection;
   - Specification translation (or circuit sizing);
   - Design verification.

(2) **Bottom-up path**
   - Layout generation;
   - Detailed design verification (after extraction).

Topology selection: Select the best topology that matches the system. For an analog-to-digital converter, for instance, topology selection could be selecting between a flash, a successive approximation, etc.

Specification translation: It is the step of mapping the specifications for the block under design at a given level (e.g. a converter) into individual specifications for each of the subblocks (e.g. a comparator) within the selected block topology, so that the complete block meets its specifications, while possibly optimizing the design toward some application-specific design objectives (e.g. minimal power consumption). The downstream specifications are verified by simulations.

Layout generation: It is the step of generating the geometrical layout of the block under design by assembling (place and route) the already generated layouts of the

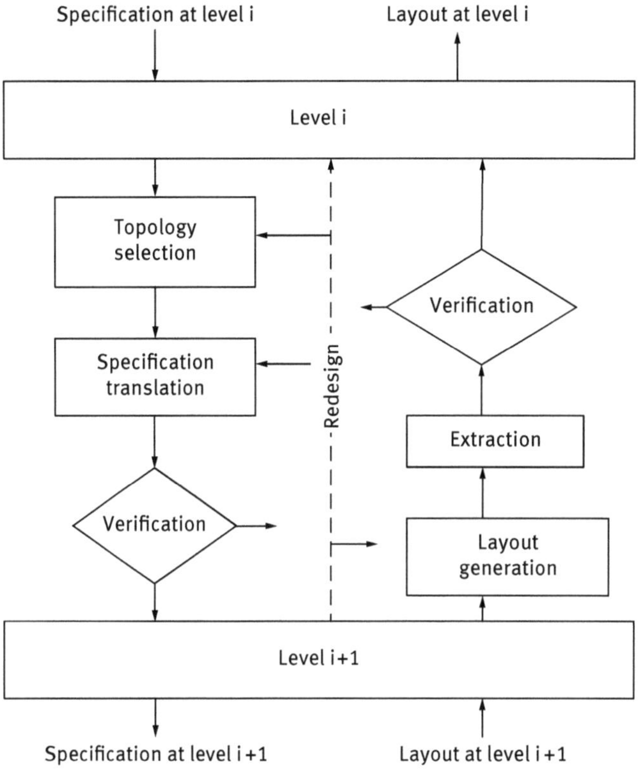

Figure 7.47: Hierarchical design strategy.

composing subblocks. At the lowest level, these subblocks are individual devices or selected device groupings, which themselves are generated by parameterized procedural device layout generators. Also, power, ground and substrate connection routing are part of the layout generation step. This step is followed by extraction and, again, detailed verification and simulation to check the impact of the layout parasitics on the overall circuit performance.

The above methodology is called ***performance-driven or constraint-driven***, as the performance specifications are the driving input to each of the steps. Throughout the design flow, constraints need to be propagated down the hierarchy to make sure that the top-level block in the end meets its target specifications. The advantage of top-down design: **it catches problems early in the design flow and, therefore, has a higher chance of first-time success, while obtaining a better overall system design**.

In the above top-down/bottom-up design flow, **redesign or backtracking iterations** may therefore be needed at any point where a design step fails to meet its input specifications. In that case, one or more of the previously executed steps will have to be redone; for example, another circuit topology can be selected instead of the failing one, or another partitioning of subblock specifications can be performed.

The following section identifies the main tasks needed for above analog design methodology.

7.9.4 Current Status for the Main Tasks in Analog Design Automation

A. Simulation

Simulation tools are provided to ensure cost-effective design correctness by means of design verification without actual fabrication of the device. Simulation is crucial and it plays a major key role in design flow. Analog ICs are typically impacted by many higher order effects that can severely degrade the circuit performance once fabricated, if the effects are not properly accounted for during the design process.

(1) *Circuit Simulation:* Circuit simulation begins with SPICE. This simulator numerically solves the system of nonlinear differential-algebraic equations (DAE) that characterize the circuit by using traditional techniques of numerical analysis. For example, to compute the evolution of the circuit's response in time, the time-derivative operator is first discretized by an implicit integration scheme such as the backward-Euler or trapezoidal rule. At each time point, the resulting set of nonlinear algebraic equations is solved iteratively via the Newton–Raphson method, with the matrix solution required to update the approximate solution computed using sparse matrix techniques.

Circuit simulation is a good design aid here, providing the capability of simulating many of the higher order effects and verifying circuit performance prior to fabrication, provided the effects are modeled properly. Many variants and descendants of the SPICE simulator are now marketed by a number of CAD vendors and many IC manufacturers have in-house versions of the SPICE simulator that have been adapted to their own proprietary processes and designs. A few examples of the many commercial SPICE-class simulators are HSPICE (Synopsys), Spectre (Cadence Design Systems) and Eldo (Mentor Graphics).

The inaccuracy of the models used in SPICE simulation is a matter of concern. However, recent models such as BSIM3 v3, Philips model 9 or EKV look more promising for analog design by providing smooth and continuous transitions between different operating regions. For RF applications, however, even these models are not accurate enough, creating a need for accurate RF modeling.

With the explosion of MS designs, the need has also arisen for simulation tools that allow not only simulation of analog or digital circuits separately, but also simulation of truly mixed analog-digital designs. Therefore, so-called **glued mixed-mode simulators** address the conversions of the signals between analog and digital signal representations. Therefore, mixed-mode simulators were developed that glue together

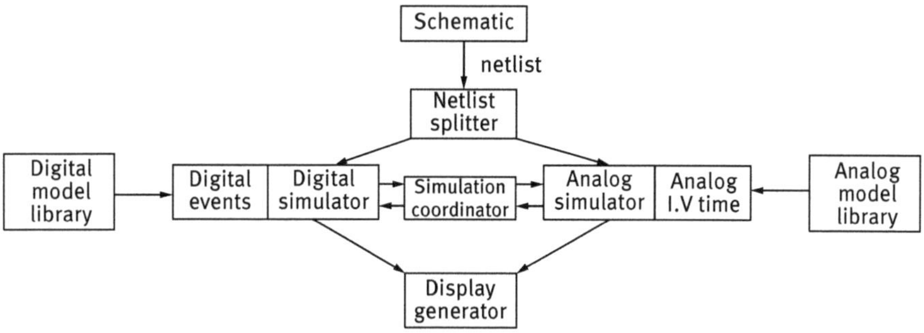

Figure 7.48: General scheme of an MS simulator, gluing an analog and a digital simulation kernel together.

an accurate SPICE-like analog simulator to an efficient digital simulator. The general scheme of these so-called glued MS simulators is shown in Figure 7.48. These simulators partition the netlist into digital and analog parts that are each simulated by a separate kernel that interfaces through a simulation "backplane": an analog kernel with adaptive time-step control, and a digital kernel with its event-driven mechanism. As representative examples of modern MS simulators, we would mention the Advance MS (Mentor Graphics) and AMS Designer tools (Cadence Design Systems).

B. Circuit Modeling

In recent years, the need has also arisen for higher levels of abstraction to describe and simulate analog circuits, because of the following reasons:

(1) In a top-down design methodology it will be possible to describe a design without knowing the lower level actual implementation details.
(2) At a higher level verification can be done to check the practicality of the system.

For a commercial simulator to be useful in current industrial MS design practice, it therefore has to be capable of simulating a system containing a mix of analog blocks described at different levels and in different domains, in combination with digital blocks. This requires a true MS multilevel, mixed-domain simulator.

To allow an easy exchange of these models across different simulators and different users, the need arose for ***standardized analog hardware description languages*** to describe these higher level models. For MS designs, the analog HDLs had to be compatible with the existing digital HDLs (such as VHDL and Verilog). Several parallel analog or MS HDL language standardization efforts, therefore, have been initiated, recently resulting in the standardized languages VHDL-AMS and Verilog-AMS. The VHDL-AMS language targets the MS domain and is a superset of the digital VHDL language. Recently, also, the standardization of an extension of VDHL-AMS toward RF has been started.

In addition, high level models also form a key part of several analog circuit synthesis and optimization systems, both at the circuit level as well as for the hierarchical synthesis of more complex blocks.

Most of the basic techniques in both circuit and layout synthesis today rely on powerful numerical optimization engines coupled to "evaluation engines" that qualify the merit of some evolving analog circuit or layout candidate. This basic scheme of optimization-based analog circuit sizing is shown in Figure 7.49. The most general but also by far the slowest solution is to call the transistor-level simulator as evaluation engine during each iteration of the optimization of a circuit. The Central Processing Unit (CPU) time can be reduced significantly by replacing the simulations by model evaluations. These models can be behavioral simulation models as described above, or they can be what are termed ***performance models***. Rather than modeling the input–output behavior, performance models directly relate the achievable performances of a circuit (e.g. gain, bandwidth or slew rate) to the design variables (e.g. device sizes and biasing). In the synthesis procedure, calls to the transistor-level simulation are then replaced by performance model evaluations, resulting in substantial speedups of the overall synthesis, once the performance models have been created and calibrated. For the synthesis of more complex analog blocks, a hierarchical approach is needed, in which higher level models are indispensable to bridge between the different levels.

Table 7.4 gives an overview of the different analog hardware description levels. The ***circuit level*** is the traditional level where a circuit is simulated as a network of physical components. In a ***macromodel***, an equivalent but computationally cheaper circuit representation is used that has approximately the same behavior as the original circuit. Equivalent sources combine the effect of several other elements that are eliminated from the netlist. The simulation speedup is roughly proportional to the number

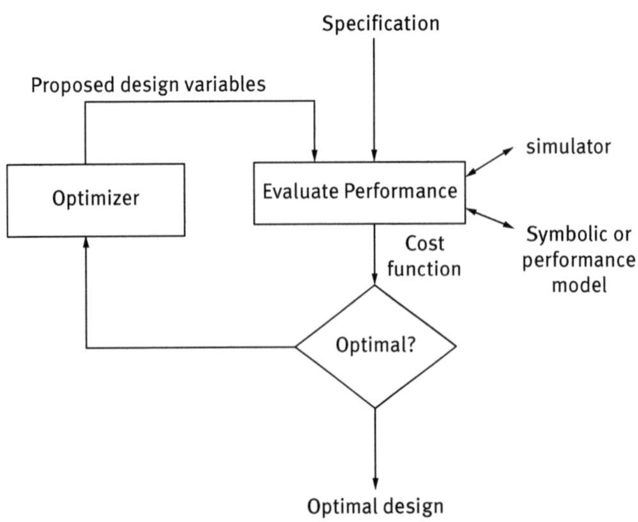

Figure 7.49: Basic flow of optimization-based analog circuit sizing.

Table 7.4: Different analog hardware description levels to describe analog circuits.

Level	Modeling primitives	Implications
Functional	Mathematical signal flow description per block, connected in signal flow diagram	No internal block structure; conservation law need not be satisfied on pins
Behavioral	Mathematical description (equation, procedures) per block	No internal block structure; conservation law must be satisfied on pins
Macromodel	Simplified circuit with controlled sources	Spatially unrelated to actual circuit; conservation law must be satisfied
Circuit	Connection of SPICE primitives	Spatially one-to-one related to actual circuit; conservation law must be satisfied

of nonlinear devices that can be eliminated. In a ***behavioral or functional model***, a purely mathematical description of the input–output behavior of the block is used. This typically will be in the form of a set of differential algebraic equations or transfer functions. At the behavioral level, conservation laws still have to be satisfied; at the functional level this is no longer the case, and the simulated system turns into a kind of signal-flow diagram.

C. Dedicated Simulation Techniques

Techniques or tools have been developed for dedicated purposes. The important types of circuits that are used in many signal processing and communication systems are the switched circuits, like switched-capacitor and, more recently, switched-current circuits. Their switched nature, with the resulting switching transients, requires many small numerical integration steps to be simulated within each clock phase if a standard SPICE simulator is used. This is exploited in several SC simulation tools like SWITCAP, SWAP and TOSCA that analyzes SC-based converters.

Another important domain where more tools are needed is RF simulation, needed for instance when developing circuits for wireless applications.

D. Analog Behavioral Model Generation Technique

One of the biggest problems today is the lack of systematic methods to create good analog behavioral or performance models. Techniques used here can roughly be divided into fitting or regression approaches and constructive approaches.

Fitting or Regression Methods. In the *fitting or regression approaches*, a parameterized mathematical model is proposed or constructed by the model developer, and the values of the parameters p are selected as to best approximate the known (simulated) circuit behavior, as schematically depicted in Figure 7.50.

Symbolic Analysis of Analog Circuits. The second class of methods, the ***constructive approaches***, tries to generate or build a model from the underlying circuit description. One approach, for instance, uses symbolic analysis techniques to generate first the exact set of algebraic/differential equations describing the circuit, which are then

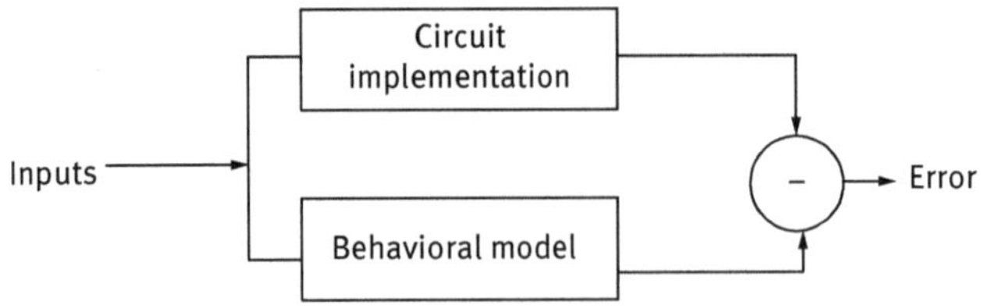

Figure 7.50: Basic flow of fitting or regression approach for analog behavioral model generation.

simplified within a given error bound of the exact response using both global and local simplifications.

A ***symbolic simulator*** is a computer tool that takes as input an ordinary (SPICE type) netlist and returns as output (simplified) analytic expressions for the requested circuit network functions in terms of the symbolic representations of the frequency variable and (some or all of) the circuit elements. They perform the same function that designers traditionally do by hand analysis (even the simplification). The difference is that the analysis is now done by the computer, which is much faster, can handle more complex circuits and does not make as many errors. An example of a complicated BiCMOS OP-AMP is shown in eq. (7.40):

$$A_{VO} = \frac{g_{m,M2}}{g_{m,M1}} \frac{g_{m,M4}}{\left(\frac{g_{o,M4}g_{o,M5}}{g_{o,M3}+g_{Mb,M3}} + \frac{G_a+g_{o,M9}+g_{oQ2}}{\beta_{Q2}}\right)}. \quad (7.40)$$

The ***symbolic expression*** gives a better insight into which small signal circuit parameters predominantly determine the gain in this OP-AMP and how the user has to design the circuit to meet a certain gain constraint. Dedicated symbolic analysis tools have been developed that use heuristic simplification. Examples of such tools are ISAAC, SYNAP and ASAP; they are successful for circuits containing only 10–15 transistors. In recent years, however, an algorithmic breakthrough in the field of symbolic circuit analysis has been realized. Tools like SYMBA and RAINIER highly reduce the computation time and, therefore, enable the symbolic analysis of large analog circuits of practical size (like the entire 741 OP-AMP).

E. Analog Circuit Synthesis and Optimization

Synthesis is a critical step since most analog designs require a custom-optimized design, and the number of (often-conflicting) performance requirements to be taken into account is large.

During synthesis, the block performance is specified and an appropriate topology to implement this block has to be decided first. This step is called topology selection. Subsequently, values for the subblock parameters have to be determined so that the final block meets the specified performance constraints. This step is called ***specification translation*** at higher levels in the design hierarchy, in which case performance specifications of subblocks have to be determined, or circuit sizing at the device level,

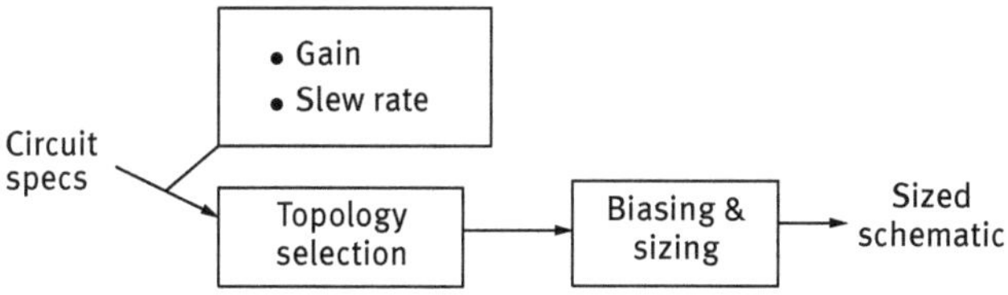

Figure 7.51: Basic flow of analog circuit synthesis for a basic cell: topology selection and circuit sizing.

in which case the sizes and biasing of all devices have to be determined. Figure 7.51 illustrates this flow for low level cells.

The circuit-level synthesis tool used in the ADA system can be optimization-based where the desired performance parameters as well as the selected topology for each block are supplied to the circuit-level synthesis tool. After optimization is carried out, the tool gives a sized netlist for the block in SPICE format. Again, with inputs and outputs defined as such, it can be used as a stand-alone tool for analog circuit designers. The most important aspect of the synthesis tool is its ***optimization algorithm***. It is well known that the performance of simulated annealing (SA) is strongly dependent on its parameters like cooling regime, initial conditions, etc. A wrong choice of these parameters can result in the algorithm to be stuck at a local minimum. Unfortunately, these parameters change from example to example and there is no way of predicting them beforehand. Evolutionary strategies (ES) suffer much less from this problem. They tend to converge to a region around the desired solution but may take a very long time or may even fail to converge to the final solution. An algorithm combines these two is superior in performance to both.

F. Analog and Mixed-signal Layout Synthesis

The next important step in the analog design flow after the circuit synthesis is the generation of the layout. The field of analog layout synthesis is more mature than circuit synthesis, in large part because it has been able to leverage ideas from the mature field of digital layout. Yet real commercial solutions are only beginning to appear in recent years. The typical analog layout hierarchy comprises two fundamentally different types of layout tasks.

Cell-level layout. The focus here is really device-level layout, placement, and routing of individual active and passive devices. At this level, many of the low level matching, symmetry, merging, reshaping and proximity management tactics for polygon-level optimization are applied. The goal is to create cells that are suitably insulated not only from fluctuations in fabrication and operating environment, but from probable coupling with neighboring pieces of layout.

System-level layout. The focus here is cell composition, arranging and interconnecting the individual cells to complete an analog subsystem. At this level, isolation is one major concern: from nearby noisy digital circuits coupled in via substrate, power grid or package. The other concern is signal integrity. Some number of digital signals from neighboring digital blocks needs to penetrate into any analog regions, and these signals may be fundamentally incompatible with some sensitive analog signals.

Programmable layout. The focus here is applying field-programmable gate array (FPGA) ideas to analog designs (***FPAA***). The idea is to bypass completely the need to do custom cells. Rather, a set of programmable active and passive elements is connected with programmable wiring to achieve low performance analog functions. This is a relatively recent implementation strategy, but we expect to see it grow.

G. Yield Estimation and Optimization

The manufacturing of ICs suffers from statistical fluctuations inherent to the fabrication process itself. These fluctuations can be local (e.g. lithographic spots) or global (e.g. gate oxide gradients). These process-induced deformations result in deviations of the actual performances of the fabricated ICs from the expected performances. Some of these deformations result in the fabricated IC not having the expected functionality. This is then called a ***structural failure***. Others cause the fabricated IC to have, despite the correct functionality, observed performances far from the targeted values. This is then called a ***hard performance failure***. Both failures are also termed ***catastrophic failures***, caused by a catastrophic fault. Other process deformations result in the observed performances deviating only slightly around the targeted values due to the statistical tolerances (inter-die variations) and mismatches (intra-die variations) of the device parameters. This is then called a ***parametric (or soft) performance failure***, caused by a parametric fault.

The ratio of the accepted total number of fabricated ICs is called the ***yield***. If the yield is significantly less than 100%, this implies a financial loss to the IC manufacturer. Therefore, it is important to already calculate and maximize the manufacturing yield during the design stage. This is called ***design for manufacturability***, which implies techniques for yield estimation and yield optimization. A simple approach often used in practice is ***worst-case analysis***. Instead of doing a true statistical analysis, some worst-case combinations (e.g. slow p-type devices, fast n-type devices, etc.) of the device parameters are used to calculate the "worst case" corners of the performances.

The most straightforward statistical approach is to use ***Monte-Carlo simulations***, in which a large number of samples of the device parameters are generated according to the given statistics.

H. Analog and Mixed-signal Testing and Design for Testability

The final step in the IC production cycle is the testing of the IC. The objective of testing is to verify whether the fabricated IC meets the specifications it was designed for or not. The cost of testing is a considerable part of the final cost of an IC, and the later in the production cycle if a certain defect is detected, the more expensive is the financial loss. The effective cost of a product is therefore related to the quality of the applied tests and the time necessary to generate and apply the tests. In MS designs, an additional complication is that, although analog circuits constitute only a small fraction of the die area, they are responsible for the largest fraction of the overall test cost. This is not only because they require more expensive MS test equipment, but also because they require longer test times than the digital circuitry and because there are no structured test waveform generation tools or structured design for test methodologies.

7.10 Field-programmable Analog Arrays

Finally we mention one very recent analog layout style which is radically different from those mentioned above. In the digital realm, FPGAs have revolutionized digital prototyping and rapid time-to-market designs. Integrating programmable logic elements and programmable interconnect, these allow rapid customization with no fabrication steps. An obvious question is: can a similar technology be adapted for analog and MS designs? The apparent answer is a qualified “yes”.

Early work directly mimicked the FPGA style of small primitive functions connectable by programmable interconnect. However, the loading that is already problematic in digital designs proved even more deleterious here. Later designs moved up to higher level building blocks (e.g. OP-AMPs, switches, capacitor arrays) and also focused new energy on sensitive analog design of the programmable interconnect. FPAAs are just beginning to be commercially available. They are currently so small however, that layout is not really a problem. It will be interesting to see if these designs become larger (e.g. larger digital blocks with smaller analog blocks), and if so, if automatic layout becomes a requirement.

A field-programmable analog array (FPAA) is an integrated device containing configurable analog blocks (CAB) and interconnects between these blocks as shown in Figure 7.52. Unlike their digital cousin, the FPGA, the devices tend to be more application driven than general purpose as they may be current mode or voltage mode devices. For voltage mode devices, each block usually contains an operational amplifier in combination with programmable configuration of passive components. The blocks can, for example, act as summers or integrators.

FPAAs usually operate in one of two modes: continuous time and discrete time.

(a) Discrete-time devices possess a system sample clock. In a SC design, all blocks sample their input signals with a sample-and-hold circuit composed of a semiconductor switch and a capacitor. This feeds a programmable OP-AMP section, which can be routed to a number of other blocks. This design requires more complex semiconductor construction. An alternative, switched-current design, offers simpler construction and does not require the input capacitor, but can be less accurate and has lower fan out; it can drive only one following block. Both discrete-time device types must compensate for switching noise, aliasing at the system sample rate, and sample-rate limited bandwidth, during the design phase.

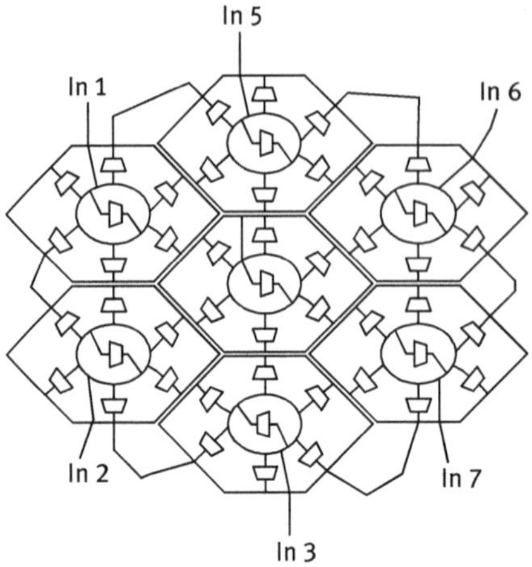

Figure 7.52: FPAA structure.

(b) Continuous-time devices work more like an array of transistors or OP-AMPs, which can operate at their full bandwidth. The components are connected in a particular arrangement through a configurable array of switches. During circuit design, the switch matrix's parasitic inductance, capacitance and noise contributions must be taken into account.

The FPAA structure consists of a two-dimensional array of CABs, which include digitally configurable transconductors (differential transconductance amplifiers). The arrangement in a hexagonal layout allows reconfigurable routing of analog signals throughout the chip. Series and parallel connection as well as feedback of any order is provided by the structure. Every CAB has six branches connecting to the respective neighbor CABs and one for self-feedback. The massive parallel connections of both parasitic input capacitances as well as parasitic output capacitances at the input nodes of each CAB sum up to capacitances in an order of magnitude such that they are suitable as integrating capacitances for the filter up to unity-gain bandwidths of 200 MHz.

It can be shown that a large number of different filter types with different orders can be synthesized on the FPAA, for example, second-order low pass (blue), lossy integrator (green) or fourth-order biquad Butterworth band pass (red). All the CAB nodes are fed to outputs so that any combination of six inputs and seven outputs is possible. The FPAA could be changed from six independent first-order low pass paths to only one path with a large number of transconductors while being embedded into the application system. Thus, the circuit can instantaneously be adapted to a wide range of varying needs in analog filtering.

7.11 Summary

The last few years have seen significant advances in both design methodology and CAD tool support for analog, MS and RF designs. Despite this enormous progress in research and commercial EDA offerings that today enable the efficient design of analog blocks for embedding in MS integrated systems (ASICs, SoCs), several problems still remain to be solved.

In order to meet the economic constraints (time to market, cost, quality) of future semiconductor products, analog design will have to be carried out in a much more systematic and structured way, supported by methodologies and tools that fit in the overall system design flow. Given the demand for MS ICs, we expect no reduction in the interest in various layout tools to speed the design of these important circuits.

References

[1] M. Pelgrom, A. Duinmaijer, A. Welbers, “Matching properties of MOS transistors”, IEEE Journal of Solid-State Circuits, 24 (1989): 1433–1439.

[2] J. Bastos, M. Steyaert, R. Roovers, P. Kinget, W. Sansen, B. Graindourze, A. Pergoot and Er Janssens, “Mismatch characterisation of small size MOS transistors.” Proceedings IEEE International Conference on Microelectronic Test Structures, March 1995, pp. 211–276.

[3] O. Erdogan, R. Gupta, D. Yee, J. Rudell, J. Ko, R. Brockenbrough, S. Lee, E. Lei, J. Tham, H. Wu, and C. Conroy, “A single-chip quad-band GSM/GPRS transceiver in 0.18 um Standard CMOS.” Digest of Technical Papers IEEE International Solid-State Circuits Conference 2005 (ISSCC 2005), pp. 318–319.

[4] H. Darabi, S. Khorram, Z. Zhou, T. Li, B. Marholev, J. Chiu, J. Castaneda, E. Chien, S. Anand, S. Wu, and M. Pan, “A fully integrated SoC for 802.11b in 0.18 um CMOS.” Digest of Technical Papers IEEE International Solid-State Circuits Conference 2005 (ISSCC 2005), pp. 96–97.

[5] W. Krenik, D. Buss and P. Rickert, ”Cellular handset integration – SIP versus SOC.” in Proceedings of IEEE 2004 Custom Integrated Circuits Conference, pp. 63–70.

[6] J.M. Rabaey, F. De Bernardinis, A.M. Niknejad, B. Nikolic and A. Sangiovanni-Vincentelli, “Embedding mixed-signal design in systems-on-chip.” Proceedings of the IEEE, 94.6 (2006): 1070–1088.2006

[7] G.E. Moore, “Cramming more components onto integrated circuits.” Electronics, 38.8 (1965): pp. 114–117.

[8] H.P. Hofstee, “Power constrained microprocessor design.” Proceedings of IEEE Conference Computer Design, 2002, pp. 14–16.

[9] R.H. Dennard, F.H. Gaensslen, H.N. Yu, V.L. Rideout, E. Bassous and A.R. LeBlanc, “Design of ion-implanted MOSFET’s with very small physical dimensions.” IEEE Journal Solid-State Circuits, SC-9.5 (1974): 256–268.

[10] The 2010 International Technology Roadmap for Semiconductors. Available at: http://www.itrs.net last accessed on 8th June 2016.

[11] Y. Uemoto, E. Fujii, A. Nakamura, K. Senda and H. Takagi, “A stacked-CMOS cell technology for high-density SRAMs.” IEEE Transactions on Electron Devices, 39.10 (1992): pp. 2359–2363.

[12] H. Qin, Y. Cao, D. Markovic, A. Vladimirescu and J. Rabaey, “SRAM leakage suppression by minimizing standby supply voltage.” Proceedings of 5th International Symposium Quality Electronic Design (ISQED ’04), 2004, pp. 55–60.

[13] E. Seevinck, F.J. List and J. Lohstoh, "Static noise margin analysis of MOS SRAM cells." IEEE Journal of Solid-State Circuits, SC-22.5 (1987): 748–754.
[14] A. Sarkar, S. De, A. Dey and C.K. Sarkar, "Analog and RF performance investigation of cylindrical surrounding-gate MOSFET with an analytical pseudo-2D model." Journal of Computational Electronics (2012), pp. 182–195, vol. 11, No. 2.
[15] A. Sarkar, S. De, A. Dey C.K. Sarkar, "1/f noise and analogue performance study of short-channel cylindrical surrounding gate MOSFET using a new subthreshold analytical pseudo-two-dimensional model." IET Circuits, Devices & Systems (2012), pp. 28–34, vol. 6, No. 1.
[16] Y. Chiu, P. Gray, and B. Nikolić, A 14-b 12-MS/s CMOS pipeline ADC with over 100-dB SFDR, IEEE Journal of Solid-State Circuits, 39.12 (2004): 2139–2151.
[17] A. Sarkar, A.K. Das, S. De, C.K. Sarkar, "Effect of gate engineering in double-gate MOSFETs for analog/RF applications." Microelectronics Journal (2012), pp. 873–882, vol. 43, No. 11.
[18] R.E. Suarez, P.R. Gray and D.A. Hodges, "All-MOS charge-redistribution analog-to digital conversion techniques." IEEE J. Solid-State Circuits Part II, SC-10.6 (1975): 379–385.
[19] M. Pelgrom, H. Tuinhout, and M. Vertregt, "Transistor matching in analog CMOS applications." Technical Digest International Electron Devices Meeting, 1998, pp. 915–918.
[20] A. Lazaro and B. Iniguez, RF and noise model of gate-all-around MOSFETs. Semiconductor Science and Technology 23.7 (2008): 075022.
[21] M. Fulde, D. Schmitt-Landsiedel and G. Knoblinger, "Analog and RF design issues in high-κ & multi-gate CMOS technologies." IEDM Technical Digest, 2009, p. 447.
[22] Nae, B., Lazaro, A. and B. Iniguez "High frequency and noise model of gate-all-around metal-oxide-semiconductor field-effect transistors." Journal of Applied Physics, 105.7 (2009): 074505–074508.
[23] K. Biswas, A. Sarkar and C.K. Sarkar, "Impact of barrier thickness on Analog, RF and Linearity performance of nanoscale DG heterostructure MOSFET." Superlattices and Microstructures, 2015.
[24] P. Allen and D. Holberg, CMOS Analog Circuit Design, ser. The Oxford Series in Electrical and Computer Engineering. Oxford University Press, USA, 2002.
[25] T. Carusone, D. Johns and K. Martin, *Analog Integrated Circuit Design*, ser. Wiley Desktop Editions. John Wiley & Sons, New York, USA, 2012.
[26] B.J. Hosticka, R.W. Brodersen, and P.R. Gray, "MOS sample data recursive filters using switched capacitor integrators." IEEE Journal Solid State Circuits, SC-12.6 (1977): 600–608.
[27] K. Martin, "Improved circuits for the realization of switched-capacitor filters." IEEE Transactions on Circuits and Systems, SC-27.4 (1980): 237–244.
[28] P. Gray and R. Meyer, *Analysis and Design of Analog Integrated Circuits*. Wiley, New York, USA, 1993.
[29] S. Rabii and B.A. Wooley, "A 1.8V, 5.4 mW, Digital-Audio XA Modulator in I .2bm CMOS." Proceedings of ISSCC, February 1996, pp. 228–229.
[30] J. Grilo, E. MacRobbie, R. Halim and G. Temes, "A I .8V 94 dB dynamic range AX modulator for voice applications." Proceedings of ISSCC, February 1996, pp. 230–231.
[31] M. Steyaert, J. Crols and S. Gogaert, "Switched opamp, a technique for realising full CMOS switched-capacitor filters at very low voltages." Proceedings ESSCIRC, 1993, pp. 178–181.
[32] J. Crols, M. Steyaert, "Switched-Opamp: An approach to realize full-CMOS switched-capacitor circuits at very low power supply voltages." IEEE Journal of Solid-State Circuits, 29 (1994): 936–942.
[33] S. Chatterjee, Y. Tsividis and P. Kinget, "0.5-V analog circuit techniques and their application in OTA and filter design." IEEE Journal of Solid-State Circuits, 40.12 (2005): 2373–2387.
[34] Paul R. Gray, Paul J. Hurst, Stephen H. Lewis and Robert G. Meyer, *Analysis and Design of Analog Integrated Circuits*. John Wiley & Sons, New York, USA, 1984.

[35] S.C. Terry, M.M. Mojarradi, B.J. Blalock and J.A. Richmond, "Adaptive gate biasing: A new solution for body-driven current mirrors. ACM Great Lakes Symposium on VLSI, April 2005, pp. 472–477.
[36] D.G. Nairn and C.A.T. Salama "Current mode algorithmic analog to digital converters." IEEE Journal of Solid State Circuits, 25 (1990): 997–1004.
[37] T. Serrano and B. Linares-Barranco "The active input regulated cascode current mirror." IEEE Transactions on Circuits and Systems-I: Fundamental Theory and Application, 41.6 (1994): 464–467.
[38] R. Carvajal, J. Ramirez-Angulo, A. Lopez Martin, A. Torralba, J. Galan, A. Carlosena, et al. "The flipped voltage follower: A useful cell for low voltage low power circuit design." IEEE Transactions on Circuits and Systems (2005), pp.1276–1291, vol. 52, No. 7.
[39] J. Ramirez Angulo, R.G. Carvajal and A. Torralba, "Low supply voltage high-performance CMOS current mirror with low input and output voltage requirements." IEEE Transactions on Circuits and Systems-II (2004).
[40] J. Ramirez-Angulo M.S. Sawant, A. Lopez Martin and R.G. Carvajal, "Compact implementation of high performance CMOS current mirror." Electronics Letters (2005), pp. 1–2, vol. 41, No. 10.
[41] B. Aggarwal and M. Gupta, "Low-voltage cascode current mirror based on bulk-driven MOSFET and FGMOS techniques." International Conference on Advances in Recent Technologies in Communication and Computing, October, 2009, pp. 473–477.
[42] X.-G. Zhang and E.I. El-Masry, "A regulated body-driven CMOS current mirror for low-voltage applications." IEEE Transactions on Circuits and Systems II: Express Briefs, 51.10 (2004): 571–577.
[43] Bradley A. Minch, "Low-voltage cascode current mirror with enhanced dynamic performance." IEEE Transactions on Circuit System II (2012).
[44] R.L. Geiger, P.E. Allen, and N.R. Strader VLSI Design *Techniques for Analog and Digital Circuits*. McGraw-Hill, New York, 1990.
[45] S.S. Rajput and S.S. Jamuar, "Current mirrors essential units for analog circuit structures." IETE, 2002.
[46] N. Tarun, S. Suman, P.K. Ghosh, "Design of low voltage improved performance current mirror", Control Theory and Informatics, 4.2 (2014), pp. 26–38.
[47] R.A. Rutenbar, "Analog design automation: Where are we? Where are we going?" Proceedings of IEEE Custom IC Conference, San Diego, CA, 1993.
[48] International Technology Roadmap for Semiconductors, http://public.itrs.net last accessed on 8th June 2016.
[49] G. Gielen and R. Rutenbar, "Computer-aided design of analog and mixed-signal integrated circuits." Proceedings of IEEE, 88 (2000): 1825–1854.
[50] IEEE 1076.1 Working Group, "Analog and mixed-signal extensions to VHDL." http://www.eda.org/ vhdl-ams/ last accessed on 8th June 2016.

Index

Bei Fragen zur Produktsicherheit wenden Sie sich bitte an:
If you have any questions regarding product safety,
please contact:

Walter de Gruyter GmbH
Genthiner Straße 13
10785 Berlin
productsafety@degruyterbrill.com